Finite Element Analysis for Building Assessment

Finite Element Analysis for Building Assessment

Advanced Use and Practical Recommendations

Paulo B. Lourenço and
Angelo Gaetani

Routledge
Taylor & Francis Group

Cover image: Computer model of Ica Cathedral, Peru, courtesy of Maria Pia Ciocci and Satyadhrik Sharma, University of Minho, Portugal

First published 2022
by Routledge
4 Park Square, Milton Park, Abingdon, Oxon OX14 4RN

and by Routledge
605 Third Avenue, New York, NY 10158

Routledge is an imprint of the Taylor & Francis Group, an informa buisness

British Library Cataloging-in-Publication Data
A catalogue record for this book is available from the British Library

ISBN: 978-0-367-35767-2 (hbk)
ISBN: 978-1-032-22839-6 (pbk)
ISBN: 978-0-429-34156-4 (ebk)

DOI: 10.1201/9780429341564

Typeset in Sabon
by codeMantra

Contents

Foreword

This is the second book inspired by the lecturing activity in the International Master Course in Structural Analysis of Monuments and Historical Constructions, SAHC (www.msc-sahc.org), which was initiated in 2007 and graduated since then more than 400 students from 72 countries across the world. This initiative offers an advanced education programme on the conservation of existing structures, focusing on the application of scientific principles and methodologies in analysis, innovation and practice.

SAHC is the only international programme that specifically addresses the conservation of structures. It effectively creates professionals who have the ability to protect existing buildings and civil engineering works from various threats, such as natural decay, past disrespectful interventions, climatic change and natural hazards. The programme creates the specialized expertise necessary to advance the protection of our built heritage, a niche area that is becoming progressively more important. These highly trained professionals have intrinsic market value and possess extensive knowledge about existing buildings, often not found, such as knowledge about past building materials, shapes and technologies, forensic engineering skills related to inspection and diagnosis in various forms of construction, understanding on how to arrest damage and deterioration, in-depth knowledge on survey techniques, the ability to use today's best available computational tools for structural analysis, as well as good writing and communication skills.

The alumni of SAHC are working across the world, creating a unique network of knowledge and friendship. As a course coordinator, I am indebted to our amazing students and all lecturers involved in making the course so successful. The authors acknowledge the collaboration of the colleagues contributing to the original development of the contents addressed in this book, Petr Kabele and Milan Jirásek from Czech Technical University in Prague (Czech Republic). The contribution of Jorge Branco and Isabel Valente for Sections 4.2 and 4.4 of the book, respectively, is also gratefully acknowledged.

The book aims at being an important aid for advanced structural analysts, professionals and students, including an introduction to the finite element method, the main concepts related to nonlinear analysis and constitutive modelling, as well as practical recommendation for adopting material properties and for carrying out successful computations. I sincerely hope that the document can be of assistance for the use of modern sophisticated structural analysis tools in profession and may contribute to the safeguard of the existing built heritage.

Paulo B. Lourenço, Professor, SAHC Course Coordinator
July 2021, Guimarães, Portugal

Preface

Existing buildings represent a very inhomogeneous category in civil engineering. This is due to aspects such as the likely presence of archaic materials, damage and disconnections, uncommon construction techniques and subsequent interventions made throughout the building history. Existing structures were usually built with readily available materials according to traditional rules of local masons and carpenters, almost exclusively focusing on gravitational loads. Later additions or alterations, such as the construction of a new floor or different space and opening distributions, may have led the building to a structural organization where the flow of forces is not as it was originally conceived. In this scenario, the definition of a specific structural category, as usually done for the design of new buildings (e.g. reinforced concrete frame with high ductility class), is hardly possible because a given building is characterized by its uniqueness.

Being substantially different from new buildings, the structural approach to analyse the building should not be limited to the assumption of linear elastic materials. This approach may lead to over-conservative results with dramatic costs associated with extensive (in lieu of moderate or limited) structural strengthening. It is worth underlining that costs must be understood not only in terms of economic resources, as related to the extent of the strengthening measures and the possible interruption of use of the building itself, but costs include also invasiveness, with likely loss in terms of aesthetics or cultural values. Advanced approaches, such as nonlinear analysis, are often invoked to provide a more accurate estimation of the structural capacity. Accordingly, the buildings' response is analysed in stages well beyond the elastic one and close to the actual inelastic deformation capacity, taking advantage of stress redistribution and reserves of capacity. The main consequence is a more rational and sustainable use of the resources needed for the strengthening measures, with an overall reduction of costs. It is worth noticing that nonlinear analysis is often employed to explain past or present observations, to predict future behaviour, and to provide some limits to the degree of confidence of the results. In this regard, it is not uncommon that nonlinear analysis procedures find applications in forensic engineering.

As far as engineering practice is concerned, finite element analysis represents the most diffused structural analysis tool. Thanks to market competition and powerful computers, many software codes based on finite elements are currently available. These are usually general purpose, employing a large variety of material models, element types and solution procedures. As an example, limiting the discussion to the material response, different approaches are typically available, namely, nonlinear elasticity, plasticity, fracture mechanics, damage continuum mechanics and other formulations. Cracking, for instance, can be modelled discretely or using smeared crack approaches. The latter may be subdivided into a rotating crack model, a fixed crack model, a model with multiple nonorthogonal cracks and hybrid models. Some formulations place emphasis on classical mechanics formulations derived from theoretical concepts, others rely on empirical data and phenomenological models derived from experimental observation. One approach may be particularly suited to describe a certain phenomenon and less suited to do so to others, making the selection of a unique approach that performs well over the entire range of structural problems encountered in practice simply impossible.

In this scenario, the analyst is asked to make the necessary assumptions to produce reliable estimates of the response with reasonable effort, based on experience and judgement. No computational model is expected to describe the reality exactly. The need of simplifying hypotheses to understand, define, quantify, visualize and simulate the most influencing features is evident, aimed at reaching a compromise between realism and costs. In this case, costs are related to the capacity of modern computers and to the availability of the data required (e.g. material properties, geometry and morphology), sometimes difficult (or expensive) to determine in practice. Without loss of generality, the selected model should be the simplest and less costly one, still able to simulate the features or phenomena intended.

Decisions made by the analyst regarding finite element mesh arrangement, type of finite element, boundary conditions, method of loading, convergence criteria and selection of material behaviour model are likely to produce some discrepancy in the results. Moreover, the possible mistakes in input may exacerbate the situation. A healthy degree of caution when analysing the results is, therefore, indispensable. At every stage of the analysis, numerical models should be verified or calibrated against an educated guess or benchmark results involving similar constructions and loading details. These processes should contribute to inform the analyst about the accuracy of the results and provide confidence in the same.

Along these lines, the goal of the book is to provide support for the analyst's decision-making processes, to ensure efficient and safe use of finite element programs for the nonlinear analysis of existing structures. Written primarily for the benefit of the practicing engineer, the book concentrates on practical aspects with moderate theoretical contents. However, the basics of solid and structural mechanics, as well as of elementary calculus and algebra, are essential for the comprehension of some topics. In this regard, the reader is encouraged to investigate more on these matters, if required. Practitioners with a deeper understanding on finite element analysis may use the book and related references to reduce the likelihood of mistakes. Further insight may help the reader to fully understand the computational aspects and the mathematical formulations involved, or to be able to create a new material model, to add a user-supplied subroutine in a software code for a special purpose or to fully develop an own structural analysis code for tailored applications, topics that fall outside the scope of the book.

The book is articulated into five chapters. Chapter 1 deals with fundamentals of the finite element theory, moving from the weak form of the governing equations in its integral form to shape functions and numerical integration. Commonly used finite elements are also discussed in the light of structural mechanics applications. Chapter 2 introduces nonlinear structural analysis. The potential sources of nonlinearities are analysed by means of examples. The chapter tackles also the main computational strategies to solve nonlinear problems. Typical incremental-iterative methods (e.g. Forward Euler method and Newton–Raphson method) are presented together with advanced solution procedures for nonlinear problems (namely, arc-length method and line search algorithm). Examples are disseminated throughout the chapter to highlight the advantages and disadvantages of different approaches or concepts, aiming at discussing analogies and differences among them. Chapter 3 deals with constitutive models, namely, elasticity, plasticity, damage and fracture. The discussion focuses on the practical relevance of each model rather than following an in-depth theoretical approach. The last part of the chapter is devoted to typical constitutive models available in commercial software codes.

Chapter 4 is aimed at providing a general overview of the nonlinear response of the main construction materials (masonry, timber, concrete and metals). Recommended properties for advanced numerical analysis are also proposed. As masonry construction is considered one of the most spread structural typologies for existing buildings and has intrinsic nonlinear response, this material is treated extensively. Many topics discussed in the chapter are the subject of current research and may require an update in a revised edition. The last part is dedicated to the introduction of safety assessment according to codes of practice, with reference to the European standard for structural design (or Eurocode), the Italian code and the American Model Code. Finally, Chapter 5 provides a step-by-step guidance for nonlinear finite element analysis, from the preprocessing, through the verification and interrogation of the results, to postprocessing. Various conclusive examples of the analysis of simple structures are also presented.

The authors hope that the reader can feel motivated to the use of advanced nonlinear structural analysis tools, understanding the complexity of nature and the need for verification of results. Experience, personal judgement and persistence, as well as proper allocation of time and computer resources, are necessary requirements for successful results.

Authors

Paulo B. Lourenço is Professor at the Department of Civil Engineering, University of Minho, Guimarães (Portugal) since 2006. He earned his degree in Civil Engineering at University of Porto (Portugal) in 1990 and his PhD in Civil Engineering at Delft University of Technology (Netherlands) in 1996. He has been Director of the Institute in Sustainability and Innovation in Structural Engineering since 2007. He is experienced in the fields of nondestructive testing, advanced experimental and numerical techniques, innovative strengthening techniques and earthquake engineering. He is a specialist in structural conservation and forensic engineering, with works on many monuments and existing buildings, including 17 UNESCO World Heritage sites. He is also a structural masonry expert, responsible for R&D projects with the clay brick, concrete block and lightweight concrete block masonry and mortar industry. He was the Project Team Leader for the revision of Part 1 of the masonry European code (EN 1996-1-1). He is the coordinator of the MSc on Structural Analysis of Monuments and Historical Constructions (SAHC) since 2007, with alumni from 70 countries, recipient of the European Heritage/Europa Nostra Award in 2017, and Editor of the *International Journal of Architectural Heritage*. He has supervised more than 60 PhD theses and coordinates an innovative training network for sustainable building lime applications with 15 PhD students across Europe. He also holds an Advanced European Research Council Grant to develop an integrated seismic assessment approach for heritage buildings.

Angelo Gaetani is a professional structural engineer and consultant in Rome (Italy), leading structural restoration and seismic retrofit projects. He earned his MSc in Building Engineering and Architecture at University of Calabria (Italy) in 2008, after being visiting student at SINTEF Building and Infrastructure research institute of Trondheim (Norway). In 2010 he received the Postgraduate Professional Degree in Seismic Design at Polytechnic University of Milan (Italy). In 2016 he earned the joint PhD in Civil Engineering at University of Sapienza in Rome and at University of Minho in Guimarães (Portugal). The thesis was awarded the "Prize for PhD Thesis 2016" and published by Sapienza University Press. At the 2018 APT conference held in Buffalo (USA), his work on masonry cross vaults was the inaugural recipient of the David Fischetti Award presented by the Preservation Technical Committee (PETC) of the Association for Preservation Technology International (APT). After earning his PhD, he was a postdoctoral research fellow at University of Minho and University of Sapienza until 2020.

Chapter 1

Fundamentals of the finite element method

INTRODUCTION

This chapter provides a brief description of the Finite Element Method (FEM) in the framework of linear elasticity. This type of problem includes two assumptions: *small displacements* (with respect to the characteristic dimensions of the body) and a *linear elastic stress–strain relationship* (meaning that the relations between stresses and strains are given by a fixed set of constants; these relations are univocal, with no dependence on the load path). In practice, the first assumption states that geometrical changes after load application to the body are so small that they may be neglected, while the second assumption means that the effects of different loads can be superimposed on each other. In this framework, a linear proportion between causes and effects holds, as first stated in 1675 by Robert Hooke (1635–1703) by means of a succinct Latin anagram standing for *"ut tensio, sic vis"*, i.e. *"as the extension, so the force"*. Basically, if the external load (understood as a point load or the full set of forces applied to a body) doubles, so do the effects, namely, displacements, strains and stresses. Although linear elasticity is a simplification with respect to the goal of the book, which is nonlinear analysis, linear elasticity allows a simple presentation of the FEM, providing the basis for further insight discussed in the next chapters. Here, applications of FEM to simple problems are also discussed and verified with respect to theoretical considerations.

1.1 EQUILIBRIUM OF A THREE-DIMENSIONAL BODY IN ELASTICITY

In the analysis of quasi-static problems (i.e. when inertia effects can be neglected and quantities are allowed to vary only slowly in time), the goal of structural engineering can be summarized as follows:

For a given system of body forces $\mathbf{b}$ and surface tractions $\mathbf{t}$, imposed displacements $\mathbf{u}_S$, initial strains $\boldsymbol{\varepsilon}_0$ (with no evident relation with stress,

DOI: 10.1201/9780429341564-1

e.g. temperature change or shrinkage) and initial (residual) stresses $\boldsymbol{\sigma}_0$ applied to a constrained structure of a certain geometry and materials, calculate the compatible displacement field $\mathbf{u}$ and strain state $\boldsymbol{\varepsilon}$, as well as the equilibrated stress state $\boldsymbol{\sigma}$ and constraint reactions $\mathbf{r}$.

This statement can be applied to the masonry bridge shown on the left of Figure 1.1. Its geometrical schematization, depicted on the right, highlights the physical quantities involved in the problem. In particular, body forces $\mathbf{b}$ are associated to the material density within the physical volume V of the bridge, whereas surface tractions $\mathbf{t}$ are associated to e.g. live loads, weight of statues on the bridge deck or hydrostatic pressure of the river water. The boundary surface includes constrained and free parts, S_u and S_t, respectively. The former represents the location where displacements are assigned (e.g. the foundation may be considered with null displacement, if sufficiently rigid) so that reaction forces $\mathbf{r}$ are obtained; the latter is generally the complementary part, which is able to freely deform. For the sake of conciseness, S_t is shown only in correspondence of the surface tractions in Figure 1.1.

According to solid mechanics, equilibrium conditions (involving forces and stresses) and kinematic conditions (involving displacements and strains) are derived locally and independently, i.e. Eqs. (1.1) and (1.2), respectively. The two sets of conditions are linked by means of the constitutive law in Eq. (1.3). The resulting combination is the **strong form of the governing equations** in solid mechanics shown in Eq. (1.4), provided that the boundary conditions remain valid. Figure 1.2 synthesizes the equations that govern the elastic problem with the relations between the physical quantities involved. Here, $\boldsymbol{\partial}$ is the differential operator, $\mathbf{D}$ is the material stiffness matrix arising from Hooke's linear elasticity law and $\mathbf{n}$ is the outward vector normal to the free boundary S_t. According to solid mechanics, the solution must satisfy this set of differential equations with additional constraints (leading to the so-called *Boundary Value Problem*). Closed-form solutions, such as

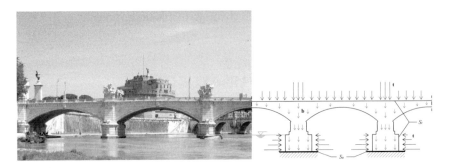

Figure 1.1 Structural analysis of a masonry bridge. Image and geometrical model with body forces **b**, surface tractions **t**, free and constrained boundary surfaces, S_t and S_u, respectively. (Photo by Sergio D'Afflitto, CC-BY-SA-4.0, edited.)

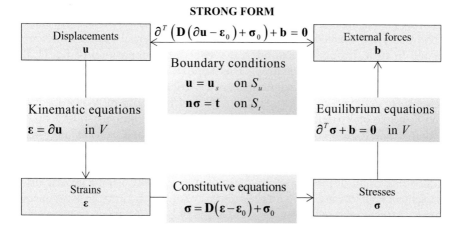

Figure 1.2 Schematic representation of the governing equations of elasticity problems and the related strong form of equilibrium.

$\mathbf{u} = [u_{(x, y, z)}, v_{(x, y, z)}, w_{(x, y, z)}]^T$ defined over the entire problem domain, are possible only when simple geometries and loadings are considered.

$$\partial^T \boldsymbol{\sigma} + \mathbf{b} = 0 \qquad \qquad \text{within the volume } V$$
$$\qquad \qquad \qquad \qquad \qquad \qquad \qquad \qquad \qquad \qquad \qquad \qquad \qquad \qquad (1.1)$$
$$\mathbf{n}\boldsymbol{\sigma} = \mathbf{t} \qquad \qquad \text{on the free boundary } S_t$$

$$\boldsymbol{\varepsilon} = \partial \mathbf{u} \qquad \qquad \text{within the volume } V$$
$$\qquad \qquad \qquad \qquad \qquad \qquad \qquad \qquad \qquad \qquad \qquad \qquad \qquad \qquad (1.2)$$
$$\mathbf{u} = \mathbf{u}_s \qquad \qquad \text{on the constrained boundary } S_u$$

$$\boldsymbol{\sigma} = \mathbf{D}(\boldsymbol{\varepsilon} - \boldsymbol{\varepsilon}_0) + \boldsymbol{\sigma}_0 \qquad \qquad \qquad \qquad \qquad \qquad \qquad (1.3)$$

$$\partial^T \left(\mathbf{D}(\partial \mathbf{u} - \boldsymbol{\varepsilon}_0) + \boldsymbol{\sigma}_0 \right) + \mathbf{b} = 0 \qquad \qquad \qquad \qquad (1.4)$$

Kinematic and equilibrium conditions are also linked by the *Principle of Virtual Work*. It must be stressed that this principle does not add equations for the problem to be solved, but it is a *consequence* of such conditions. In fact, for an equilibrated stress field and system of forces producing work on any compatible strain and displacement field, the internal virtual work W_i is equal to the external virtual work W_e, as given by Eq. (1.5). The word *virtual* means that the displacement field and the external loads (as well as stresses $\boldsymbol{\sigma}$ and strains $\boldsymbol{\varepsilon}$) are not linked by any law of cause and effect, i.e. the displacements are not a consequence of the loads. Hence, the consequent work is virtual and not real.

$$\underbrace{\int_V \boldsymbol{\varepsilon}^T \boldsymbol{\sigma} \, dV}_{\text{internal virtual work } W_i} = \underbrace{\int_V \mathbf{u}^T \mathbf{b} \, dV + \int_{S_t} \mathbf{u}^T \mathbf{t} \, dS + \int_{S_u} \mathbf{u}_s^T \mathbf{r} \, dS}_{\text{external virtual work } W_e} \qquad (1.5)$$

More concisely, if equilibrium and compatibility (the latter defined in a simple way as a deformed shape of the structure compatible with the kinematic conditions) are satisfied, then the external and the internal virtual works are equal. In turn, if only compatibility is satisfied, the imposition of Eq. (1.5) becomes an equilibrium equation. This approach goes under the name of *Principle of Virtual Displacements* because compatible displacements and strains are assumed as the primary variables of the problem. Consequently, the Principle of Virtual Displacements states that the equilibrium of a body requires that, for any compatible displacement field $\delta\mathbf{u}$ (and strain $\delta\varepsilon$) imposed on the body and its surfaces S_u and S_t, the internal virtual work is equal to the external virtual work given by Eq. (1.6). In this case, δ is introduced for addressing small displacements and strains. Operatively, using a generic displacement field (compatible with the boundary conditions) and the consequent strains, stresses and reactions, the satisfaction of Eq. (1.6) means that stresses and forces are equilibrated, and the displacement field chosen is the exact solution. This approach goes under the name of the **weak form of the governing equations** in its *integral form*. Upon integration, equilibrium is sought over the entire body and surfaces rather than locally, as enforced in Eq. (1.4).

$$\int_V \delta\varepsilon^T \sigma \, dV = \int_V \delta\mathbf{u}^T \mathbf{b} \, dV + \int_{S_t} \delta\mathbf{u}^T \mathbf{t} \, dS + \int_{S_u} \delta\mathbf{u}_S^T \mathbf{r} \, dS$$

$$\forall \left(\delta\mathbf{u}, \delta\varepsilon\right) \big| \delta\varepsilon = \partial\delta\mathbf{u} \text{ in } V$$

(1.6)

1.2 BASICS

As already pointed out, a closed-form solution of the Boundary Value Problem is not achievable in practical engineering problems, for which a numerical approach is generally pursued. With this goal, any real structure is subdivided into a *finite* number of *discrete elements* of given dimensions (provided with a finite number of Degrees of Freedom – DOFs), whose behaviour and properties are readily understood. Once these elements are chosen and defined, it is possible to (mathematically) rebuild the original structure and evaluate its overall behaviour. How to move from a real/continuum problem to a discrete problem is critical in structural engineering. This process presumes a mathematical idealization (*discretization*) of the reality and the creation of a *mathematical model*. Finite Element Analysis is aimed at solving this mathematical model, which necessarily reflects the incorporated assumptions.

1.2.1 Weak form of governing equations in elasticity

Although the analysis of natural phenomena is marked by a more or less pronounced level of approximation, the intrinsic approximation of finite element method (FEM) is explicitly assumed in the description of the displacement field. In this regard, the following assumptions are employed:

1. Elements are interconnected at a discrete number of joints called nodes, generally placed on the element boundaries. The complete set, or assemblage of elements, is called mesh.
2. The displacements within and on the boundaries of each element are uniquely defined according to the nodal displacements by means of well-suited algebraic shape functions (generally, but not necessarily, polynomial) that work as interpolation functions for the full continuous field.
3. The internal state of strain depends on the displacement field, thus on the shape functions and nodal displacements; the state of stress is calculated from the strain field accordingly.
4. Body forces and external tractions (either loads or reactions) are lumped in the nodes where equilibrium is sought.

Accordingly, once the displacement functions are selected for the chosen finite elements, the goal of the analysis is the calculation of nodal displacements, which are the state variables of the problem. These are calculated by imposing the simultaneous equilibrium of all nodes: the equilibrium of the entire body is guaranteed together with the equilibrium of its individual elements. In terms of the integral approach of the weak form of governing equations, i.e. Eq. (1.6), this method is allowed by the fact that the integral over the entire volume of the body is the summation of the integrals over its subvolumes.

In this regard, let the displacement field $\mathbf{u}_e$ of a generic element be approximated (note the usage of the symbol $\approx$ and the subscript "e") according to Eq. (1.7), where $\mathbf{N}_e$ is the matrix of the shape functions and $\mathbf{d}_e$ is the vector of the nodal displacements, hereinafter referred to as degrees of freedom or DOFs. As stated, DOFs are the principal unknowns of the problem, and their number is defined according to kinematic considerations. A three-dimensional beam may be provided with six DOFs per node (i.e. three translations and three rotations along the axes), or three DOFs in two-dimensional beam problems (i.e. two translations and one rotation); with the same reasoning, a three-dimensional volume problem considers three DOFs per node (i.e. translations along the axes) and a membrane problem may consider only two DOFs per node (i.e. translations along the axes corresponding to a plane stress state). With no doubt, a reduction of DOFs necessarily implies a reduction of the computational demand.

With simple mathematical processing, in line with the previous section, strains and stresses are derived in Eqs. (1.8) and (1.9), respectively. Here, $\mathbf{B}_e$ is the strain-displacement transformation matrix given by $\partial\mathbf{N}_e$. It is worth noticing that the differential operator ∂ collects the partial derivatives and, provided the shape functions are polynomial expressions of a suitable order, the strains $\boldsymbol{\varepsilon}_e$ and stresses $\boldsymbol{\sigma}_e$ are also polynomial expressions.

$$\mathbf{u}_e \approx \mathbf{N}_e \mathbf{d}_e \tag{1.7}$$

$$\boldsymbol{\varepsilon}_e = \partial \mathbf{u}_e \approx \partial \mathbf{N}_e \mathbf{d}_e = \mathbf{B}_e \mathbf{d}_e \tag{1.8}$$

$$\boldsymbol{\sigma}_e = \mathbf{D}_e \left(\boldsymbol{\varepsilon}_e - \boldsymbol{\varepsilon}_{0_e} \right) + \boldsymbol{\sigma}_{0_e} \approx \mathbf{D}_e \left(\mathbf{B}_e \mathbf{d}_e - \boldsymbol{\varepsilon}_{0_e} \right) + \boldsymbol{\sigma}_{0_e} = \mathbf{D}_e \mathbf{B}_e \mathbf{d}_e - \mathbf{D}_e \boldsymbol{\varepsilon}_{0_e} + \boldsymbol{\sigma}_{0_e} \tag{1.9}$$

According to the weak form of the governing equations, equilibrium is guaranteed by means of the Principle of Virtual Work in Eq. (1.10), provided $\delta \mathbf{u}_e = \mathbf{N}_e \delta \mathbf{d}_e$, $\delta \boldsymbol{\varepsilon}_e = \mathbf{B}_e \delta \mathbf{d}_e$, and $\mathbf{r}_e$ is the vector of the external nodal forces in lieu of the external reactions. Here, for the sake of clarity, the subscript "e" is dropped out. Expanding all the terms of Eq. (1.10) and isolating $\delta \mathbf{d}$, the result is reported in Eq. (1.11). As this relation is valid for any compatible virtual displacement, the multipliers must be equal. Additionally, $\mathbf{d}$ can be taken out of the integral and, after Eq. (1.12), the new relation can be written as in Eq. (1.13), where the subscript "e" is added back.

$\mathbf{K}_e$ is defined as the stiffness matrix of the element, i.e. the matrix (containing the elastic constants) that allows calculating the nodal reactions of the element due to its nodal displacements $\mathbf{d}_e$. These nodal reactions must equate the nodal forces $\mathbf{f}_e$ due to body forces $\mathbf{b}$, surface traction $\mathbf{t}$, initial strains $\boldsymbol{\varepsilon}_0$, initial stresses $\boldsymbol{\sigma}_0$ and external nodal forces $\mathbf{r}$. In turn, if displacements are imposed on S_t, these can be satisfied by specifying some of the nodal parameters of $\mathbf{d}_e$. Once the nodal displacements are calculated, the other mechanical quantities can be evaluated according to Eqs. (1.7–1.9).

$$\int_V \left(\mathbf{B} \delta \mathbf{d} \right)^T \left(\mathbf{DBd} - \mathbf{D} \boldsymbol{\varepsilon}_0 + \boldsymbol{\sigma}_0 \right) dV$$
$$= \int_V \left(\mathbf{N} \delta \mathbf{d} \right)^T \mathbf{b} \, dV + \int_{S_u} \left(\mathbf{N} \delta \mathbf{d} \right)^T \mathbf{t} \, dS + \delta \mathbf{d}^T \mathbf{r} \tag{1.10}$$

$$\delta \mathbf{d}^T \left(\int_V \mathbf{B}^T \mathbf{DBd} \, dV - \int_V \mathbf{B}^T \mathbf{D} \boldsymbol{\varepsilon}_0 \, dV + \int_V \mathbf{B}^T \boldsymbol{\sigma}_0 \, dV \right)$$
$$= \delta \mathbf{d}^T \left(\int_V \mathbf{N}^T \mathbf{b} \, dV + \int_{S_u} \mathbf{N}^T \mathbf{t} \, dS + \mathbf{r} \right) \tag{1.11}$$

$$\int_V \mathbf{B}^T \mathbf{DB} \, dV = \mathbf{K}$$
$$\int_V \mathbf{B}^T \mathbf{D} \boldsymbol{\varepsilon}_0 \, dV - \int_V \mathbf{B}^T \boldsymbol{\sigma}_0 \, dV + \int_V \mathbf{N}^T \mathbf{b} \, dV + \int_{S_u} \mathbf{N}^T \mathbf{t} \, dS + \mathbf{r} = \mathbf{f} \tag{1.12}$$

$$\mathbf{K}_e \mathbf{d}_e = \mathbf{f}_e \tag{1.13}$$

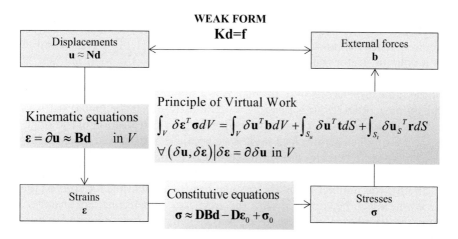

Figure 1.3 Schematic representation of the governing equations of elasticity problems and the relative weak form of equilibrium in finite element analysis (with **d** nodal displacements).

The same reasoning can be followed for all the finite elements of the structure, summing up all their contributions. Eq. (1.14) describes the equilibrium equations of the entire structure (see Section 1.3.2 for details) where the quantities involved have the same shape of Eq. (1.13), but they embrace all the nodal variables. Accordingly, the unknown nodal displacements **d** can be calculated with Eq. (1.15). Although easy in theory, the inversion of the stiffness matrix **K** is not trivial in practice because of its order, namely, the number of DOFs, thus equations, involved (it is not uncommon to have structural problems larger than hundreds of thousands or millions DOFs). Several techniques have been developed to solve this problem, as mentioned in Section 1.3.2.

$$\mathbf{Kd} = \mathbf{f} \tag{1.14}$$

$$\mathbf{d} = \mathbf{K}^{-1}\mathbf{f} \tag{1.15}$$

As a counterpart of Figure 1.2, and in line with the symbols already used, the schematic representation of the governing equations in finite element analysis is shown in Figure 1.3. As it is possible to notice, the Principle of Virtual Work is used to obtain the equilibrium equations, leading, ultimately, to Eq. (1.14). Basically, if compared to Eq. (1.4), the boundary value problem (i.e. the set of differential equations and boundary conditions) is substituted by a set of linear equations (applicable only in the case of linear elastic problems).

It is worth focusing on the dimensions of the matrix and vectors involved. Looking at Eq. (1.13), for a two-node finite element where only the longitudinal displacement is considered, $\mathbf{d}_e$ is a 2×1 column vector of the nodal displacements, and $\mathbf{f}_e$ is a 2×1 column vector of the nodal forces. Consequently, $\mathbf{K}_e$ must be square and 2×2. If Eq. (1.14) is considered, $\mathbf{d}$ is a column vector with a number of rows equal to the total amount of DOFs. Consequently, being $\mathbf{f}$ the column vector of all the nodal forces, it is of the same dimensions of $\mathbf{d}$. In turn, $\mathbf{K}$ is a square matrix of order equal to the number of DOFs. It is possible to demonstrate that the global stiffness matrix $\mathbf{K}$ and the vector $\mathbf{f}$ are obtained from the superposition of the single $\mathbf{K}_e$ and $\mathbf{f}_e$ of all elements. As a node is generally shared by several elements, the superposition creates interconnectivity. In turn, the vector $\mathbf{d}$ contains all the unknown DOFs and no superposition applies.

Finally, the number of unknowns defines the dimension (or size) of the problem, not to be confused with the geometric dimensions. For instance, from the geometrical point of view, the axial behaviour of a column is a one-dimensional problem (in the case of linear elastic analysis, only the longitudinal displacement needs to be considered) but the unknowns are related to the number of subdivisions (finite elements) and the nodes defined between the subdivisions, making the problem multidimensional. The examples provided in the next sections will make the concepts just introduced clearer.

1.2.2 Example of an axially loaded column

Let us consider a column with a rectangular cross section $0.3 \times 0.5\,\mathrm{m}^2$ and a height $h = 3\,\mathrm{m}$ (Figure 1.4a). The column is subjected to its axial compressive body force (specific weight) $b = -25\,\mathrm{kN/m}^3$ and an external surface

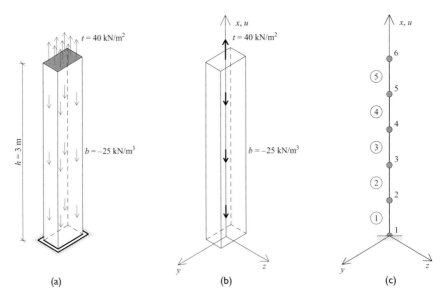

Figure 1.4 Axially loaded column. (a) Overview, (b) structural model and (c) FEM discretization.

tensile action (applied stress) $t = 40$ kN/m² at the free extremity (minus sign means downward, as shown in Figure 1.4b). The tensile force can be the result of e.g. an earthquake or wind load which may introduce tension in a column in the exterior part of a building. The governing equations and the strong form of the equilibrium are described first, followed by the finite element discussion (Figure 1.4c).

As the problem is one-dimensional, matrices and vectors are substituted with the equivalent scalar quantities along the longitudinal direction x, as in Eqs. (1.16–1.19), where the volume V is substituted by the height h. Provided that Young's modulus $E = 30 \times 10^6$ kN/m² is constant along the beam, by means of subsequent integrations and imposing the boundary conditions, the results are shown in Eq. (1.20). As $\sigma = -35$ kN/m² at the bottom of the column, the constraint reaction (or reaction at the base) is equal to its opposite, with a resultant force of $r = + 35$ kN/m² $\times$ (0.3 $\times$ 0.5) m² $= + 5.25$ kN.

$$\frac{\partial \sigma}{\partial x} + b = 0 \qquad \text{along the height } h$$

$$\sigma = t \qquad \text{on the free boundary } (x = 3\,\text{m}) \tag{1.16}$$

$$\varepsilon = \frac{\partial u}{\partial x} \qquad \text{along the height } h$$

$$u = 0 \qquad \text{on the constrained boundary } (x = 0\,\text{m}) \tag{1.17}$$

$$\sigma = E\varepsilon \tag{1.18}$$

$$\frac{\partial}{\partial x} E \frac{\partial u}{\partial x} + b = 0 \tag{1.19}$$

$$u = -\frac{b}{2E}x^2 + \frac{t + bh}{E}x = \frac{4.17 \times 10^{-7}}{\text{m}}x^2 - 1.17 \times 10^{-6}x$$

$$\varepsilon = -\frac{b}{E}x + \frac{t + bh}{E} = \frac{8.33 \times 10^{-7}}{\text{m}}x - 1.17 \times 10^{-6} \tag{1.20}$$

$$\sigma = -bx + t + bh = 25\frac{\text{kN}}{\text{m}^3}x + 40\frac{\text{kN}}{\text{m}^2} - 75\frac{\text{kN}}{\text{m}^2}$$

The column is now studied according to FEM as described in Figure 1.4c, considering six nodes (as labelled) and five equal elements (encircled numbers) of length $l_e = 0.6$ m. The nodal displacements are the main unknowns of the problem. Being the problem one-dimensional in terms of geometry and loading, each node is assumed to move only longitudinally, thus having six DOFs in total (namely, the vertical displacements of each node). Additionally, the shape functions are calculated assuming that displacements vary linearly within the element. As highlighted below in Section 1.4,

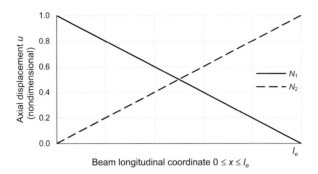

Figure 1.5 Shape functions for axial elongation in beam.

the shape functions of each element must be chosen so that the unit displacement of one node corresponds to null displacement of the others. Considering the first element between nodes 1 and 2, the following relationships hold, where $0 \leq x \leq l_e$ ($x = 0$ and $x = l_e$ correspond to the node 1 and 2, respectively), and d_1 and d_2 are the nodal displacements of nodes 1 and 2. In particular, the shape functions for the analysed case, i.e. axially loaded column with a pure longitudinal displacement u, are shown in Eq. (1.21) and drawn in Figure 1.5, where N_1 and N_2 are the two components of vector $\mathbf{N}_1$ of Eq. (1.21). The full set of Eqs. (1.7–1.9) are recast in Eqs. (1.21–1.23).

$$u_1 \approx \mathbf{N}_1 \mathbf{d}_1 = \begin{bmatrix} 1 - \dfrac{x}{l_e} & \dfrac{x}{l_e} \end{bmatrix} \begin{bmatrix} d_1 \\ d_2 \end{bmatrix} \tag{1.21}$$

$$\varepsilon_1 = d\mathbf{u}_1 \approx \dfrac{d}{dx} \begin{bmatrix} 1 - \dfrac{x}{l_e} & \dfrac{x}{l_e} \end{bmatrix} \begin{bmatrix} d_1 \\ d_2 \end{bmatrix} = \begin{bmatrix} -\dfrac{1}{l_e} & \dfrac{1}{l_e} \end{bmatrix} \begin{bmatrix} d_1 \\ d_2 \end{bmatrix} \tag{1.22}$$

$$\mathbf{B}_1 = \begin{bmatrix} -\dfrac{1}{l_e} & \dfrac{1}{l_e} \end{bmatrix}$$

$$\sigma_1 = E\varepsilon_1 \approx E \begin{bmatrix} -\dfrac{1}{l_e} & \dfrac{1}{l_e} \end{bmatrix} \begin{bmatrix} d_1 \\ d_2 \end{bmatrix} = \begin{bmatrix} -\dfrac{E}{l_e} & \dfrac{E}{l_e} \end{bmatrix} \begin{bmatrix} d_1 \\ d_2 \end{bmatrix} \tag{1.23}$$

The Principle of Virtual Work can be applied according to Eqs. (1.12) and (1.13), yielding Eqs. (1.24) and (1.25). As the mechanical and geometrical parameters are constant within the cross section A, the integral over the volume is limited to the longitudinal direction x.

$$\mathbf{K}_1 = \int_V \mathbf{B}_1^T \mathbf{D}_1 \mathbf{B}_1 \, dV = A \int_0^{l_e} \begin{bmatrix} -\dfrac{1}{l_e} \\[2mm] \dfrac{1}{l_e} \end{bmatrix} E \begin{bmatrix} -\dfrac{1}{l_e} & \dfrac{1}{l_e} \end{bmatrix} dx = \begin{bmatrix} \dfrac{EA}{l_e} & -\dfrac{EA}{l_e} \\[3mm] -\dfrac{EA}{l_e} & \dfrac{EA}{l_e} \end{bmatrix}$$

$$\mathbf{f}_1 = A \int_0^{l_e} \begin{bmatrix} 1 - \dfrac{x}{l_e} \\[2mm] \dfrac{x}{l_e} \end{bmatrix} b \, dx = Ab \begin{bmatrix} \dfrac{l_e}{2} \\[3mm] \dfrac{l_e}{2} \end{bmatrix} = \begin{bmatrix} \dfrac{Abl_e}{2} \\[3mm] \dfrac{Abl_e}{2} \end{bmatrix} \tag{1.24}$$

$$\mathbf{K}_1 \mathbf{d}_1 = \mathbf{f}_1 \quad \rightarrow \quad \begin{bmatrix} \dfrac{EA}{l_e} & -\dfrac{EA}{l_e} \\[3mm] -\dfrac{EA}{l_e} & \dfrac{EA}{l_e} \end{bmatrix} \begin{bmatrix} d_1 \\[3mm] d_2 \end{bmatrix} = \begin{bmatrix} \dfrac{Abl_e}{2} \\[3mm] \dfrac{Abl_e}{2} \end{bmatrix} \tag{1.25}$$

With the aim of assembling the global structural matrix, it is needed to expand the quantities in Eq. (1.25) to the global form. In other words, the quantities of each element must be rearranged within the global matrix and vectors by adequate compartmentalization. Since the problem has six DOFs, the stiffness matrix $\mathbf{K}_1$ must be transformed into a 6×6 matrix, whereas the full vector of nodal displacements $\mathbf{d}$ (including nodal displacements $\mathbf{d}_1$) and forces $\mathbf{f}_1$ become 6×1. Again, it is worth noticing that the vector $\mathbf{d}$ contains all the unknown DOFs, and there is no need to differentiate it according to each element. The new relation for the first element is, thus, reported in Eq. (1.26) where the matrices assume nonzero values only in correspondence of the cells K_{ij} related to the nodes of the element (in this case, $i, j = 1, 2$). Analogously, the equilibrium relation for the second element (connecting node 2 and 3) is shown in Eq. (1.27).

$$\mathbf{K}_1 \mathbf{d} = \mathbf{f}_1 \rightarrow \begin{bmatrix} \dfrac{EA}{l_e} & -\dfrac{EA}{l_e} & \cdot & \cdot & \cdot & \cdot \\[3mm] -\dfrac{EA}{l_e} & \dfrac{EA}{l_e} & & & & \\[3mm] \cdot & & \cdot & & & \\ \cdot & & & \cdot & & \\ \cdot & & & & \cdot & \\ \cdot & & & & & \cdot \end{bmatrix} \begin{bmatrix} d_1 \\ d_2 \\ d_3 \\ d_4 \\ d_5 \\ d_6 \end{bmatrix} = \begin{bmatrix} \dfrac{Abl_e}{2} \\[3mm] \dfrac{Abl_e}{2} \\[2mm] \cdot \\ \cdot \\ \cdot \\ \cdot \end{bmatrix}$$

$$\tag{1.26}$$

$$K_2 d = f_2 \rightarrow \begin{bmatrix} \cdot & \cdot & & \cdot & \cdot & \cdot & \cdot \\ \cdot & \dfrac{EA}{l_e} & -\dfrac{EA}{l_e} & & & \\ \cdot & -\dfrac{EA}{l_e} & \dfrac{EA}{l_e} & & & \\ \cdot & & & & \cdot & \\ \cdot & & & & \cdot & \\ \cdot & & & & & \cdot \end{bmatrix} \begin{bmatrix} d_1 \\ d_2 \\ d_3 \\ d_4 \\ d_5 \\ d_6 \end{bmatrix} = \begin{bmatrix} \cdot \\ \dfrac{Abl_e}{2} \\ \dfrac{Abl_e}{2} \\ \cdot \\ \cdot \\ \cdot \end{bmatrix}$$

(1.27)

Recalling the surface traction on the extremity of the element 5, Eq. (1.28) shows the global equilibrium of the entire column by adding the contribution of the different finite elements. Nonetheless, as the boundary conditions impose a null displacement for d_1, the equilibrium equation of node 1 can be neglected (i.e. it is possible to delete the first row of K, d and f) and all the products with d_1 are identically null (i.e. it is possible to delete the first column of K). Accordingly, the set of equilibrium equations for the entire column is reported in Eq. (1.29). It is worth stressing that, in the case of lack of boundary conditions, rigid movement would not be prevented, and no solution is possible (see Section 1.3.2).

$$Kd = f$$

$$K = \sum_{e=1}^{5} K_e$$

$$f = \sum_{e=1}^{5} f_e$$

$$\begin{bmatrix} \dfrac{EA}{l_e} & -\dfrac{EA}{l_e} & \cdot & \cdot & \cdot & \cdot \\ -\dfrac{EA}{l_e} & \dfrac{2EA}{l_e} & -\dfrac{EA}{l_e} & & & \\ \cdot & -\dfrac{EA}{l_e} & \dfrac{2EA}{l_e} & -\dfrac{EA}{l_e} & & \\ \cdot & & -\dfrac{EA}{l_e} & \dfrac{2EA}{l_e} & -\dfrac{EA}{l_e} & \\ & & & -\dfrac{EA}{l_e} & \dfrac{2EA}{l_e} & -\dfrac{EA}{l_e} \\ \cdot & & & & -\dfrac{EA}{l_e} & \dfrac{EA}{l_e} \end{bmatrix} \begin{bmatrix} d_1 \\ d_2 \\ d_3 \\ d_4 \\ d_5 \\ d_6 \end{bmatrix} = \begin{bmatrix} \dfrac{Abl_e}{2} \\ Abl_e \\ Abl_e \\ Abl_e \\ Abl_e \\ \dfrac{Abl_e}{2} + At \end{bmatrix}$$

(1.28)

$$
\begin{bmatrix}
\dfrac{2EA}{l_e} & -\dfrac{EA}{l_e} & \cdot & & \cdot \\
-\dfrac{EA}{l_e} & \dfrac{2EA}{l_e} & -\dfrac{EA}{l_e} & & \\
\cdot & -\dfrac{EA}{l_e} & \dfrac{2EA}{l_e} & -\dfrac{EA}{l_e} & \\
& & -\dfrac{EA}{l_e} & \dfrac{2EA}{l_e} & -\dfrac{EA}{l_e} \\
\cdot & & & -\dfrac{EA}{l_e} & \dfrac{EA}{l_e} \\
\cdot & & & &
\end{bmatrix}
\begin{bmatrix}
d_2 \\ d_3 \\ d_4 \\ d_5 \\ d_6
\end{bmatrix}
=
\begin{bmatrix}
Abl_e \\ Abl_e \\ Abl_e \\ Abl_e \\ \dfrac{Abl_e}{2} + At
\end{bmatrix}
$$

$$(1.29)$$

Finally, according to Eq. (1.15), the nodal displacements are obtained in Eq. (1.30).

$$
\mathbf{d} = \mathbf{K}^{-1}\mathbf{f} =
\begin{bmatrix}
d_2 \\ d_3 \\ d_4 \\ d_5 \\ d_6
\end{bmatrix}
=
\begin{bmatrix}
-5.5 \\ -8.0 \\ -7.5 \\ -4.0 \\ +2.5
\end{bmatrix}
\times 10^{-7}\,\mathrm{m}
$$

$$(1.30)$$

Strains and stresses within the elements can be calculated according to Eqs. (1.22) and (1.23). The comparisons between the closed-form (continuum) solution and the FEM (discrete) solution are shown in Figure 1.6. The comments on these results are provided in the following section.

1.3 GENERALIZATION

The strong form solution of the example discussed in the previous section was apparently easier than FEM. This would not be the case if additional features were considered. For instance, an additional axial point load in the central part of the column would have requested the study of two sets of differential equations (each set relative to the single portion of the column defined by the point load) imposing mutual boundary conditions. In turn, with FEM, the assemblage of the equilibrium equations follows a systematic procedure that can be easily implemented in software codes, and the additional point load would only change the vector of external forces locally. The stiffness matrix depends exclusively on the geometrical and mechanical parameters of the discrete model, and any additional load is simply lumped in the nodal force vector.

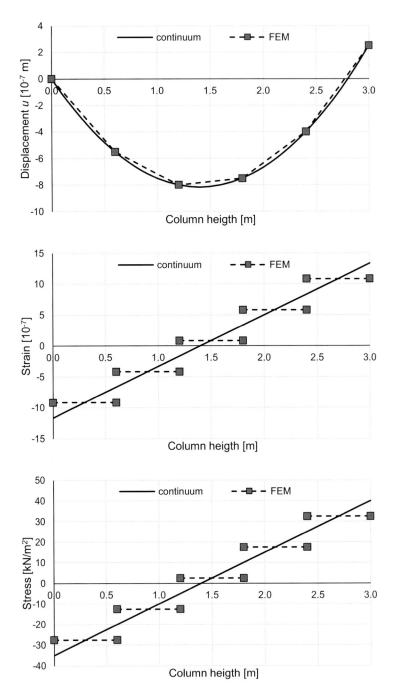

Figure 1.6 Axially loaded column: comparison between closed-form and FEM solutions.

In this regard, the shape functions play a crucial role in approximating the real behaviour of the structure to be analysed. Although the solution of the continuum problem suggested a quadratic variation of u along the column height, the adoption of linear shape functions resulted accurate in describing the displacement field exactly at the nodes (upper diagram of Figure 1.6). Expectedly, the finer is the discretization (i.e. the larger the number of DOFs), the more precise is the approximation. However, the first derivative of linear displacements leads to constant strains and stresses within each element (thus, piecewise along the column), evidently not real, see central and lower diagrams of Figure 1.6.

In this regard, it is worth noticing that, according to the integral form of the governing equations, equilibrium and compatibility are guaranteed for the element as a whole but not locally, as given by Eq. (1.6). The stress at the upper surface of the column is not in agreement with the external traction t, and it is lower (in absolute value) than expected at the bottom of the column. Stress and strain distributions are discontinuous (with jumps at each node), and their approximation is accurate only in the central part of the element. Also in this case, a finer discretization (or averaging the values at the nodes) would reduce these drawbacks.

1.3.1 Convergence criteria on shape functions

Being FEM analysis a numerical approximation, it is important to focus on two important concepts, namely, *accuracy* and *convergence*. As shown in the previous section, shape functions play the most important role because, through mesh discretization, they limit the infinite DOFs of the continuum body to a finite number. An appropriate choice of the interpolation functions can lead the numerical solution to approach the exact one. In turn, despite the fineness of the discretization, improper shape functions might prevent such convergence to be reached.

Limiting the discussion to linear elastic problems, for the numerical solution to monotonically converge to the exact solution as the number of finite elements increases (so-called *mesh refinement*), the elements must be *complete*, and the elements and mesh must be *compatible*. The *completeness* requirement is fulfilled when a shape function is able to represent the rigid body motion and a constant strain state. This means that straining of an element does not occur when the nodal displacements are caused by a rigid body motion; in turn, if nodal displacements are compatible with a constant strain condition, such constant strain will in fact be obtained. The requirement of *compatibility* means that the displacements within the elements and across the element boundaries must be continuous up to a certain derivative order. For instance, if the equilibrium equations (like the ones seen earlier) contain only the values of $\mathbf{N}$ and its first derivatives, $\mathbf{N}$ has to be C^0 continuous (i.e. the first derivatives exist but they are not continuous). If second derivatives are involved, C^1 continuity is requested (i.e. its first derivatives are continuous). This requirement guarantees that strains are finite (even though they may be discontinuous across neighbouring elements) and that the integrals of the weak form of equilibrium equations can be evaluated.

With respect to the compatibility of a finite element assemblage, the geometry and the displacements of edges/faces of adjacent elements must be the same. The geometric aspect is guaranteed if the elements share the same nodes, while the displacement compatibility along the shared edge or face is ensured if the displacements are described by equal shape functions. In addition to the lack of compatibility, adjacent elements not interconnected make the structure lose its continuity, affecting the correctness of the numerical solution of the entire problem.

Once convergence is assured, it is important to evaluate the adequacy of the numerical approximation and how this approximation can be improved, if required. For simple problems, the term of comparison can be deduced from theoretical models such as Bernoulli–Euler or Timoshenko beam. It is worth noticing that, even if this solution can be considered the *exact* solution, the word "exact" is referred to the mathematical/mechanical model adopted, which is still an approximation of the reality. Looking at the example of the axially loaded column in Section 1.2.2, the numerical solution could be improved by: (1) increasing the polynomial order p of the shape function (*p-convergence*) and (2) reducing the general dimension h (or size) of the elements (*h-convergence*). According to the former, as a quadratic distribution of displacement is expected, a unique finite element with shape function of order two or higher would be sufficient to model the entire column. The latter can be demonstrated according to Taylor's expansion of a function as an infinite sum of terms calculated from the values of the function and its derivatives at a single point/node i. Eq. (1.31) shows the cases of a two-dimensional problem in xy plane.

$$
\mathbf{u} = \mathbf{u}_i + \left(\frac{\partial \mathbf{u}}{\partial x}\right)_i (x - x_i) + \left(\frac{\partial \mathbf{u}}{\partial y}\right)_i (y - y_i) +
$$

$$
+ \frac{1}{2}\left[\left(\frac{\partial^2 \mathbf{u}}{\partial x^2}\right)_i (x - x_i)^2 + 2\left(\frac{\partial^2 \mathbf{u}}{\partial x \partial y}\right)_i (x - x_i)(y - y_i) + \left(\frac{\partial^2 \mathbf{u}}{\partial y^2}\right)_i (y - y_i)^2\right] + \dots
$$

$$
(1.31)
$$

If within an element of size h a polynomial expansion of degree p is employed, being $x - x_i$ and $y - y_i$ of the order of h, the error (i.e. the divergence) of $\mathbf{u}$ with respect to the exact solution is of the order $O(h^{p+1})$. According to the axially loaded column in Section 1.2.2, a linear expansion means $p = 1$, thus an error of $O(h^2)$ is found. This means that the error in displacement is reduced to ¼ when halving the element size. Analogously, if the strain is the mth derivative of the displacement, its error is of the order $O(h^{p+1-m})$, or

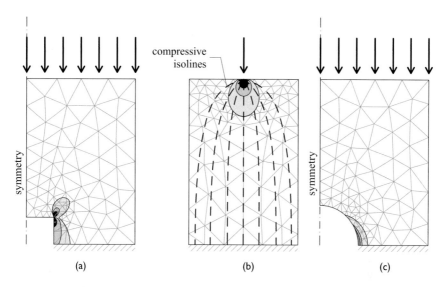

Figure 1.7 Stress concentration given by contour levels in different meshes. Impossible convergence with mesh refinement for (a) sharp re-entrant corners and (b) point loads vs. (c) full convergence with mesh refinement (semi-circular hole).

$O(h)$ for the mentioned example. This means that halving the element size reduces ½ the error in the strains. These results can be intuitively grasped by looking at Figure 1.6. It must be stressed that the convergence issues discussed in this section are not associated to any source of nonlinearity, for which the reader is referred to Chapter 2.

If the problem involves stress and strain singularities, convergence in stress and strain at such locations may be impossible, i.e. they do not converge towards a specific value. Upon mesh refinement, stress (and strain) at this point keeps increasing tending to infinite. This usually occurs when the mesh is characterized by concentrated forces or other load constraints, sharp edges, openings, cut-outs, sudden changes on cross-sectional areas or other geometrical constraints, and other features such as an abrupt change of material properties or crack tips. As an example, the stress in a sharp re-entrant corner or next to a point load cannot be defined, see Figure 1.7a and b. In a simplistic way, the stress for the elements around the sharp corner or under the point load can be evaluated as load over area. As a consequence, the smaller is the finite element, the larger is the stress calculated: upon mesh refinement, the convergence to a unique value is simply impossible, because stress is locally undefined and depends on the element size. In these cases, it is possible to recall the Saint-Venant's Principle, according to which the difference between the effects of two different but statically equivalent loads becomes very small at sufficiently large distances from the load. This means that the stress at the bottom of the element in Figure 1.7b can be evaluated as the point load divided by the area, regardless of the singularity at the upper edge. In other words, the stress inaccuracy under the point load is not crucial for assessing the behaviour of the structure at a sufficient distance from it, even if local failure can be incorrectly assessed in the case of nonlinear analysis.

This type of singularities is often not much relevant as reality overcomes the discrete model. It is rather evident that, in real life, no applied force can be considered a point load as there is always some contact area. Likewise, sharp corners tend to have a small radius, or a small crack will appear in the edge, smoothing the stress distribution shown in Figure 1.7a. For a similar plate with a nonsharp edge hole as in Figure 1.7c, the stress can be calculated theoretically according to a stress concentration factor, and FEM will provide the correct solution upon mesh refinement. Additionally, optimal meshes include a finer discretization in regions where high gradients of stress (and strain) are expected and a coarser discretization elsewhere. Typical examples are discussed in Chapter 5, to which the reader is referred.

If discretization and interpolation are the most evident sources of discrepancy between exact and numerical solution, other errors are possible, and the analyst must carefully interpret FEM outcomes. Besides obvious mistakes, such as mistyping geometrical or mechanical characteristics, or incorrect definition of boundary and loading conditions, other common sources of errors are due to: (1) *round-off processes* in calculations (a reduction of accuracy occurs every time a number is rounded off to a finite number of digits, even when using a double-precision floating-point format or higher); (2) *iterative procedures* for the solution of equilibrium equations (the threshold for measuring convergence is usually small but it is not zero, see Chapter 2); (3) evaluation of *nonlinear constitutive relations* (according to constitutive models) and their subsequent linearization and integration (Chapter 3). Whereas the first aspect is generally shared between linear and nonlinear analyses, the last two aspects are typical of the latter.

1.3.2 Matrix formulation

To consolidate the concepts described hitherto, and to show the automatism of FEM, a general example will be discussed considering the structure schematized in Figure 1.8a. The first step regards the labelling of the elements and of the associated nodes (Figure 1.8b). Once geometrical and mechanical data are set, together with the load cases, it is possible to assemble the stiffness matrix as well as the vector of nodal forces. In particular, the stiffness matrix is square of the same order of the DOFs of the system.

For the sake of simplicity, let us consider only horizontal displacements, i.e. eight DOFs totally (i.e. u_1, ..., u_8). As shown in Eqs. (1.26) and (1.27), the stiffness matrix of a single element has nonzero coefficients only in correspondence of its nodes, being zero elsewhere. Considering the first three elements, the nonzero coefficients are placed in the marked cells of the matrices in Figure 1.9. In particular, the small square cells are the ones that are shared among various elements (such as K_{44}, K_{45}, K_{54} and K_{55} between $\mathbf{K}_1$ and $\mathbf{K}_3$).

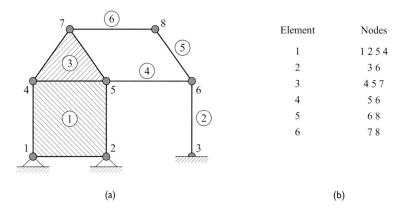

Element	Nodes
1	1 2 5 4
2	3 6
3	4 5 7
4	5 6
5	6 8
6	7 8

(a)   (b)

Figure 1.8 Finite element discretization of a general structure. (a) Element assemblage (mesh) and (b) element definition.

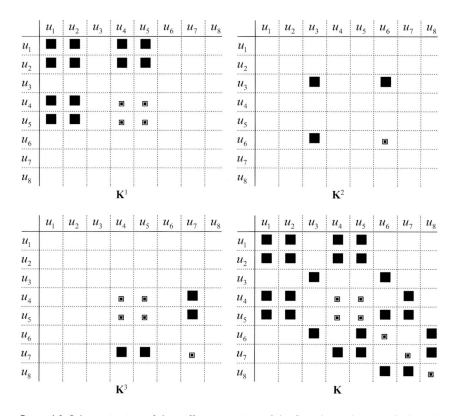

Figure 1.9 Schematic view of the stiffness matrices of the first three elements (indicated by the superscript) of Figure 1.8 and of the system as a whole (small square cells are shared between different elements).

As stated earlier, provided the coefficients are established consistently in the same coordinate system, the final matrix stiffness **K** is a result of the superposition of the single contributions. For the sake of intelligibility, DOFs are indicated in the first row and the first column, so that the compiling process may result easier. However, the DOFs are not a part of the stiffness matrix. The vector of nodal forces can be compiled accordingly.

The same reasoning adopted for the horizontal displacement can be followed in the case each node is provided with additional DOFs. The schematic view presented in Figure 1.9 still holds, but the cells become submatrices 2×2 (two translations for two-dimensional membrane elements) up to 6×6 (three translations and three rotations for three-dimensional beam or shell elements), according to the number of DOFs per node.

As it is possible to observe, the nonnull coefficients of **K** are stored in a diagonal *band* whose horizontal dimension can be calculated a priori. This is a direct consequence of the numbering of the nodes: in Figure 1.8a lower nodes have less and less connections with higher nodes (e.g. nodes 1, 2 and 3 are not connected with nodes 7 and 8). From the computational point of view, this band matrix configuration makes the numerical solution faster and the requirements for data storage lower. In the adopted example, better numbering would be possible to reduce the band of the stiffness matrix, as shown in Figure 1.10. Usually, FEM codes have built-in automatic routines to minimize the bandwidth without any necessary action by the user.

The matrix formulation is rather intuitive also in the case some nodal displacements are assigned according to prescribed boundary conditions. They affect the response of the entire structure but they are not among the unknowns of the analysis. Considering the difference between free and constrained boundaries as seen in Eqs. (1.1) and (1.2), it is possible to disassemble the equilibrium equations and write Eq. (1.32). In particular, the

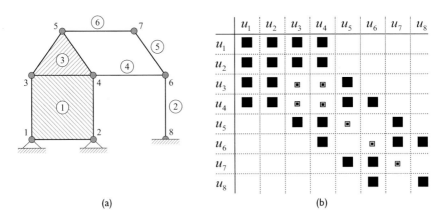

(a)   (b)

Figure 1.10 Finite element discretization of a general structure. (a) New numbering and (b) relative stiffness matrix of the system (small square cells are shared between different elements).

subscripts t and u indicate the submatrices and subvectors related to free nodes and nodes with constrained displacements, respectively.

According to Eq. (1.32) and to the rules for matrix multiplication, two sets of equations can be written in Eq. (1.33). Considering the goal of the analysis, only the free nodes submatrix is of interest and the second product of the first term represents the nodal forces on the entire structure due to the assigned displacements. Finally, the equilibrium equations are given in Eq. (1.34), and it is clear that the submatrix $\mathbf{K}_{uu}$ is not necessary to obtain the solution.

$$
\begin{bmatrix}
\mathbf{K}_{tt} & \mathbf{K}_{tu} \\
\mathbf{K}_{ut} & \mathbf{K}_{uu}
\end{bmatrix}
\begin{bmatrix}
\mathbf{d}_t \\
\mathbf{d}_u
\end{bmatrix}
=
\begin{bmatrix}
\mathbf{f}_t \\
\mathbf{f}_u
\end{bmatrix}
\tag{1.32}
$$

$$
\mathbf{K}_{tt}\mathbf{d}_t + \mathbf{K}_{tu}\mathbf{d}_u = \mathbf{f}_t
$$
$$
\mathbf{K}_{ut}\mathbf{d}_t + \mathbf{K}_{uu}\mathbf{d}_u = \mathbf{f}_u
\tag{1.33}
$$

$$
\mathbf{K}_{tt}\mathbf{d}_t = \mathbf{f}_t - \mathbf{K}_{tu}\mathbf{d}_u
\tag{1.34}
$$

Further examination of the global matrix $\mathbf{K}$ allows stating that the same is symmetric, usually sparse (i.e. with many zero elements) and banded (i.e. the nonzero elements are concentrated around the diagonal when the nodes are adequately sorted). In nonlinear problems, the tangent stiffness matrix (i.e. the one providing a relation between incremental changes in nodal displacements and incremental changes in nodal forces) may be nonsymmetric due to material nonlinearity (e.g. nonassociated plasticity) or geometrical nonlinearity. In a well-posed problem, the matrix $\mathbf{K}$ is positive definite, as defined in Eq. (1.35), but this will not be the case if (see Figure 1.11): (a) the structure or some part of it is not sufficiently constrained to prevent

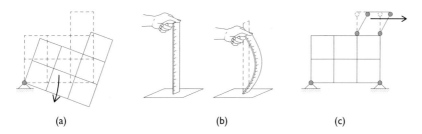

(a)   (b)   (c)

Figure 1.11 Possible causes for a nonpositive definite stiffness matrix $\mathbf{K}$. (a) Mechanism or rigid movement not prevented; degrees of freedom with zero or negative stiffness due to (b) buckling or (c) yielding of the material.

a rigid body motion, so that a mechanism is possible; the structural model develops DOFs with zero or negative stiffness as a result of (b) geometrical (e.g. buckling) or (c) material nonlinear effects (e.g. cracking or yielding), so that a portion of the structure or the full structure become unstable.

$$\mathbf{d}^T \mathbf{K} \mathbf{d} > 0 \quad \forall \mathbf{d} \neq 0 \tag{1.35}$$

As stated above, a simple and direct way to obtain a solution for the unknown nodal displacements $\mathbf{d}$ to a given set of nodal forces $\mathbf{f}$ would be to invert the stiffness matrix $\mathbf{K}$, as in Eq. (1.15). However, an inverse is almost never calculated due to the computational cost. In turn, the solution of the set of linear equations is sought by means of *direct* or *indirect/iterative* methods. Without entering into the merits, direct methods are based on a number of steps and operations that are predetermined. They include Gaussian elimination or LU, Cholesky and QR decompositions, whose effectiveness is enhanced by the peculiar characteristics of the stiffness matrix, such as symmetry and positive definiteness. However, due to the large amount of memory storage and computer time requested, direct methods are generally not recommended for large finite element systems.

It must be noted that the definition of large problems is not univocal, mainly depending on the advances in computer architecture. For instance, a few hundreds of unknowns were the maximum a computer could handle in the 1970s. Nowadays, the concept of large is often associate to problems with more than a million of unknowns, such as in many engineering three-dimensional problems. In these cases, iterative methods can be more effective. The common idea in iterative methods is that the solution is approached gradually, rather than in one large computational step. Conceptually, the equilibrium equations of linear problem are solved by means of a sequence of approximations to the solution vector $\mathbf{d}$ via a recursive operation. As expected, each iteration is achieved with a limited computational effort. Several methods (e.g. Jacobi, conjugate gradient and generalized minimal residual methods) have been developed in the last decades to optimize and speedup the solution procedure, but they are outside the scope of this book.

1.3.3 Example of a beam subjected to bending

Let us consider the beam constrained as in Figure 1.12a, with a rectangular cross section of $0.3 \times 0.5 \, \text{m}^2$ and a span of $l = 6 \, \text{m}$. The beam is subjected to a body force (specific weight) $b = 25 \, \text{kN/m}^3$, an external surface traction $t = 133 \, \text{kN/m}^2$ on the upper face (both downward) and an imposed rotation on the right-hand extremity $\theta_4 = -0.001 \, \text{rad}$. The finite element (FE) discretization is shown in Figure 1.12b where the beam is subdivided into three equal elements with length $l_e = 2 \, \text{m}$ and four nodes. Being a two-dimensional problem, each node is assumed to have three

DOFs, namely, two translations u and w (along x and z axes, respectively), and one rotation θ (within the zx plane, about y axis). The vertical displacement w is also called *deflection*. Considering the problem in the framework of small displacements and in the absence of axial loads, the longitudinal displacements u can be neglected. Consequently, the problem has eight DOFs in total.

The shape function is chosen as a polynomial of third order that, despite the high order of approximation, does not match with the Bernoulli–Euler beam theory. As shown below, considering a constant distributed load along the beam development, a polynomial of fourth order would be needed. Additionally, according to the current reference system, the first derivative of the displacement field is equal to the opposite of the rotation or *slope* (Figure 1.12c), i.e. $w' = dw/dx = -\theta$. In the following, the symbol w' will be preferred to indicate the rotational unknowns. The only components of strain and stress are ε_x and σ_x, respectively. Conversely to the axially loaded column in Section 1.2.2, stress and strain are not constant within the cross section, as they vary along the height of the beam.

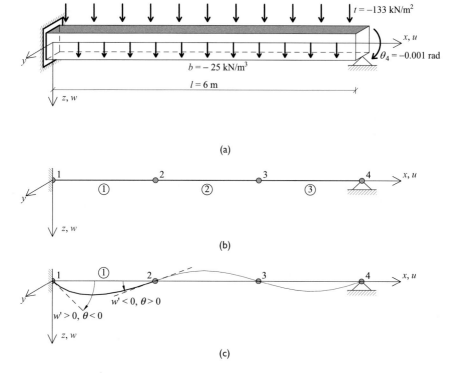

Figure 1.12 Bending beam. (a) Problem overview, (b) FEM discretization and (c) relationship between the derivative of the deflection (w') and the rotation (θ).

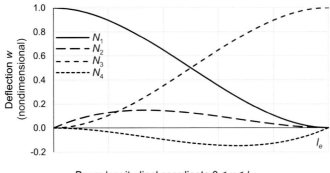

Figure 1.13 Shape functions for a beam subjected to bending.

Considering the first element, $0 \leq x \leq l_e$, where $x = 0$ and $x = l_e$ correspond to nodes 1 and 2, respectively. The shape functions for the analysed case are shown in Eq. (1.36) and drawn in Figure 1.13, where N_1, N_2, N_3 and N_4 are the components of the first line of $\mathbf{N}_1$ of Eq. (1.36). The second line of $\mathbf{N}_1$ includes the derivative of its first line. According to Eqs. (1.8) and (1.9), strains and stresses are derived in Eqs. (1.37) and (1.38), respectively.

$$\mathbf{u}_1 = \begin{bmatrix} w \\ w' \end{bmatrix} \approx \mathbf{N}_1 \mathbf{d}_1$$

$$\mathbf{N}_1 = \begin{bmatrix} \frac{1}{l_e^3}\left(2x^3 - 3x^2 l_e + l_e^3\right) & \frac{1}{l_e^2}\left(x^3 - 2x^2 l_e + x l_e^2\right) & \frac{1}{l_e^3}\left(-2x^3 + 3x^2 l_e\right) & \frac{1}{l_e^2}\left(x^3 - x^2 l_e\right) \\ \frac{1}{l_e^3}\left(6x^2 - 6x l_e\right) & \frac{1}{l_e^2}\left(3x^2 - 4x l_e + l_e^2\right) & \frac{1}{l_e^3}\left(-6x^2 + 6x l_e\right) & \frac{1}{l_e^2}\left(3x^2 - 2x l_e\right) \end{bmatrix}$$

$$\mathbf{d}_1^T = \begin{bmatrix} w_1 & w'_1 & w_2 & w'_2 \end{bmatrix} = \begin{bmatrix} d_1 & d_2 & d_3 & d_4 \end{bmatrix} \qquad (1.36)$$

$$\varepsilon_1 \equiv \varepsilon_x = -z\frac{d^2 w}{dx^2} = \begin{bmatrix} -z\frac{d^2}{dx^2} & 0 \end{bmatrix} \begin{bmatrix} w \\ w' \end{bmatrix} \approx \begin{bmatrix} -z\frac{d^2}{dx^2} & 0 \end{bmatrix} \mathbf{N}_1 \mathbf{d}_1 = \mathbf{B}_1 \mathbf{d}_1$$

$$\mathbf{B}_1 = -\frac{z}{l_e^3}\begin{bmatrix} 12x - 6l_e & 6x l_e - 4l_e^2 & -12x + 6l_e & 6x l_e - 2l_e^2 \end{bmatrix} \qquad (1.37)$$

$$\sigma_1 \equiv \sigma_x = E\varepsilon_x \approx E\mathbf{B}_1 \mathbf{d}_1 \qquad (1.38)$$

The Principle of Virtual Work can be exploited according to Eqs. (1.12) and (1.13). Conversely to the example of Section 1.2.2, the mechanical parameters are not constant along the height of the beam, and the integral over the volume can be written as the double integral over the cross-section A and the longitudinal direction x. The stiffness matrix of the first element is calculated in Eq. (1.39) where the moment of inertia I about the y axis is highlighted. The vector of nodal forces and the equilibrium equations of the first element are shown in Eqs. (1.40) and (1.41), respectively.

$$\mathbf{K}_1 = \int_V \mathbf{B}_1{}^T \mathbf{D}_1 \mathbf{B}_1 \, dV =$$

$$\int_0^{l_e} \frac{E}{l_e{}^6} \underbrace{\left(\int_A z^2 \, dA \right)}_{\text{Moment of inertia } I} \begin{bmatrix} 12x - 6l_e \\ 6xl_e - 4l_e{}^2 \\ -12x + 6l_e \\ 6xl_e - 2l_e{}^2 \end{bmatrix} \begin{bmatrix} 12x - 6l_e \\ 6xl_e - 4l_e{}^2 \\ -12x + 6l_e \\ 6xl_e - 2l_e{}^2 \end{bmatrix}^T dx =$$

$$\frac{EI}{l_e{}^3} \begin{bmatrix} 12 & 6l_e & -12 & 6l_e \\ 6l_e & 4l_e{}^2 & -6l_e & 2l_e{}^2 \\ -12 & -6l_e & 12 & -6l_e \\ 6l_e & 2l_e{}^2 & -6l_e & 4l_e{}^2 \end{bmatrix} \tag{1.39}$$

$$\mathbf{f}_1 = \int_V \mathbf{N}_1{}^T \mathbf{b}_1 \, dV + \int_{S_u} \mathbf{N}_1{}^T \mathbf{t}_1 \, dS = A \int_0^{l_e} \mathbf{N}_1{}^T \begin{bmatrix} b \\ 0 \end{bmatrix} dx + \int_0^{l_e} \mathbf{N}_1{}^T \begin{bmatrix} t \\ 0 \end{bmatrix} dx =$$

$$A \begin{bmatrix} \dfrac{bl_e}{2} \\[8pt] \dfrac{bl_e{}^2}{12} \\[8pt] \dfrac{bl_e}{2} \\[8pt] -\dfrac{bl_e{}^2}{12} \end{bmatrix} + \begin{bmatrix} \dfrac{tl_e}{2} \\[8pt] \dfrac{tl_e{}^2}{12} \\[8pt] \dfrac{tl_e}{2} \\[8pt] -\dfrac{tl_e{}^2}{12} \end{bmatrix} = (Ab + t) \begin{bmatrix} \dfrac{l_e}{2} \\[8pt] \dfrac{l_e{}^2}{12} \\[8pt] \dfrac{l_e}{2} \\[8pt] -\dfrac{l_e{}^2}{12} \end{bmatrix} \tag{1.40}$$

$$\mathbf{K}_1 \mathbf{d}_1 = \mathbf{f}_1 \rightarrow \frac{EI}{l_e{}^3} \begin{bmatrix} 12 & 6l_e & -12 & 6l_e \\ 6l_e & 4l_e{}^2 & -6l_e & 2l_e{}^2 \\ -12 & -6l_e & 12 & -6l_e \\ 6l_e & 2l_e{}^2 & -6l_e & 4l_e{}^2 \end{bmatrix} \begin{bmatrix} d_1 \\ d_2 \\ d_3 \\ d_4 \end{bmatrix} = (Ab + t) \begin{bmatrix} \dfrac{l_e}{2} \\[8pt] \dfrac{l_e{}^2}{12} \\[8pt] \dfrac{l_e}{2} \\[8pt] -\dfrac{l_e{}^2}{12} \end{bmatrix}$$

$$\tag{1.41}$$

Analogously to Eq. (1.28), the equilibrium equations of the entire beam are combined in matrix form in Eq. (1.42). Introducing the boundary conditions ($w_1 = d_1 = 0$, $w'_1 = -\theta_1 = d_2 = 0$ and $w_4 = d_7 = 0$), the equilibrium equations can be simplified by erasing the information about these DOFs, as in Eq. (1.43). The matrices are disassembled according to Eq. (1.32) to account for the imposed displacement $w'_4 = -\theta_4 = 0.001\,\text{rad}$. The result is shown in Eq. (1.44), in line with Eq. (1.34).

$$\mathbf{K}\mathbf{d} = \mathbf{f}$$

$$\mathbf{K} = \sum_{e=1}^{3} \mathbf{K}_e$$

$$\mathbf{d}^T = \begin{bmatrix} w_1 & w'_1 & w_2 & w'_2 & w_3 & w'_3 & w_4 & w'_4 \end{bmatrix} = \qquad (1.42)$$

$$\begin{bmatrix} d_1 & d_2 & d_3 & d_4 & d_5 & d_6 & d_7 & d_8 \end{bmatrix}$$

$$\mathbf{f} = \sum_{e=1}^{3} \mathbf{f}_e$$

$$\frac{EI}{l_e^3}\begin{bmatrix} 24 & 0 & -12 & 6l_e & \vdots & \cdot \\ & 8l_e^2 & -6l_e & 2l_e^2 & \vdots & \cdot \\ & & 24 & 0 & \vdots & 6l_e \\ & \text{Symm} & & 8l_e^2 & \vdots & 2l_e^2 \\ \hdashline & & & & \vdots & 4l_e^2 \end{bmatrix}\begin{bmatrix} d_3 \\ d_4 \\ d_5 \\ d_6 \\ \hdashline -\theta_4 \end{bmatrix} = (Ab+t)\begin{bmatrix} l_e \\ 0 \\ l_e \\ 0 \\ \hdashline -\dfrac{l_e^2}{12} \end{bmatrix}$$

$$(1.43)$$

$$\frac{EI}{l_e^3}\begin{bmatrix} 24 & 0 & -12 & 6l_e \\ & 8l_e^2 & -6l_e & 2l_e^2 \\ & & 24 & 0 \\ & \text{Symm} & & 8l_e^2 \end{bmatrix}\begin{bmatrix} d_3 \\ d_4 \\ d_5 \\ d_6 \end{bmatrix} = (Ab+t)\begin{bmatrix} l_e \\ 0 \\ l_e \\ 0 \end{bmatrix} + \frac{EI}{l_e^3}\theta_4\begin{bmatrix} 0 \\ 0 \\ 6l_e \\ 2l_e^2 \end{bmatrix}$$

$$(1.44)$$

Finally, according to Eq. (1.15), the unknown displacements are calculated in Eq. (1.45). The comparison between the closed-form (continuum) solution, obtained by integration of the beam differential equation (not derived in the present text), which is shown in Eq. (1.46), and the FEM solution

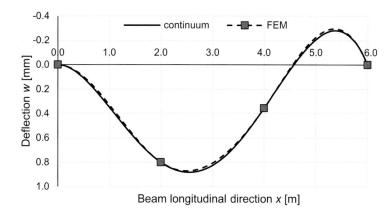

Figure 1.14 Bending beam: comparison in terms of deflection between closed-form and FEM solutions.

is displayed in Figure 1.14. The correspondence between continuum and FEM solution is *almost* exact. Despite the perfect match in terms of nodal displacements, the FE model does not approximate exactly the displacement field within the finite elements due to the adopted shape functions. The difference in terms of polynomial orders (third and fourth for FEM and continuum solutions, respectively) is evident once the displacement field is calculated.

$$\mathbf{d} = \mathbf{K}^{-1}\mathbf{f} = \begin{bmatrix} d_3 \\ d_4 \\ d_5 \\ d_6 \end{bmatrix} = \begin{bmatrix} 8.0 \text{ m} \\ 2.9 \text{ rad} \\ 3.6 \text{ m} \\ -6.2 \text{ rad} \end{bmatrix} 10^{-4} \qquad (1.45)$$

$$w = \frac{q}{24EI}x^4 + \left(\frac{\theta}{l^2} - \frac{ql}{12EI}\right)x^3 + \left(\frac{ql^2}{24EI} - \frac{\theta}{l}\right)x^2$$

$$= \frac{1.94 \times 10^{-5}}{\text{m}^3}x^4 - \frac{2.06 \times 10^{-4}}{\text{m}^2}x^3 + \frac{5.33 \times 10^{-4}}{\text{m}}x^2 \qquad (1.46)$$

Additionally, recalling the boundary conditions, strains and stresses within the elements can be calculated according to Eqs. (1.37) and (1.38). As clearly noticeable, they are not constant within the cross section because they depend on the z coordinate. This aspect is stressed in Section 1.4.1 where *curvature* and *moment*, the so-called *generalized strain* and *generalized*

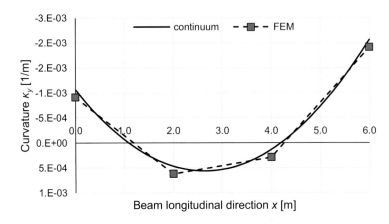

Figure 1.15 Bending beam: comparison in terms of curvature κ_y between closed-form and FEM solutions.

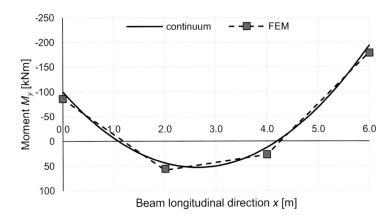

Figure 1.16 Bending beam: comparison in terms of moment M_y between closed-form and FEM solutions.

stress, are introduced. For the sake of completeness, the curvature and moment evolution along the beam are shown in Figures 1.15 and 1.16, respectively. As explained below in Section 1.4.4, curvature κ_y and moment M_y match perfectly the closed-form solution in two particular points within the elements, but not at their ends.

1.4 SHAPE FUNCTIONS AND NUMERICAL INTEGRATION

In the present section, the main features of typical finite elements are discussed. One-, two- and three-dimensional elements are named line, plane and solid elements, respectively.

1.4.1 Hermitian beam elements

The examples described in Sections 1.2.2 (axially loaded column) and 1.3.3 (beam subjected to bending) introduced the fundamentals of the line beam element (based on the Bernoulli–Euler beam theory). According to Figure 1.17, if the beam local system is collinear with the axes x and z, the mechanical behaviour of the element in the xz plane is fully described by means of the equations in Table 1.1.

Based on the continuum model, bending and axial behaviour are decoupled, and they can be considered as such also in finite element analysis. In this regard, it is preferable to introduce two *generalized stresses*, namely, N_x and M_y (axial force and bending moment, respectively), and two *generalized strains*, namely ε_x and κ_y (longitudinal elongation due to axial action and curvature due to bending moment, respectively). Consequently, rather than stresses and strains, the constitutive relations apply to *generalized stresses and strains*. Besides the engineering meaning of these quantities, the integration over the beam height shown in Eq. (1.39) is now avoided, as it is already included in the constitutive relationship. This approach is correct as long as the mechanical properties are constant throughout the element, and linear elasticity is adopted.

As it is possible to observe in Table 1.1, the generalized strain κ associated with bending involves the second derivative of the field variable w.

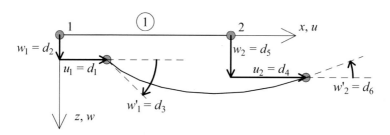

Figure 1.17 Beam element in the plane xz: indication and geometrical description of the nodal displacements.

Table 1.1 Kinematic and static properties of a beam element

Kinematics	Statics	Relationships
$\mathbf{u} = \begin{bmatrix} u & w & w' \end{bmatrix}^T$	$\mathbf{f} = \begin{bmatrix} f_x & f_z & m_y \end{bmatrix}^T$	$\mathbf{Ku} = \mathbf{f}$ (see Eqs. (1.25) and (1.39))
$\varepsilon = \begin{bmatrix} \varepsilon_x & \kappa_y \end{bmatrix}^T$	$\sigma = \begin{bmatrix} N_x & M_y \end{bmatrix}^T$	$\sigma = \mathbf{D}\varepsilon$
$\varepsilon_x = \dfrac{du}{dx}$	$N_x = \displaystyle\int_A \sigma_x \, dA$	$\mathbf{D} = \begin{bmatrix} EA & 0 \\ 0 & EI \end{bmatrix}$
$\kappa_y = -\dfrac{d^2 w}{dx^2} = -w''$	$M_y = \displaystyle\int_A z\sigma_x \, dA$	

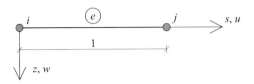

Figure 1.18 Nondimensional coordinate s in beam element.

Accordingly, C¹ continuity for the deflection is required, i.e. the *w* field and at least its first derivative must be continuous. The shape functions for the deflection must fulfil these requirements and interpolate the displacements within an element starting from its nodal displacements and rotations (first derivative of deflection). In this regard, it is possible to implement the *Hermite interpolation* (named after Charles Hermite, 1822–1901), which is a method of interpolating data points and derivatives with a polynomial function. In the case only the first derivative is considered (i.e. the slope), the Hermite polynomial is of the third order, resulting in a smooth continuous spline whose first derivative is still smooth (being a polynomial of the second order).

The resulting shape functions have already been introduced in Eq. (1.36). However, let us consider here a more rigorous approach and the longitudinal (nondimensional) coordinate *s*, where *s* = 0 and *s* = 1 are representative of node *i* and *j*, respectively (Figure 1.18). The first line of Eq. (1.47) describes the relations between *s* and *x* (the latter being the local coordinate of Figure 1.17). In particular, $s \in [0, 1]$ is a nondimensional value, and it is possible to generalize the following findings to beams of any length. The other lines of Eq. (1.47) provide the first and second derivatives of the deflection.

The deflection along the element is approximated by the Hermite interpolation polynomial function shown in Eq. (1.48); the equation illustrates also the additional simplifications once the nondimensional coordinate *s* is included. A closer inspection of Eq. (1.48) indicates that the deflection is a linear combination of the four *shape functions*, whose coefficients are the displacements *w* and the derivatives *w'* at the nodes *i* and *j*. The four shape functions were already shown in Figure 1.13, where it is evident that they impose a unit displacement/rotation of one node while keeping null all other quantities.

$$s = \frac{x}{l_e} \qquad\qquad \frac{ds}{dx} = \frac{1}{l_e} \qquad\qquad dx = l_e ds$$

$$w' = \frac{dw}{dx} = \frac{dw}{ds}\frac{ds}{dx} = \frac{1}{l_e}\frac{dw}{ds} \qquad\qquad \frac{dw}{ds} = l_e w' \qquad (1.47)$$

$$w'' = \frac{d^2 w}{dx^2} = \frac{d}{dx}\left(\frac{dw}{dx}\right) = \frac{d}{dx}\left(\frac{1}{l_e}\frac{dw}{ds}\right) = \frac{d}{ds}\frac{ds}{dx}\left(\frac{1}{l_e}\frac{dw}{ds}\right) = \frac{1}{l_e^2}\frac{d^2 w}{ds^2}$$

$$w \approx \left(2s^3 - 3s^2 + 1\right)w_i + \left(s^3 - 2s^2 + s\right)\frac{dw}{ds}\bigg|_i +$$

$$\left(-2s^3 + 3s^2\right)w_j + \left(s^3 - s^2\right)\frac{dw}{ds}\bigg|_j$$

thus   (1.48)

$$w \approx \left(2s^3 - 3s^2 + 1\right)w_i + \left(s^3 - 2s^2 + s\right)l_e w'_i +$$

$$\left(-2s^3 + 3s^2\right)w_j + \left(s^3 - s^2\right)l_e w'_j$$

Analogously, for the axial behaviour, the shape functions must be consistent with the nodal displacements, and the elongation u can be expressed through Eq. (1.49). According to Table 1.1, the approximation by linear functions is sufficient for C^0 continuity. The graphical representation of the shape functions of Eq. (1.49) was already shown in Figure 1.5.

$$u \approx \left(1 - s\right)u_i + \left(s\right)u_j$$ (1.49)

These findings can be expressed in matrix notation and, given the clear meaning, the subscript "e" is here avoided. For a single beam element between node i and j, Eq. (1.50) shows the vector of displacements $\mathbf{u}$, the vector of nodal displacements $\mathbf{d}$ and the matrix of shape functions $\mathbf{N}$. The generalized strains are calculated in Eq. (1.51), whereas generalized stresses have been defined in Table 1.1. According to Section 1.2, it is possible to derive the equilibrium equations for the single element. In particular, the stiffness matrix and the related calculations are given in Eq. (1.52).

$$\mathbf{u} = \begin{bmatrix} u & w \end{bmatrix}^T \approx \mathbf{Nd}$$

$$\mathbf{d} = \begin{bmatrix} u_i & w_i & w'_i & u_j & w_j & w'_j \end{bmatrix}^T$$

$$\mathbf{N} = \begin{bmatrix} 1-s & \cdot & \cdot & s & \cdot & \cdot \\ \cdot & 2s^3 - 3s^2 + 1 & \left(s^3 - 2s^2 + s\right)l_e & \cdot & -2s^3 + 3s^2 & \left(s^3 - s^2\right)l_e \end{bmatrix}$$

(1.50)

$$\boldsymbol{\varepsilon} = \begin{bmatrix} \varepsilon_x & \kappa \end{bmatrix}^T = \begin{bmatrix} \dfrac{du}{dx} & -\dfrac{d^2w}{dx^2} \end{bmatrix}^T = \begin{bmatrix} \dfrac{1}{l_e}\dfrac{du}{ds} & -\dfrac{1}{l_e^2}\dfrac{d^2w}{ds^2} \end{bmatrix}^T \approx \mathbf{B}\mathbf{d}$$

$$\mathbf{B} = \begin{bmatrix} -\dfrac{1}{l_e} & . & . & \dfrac{1}{l_e} & . & . \\[2mm] . & -\dfrac{12s-6}{l_e^2} & \dfrac{6s-4}{l_e} & . & \dfrac{12s-6}{l_e^2} & -\dfrac{6s-2}{l_e} \end{bmatrix} \tag{1.51}$$

$$\mathbf{K} = \int_V \mathbf{B}^T \mathbf{D}\mathbf{B}\, dV = \int_0^{l_e} \mathbf{B}^T \mathbf{D}\mathbf{B}\, dx = \int_0^1 \mathbf{B}^T \mathbf{D}\mathbf{B}\, l_e \, ds$$

$$\mathbf{K} = \begin{bmatrix}
\dfrac{EA}{l_e} & . & . & -\dfrac{EA}{l_e} & . & . \\[2mm]
 & \dfrac{12EI}{l_e^3} & \dfrac{6EI}{l_e^2} & . & -\dfrac{12EI}{l_e^3} & \dfrac{6EI}{l_e^2} \\[2mm]
 & & \dfrac{4EI}{l_e} & . & -\dfrac{6EI}{l_e^2} & \dfrac{2EI}{l_e} \\[2mm]
 & & & \dfrac{EA}{l_e} & . & . \\[2mm]
 & \text{Symm} & & & \dfrac{12EI}{l_e^3} & -\dfrac{6EI}{l_e^2} \\[2mm]
 & & & & & \dfrac{4EI}{l_e}
\end{bmatrix} \tag{1.52}$$

As a final remark, it is worth mentioning that the beam element has been considered in its local reference system xz. In the case of assembling the finite element matrices of a global structure, it is necessary to consider the global reference system $x_g\, z_g$. As it is possible to notice in Figure 1.19, nodal forces and DOFs are rotated by the angle ω and, intuitively, so is the stiffness matrix.

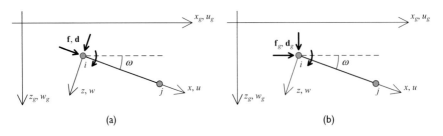

Figure 1.19 Nodal forces and DOFs in (a) local and (b) global reference system.

Knowing the nodal coordinates in the global reference system and the length of the element, it is possible to calculate the rotation (or transformation) matrix $\mathbf{T}$ in Eq. (1.53). According to Figure 1.19, the relation between nodal displacement and force vectors in global and local coordinate systems is shown in Eq. (1.54). Looking at Eq. (1.55), with due substitutions, premultiplying by $\mathbf{T}^T$ and reminding that for a rotation matrix $\mathbf{T}^T\mathbf{T} = \mathbf{I}$ (identity matrix), the equilibrium equations are written in Eq. (1.56) in terms of global coordinates. The matrix $\mathbf{K}_g$ is the stiffness matrix of the element according to the global coordinate system, i.e. this matrix describes the relationship between nodal displacements and elastic reactions of the beam in a system where these quantities are not collinear with the beam axes.

$$\cos\omega = \frac{x_{gj} - x_{gi}}{l_e} \qquad \sin\omega = \frac{y_{gj} - y_{gi}}{l_e}$$

$$\mathbf{T} = \begin{bmatrix} \cos\omega & \sin\omega & 0 & & & \\ -\sin\omega & \cos\omega & 0 & & 0 & \\ 0 & 0 & 1 & & & \\ & & & \cos\omega & \sin\omega & 0 \\ & 0 & & -\sin\omega & \cos\omega & 0 \\ & & & 0 & 0 & 1 \end{bmatrix} \tag{1.53}$$

$$\mathbf{d} = \mathbf{T}\mathbf{d}_g, \quad \mathbf{f} = \mathbf{T}\mathbf{f}_g \tag{1.54}$$

$$\mathbf{K}\mathbf{d} = \mathbf{f} \quad \rightarrow \quad \mathbf{K}\mathbf{T}\mathbf{d}_g = \mathbf{T}\mathbf{f}_g$$

$$\mathbf{T}^T\mathbf{K}\mathbf{T}\,\mathbf{d}_g = \mathbf{T}^T\mathbf{T}\mathbf{f}_g \tag{1.55}$$

$$\mathbf{T}^T\mathbf{K}\mathbf{T} = \mathbf{K}_g$$

$$\mathbf{K}_g\mathbf{d}_g = \mathbf{f}_g \tag{1.56}$$

1.4.2 Isoparametric elements

The previous section introduced the stiffness matrix of a (Bernoulli–Euler) beam element calculated in the local system, subsequently rotated to obtain the equilibrium equations in the global coordinate system. This process is straightforward for a beam (and truss) element but it may result complicated for plane elements, such as quadrilateral elements. A particular shape of given dimensions would request unique shape functions for each element, with a significant computational effort. It is possible to avoid this drawback by introducing a new and more effective type of elements, the so-called *isoparametric* elements.

Accordingly, the finite element matrices are written in the *natural coordinate system* and then transformed into a local coordinate system. The natural coordinate system is a new system whose variables range from −1 to +1 (where the positive symbol will be dropped for simplicity). As suggested by the name itself (i.e. same parameters), the transformation between natural and local coordinate systems is described by functions that coincide with the shape functions. Furthermore, the shape functions are independent from the dimensions and overall shape of the element, with an overall computational simplification. The isoparametric derivation of the stiffness matrix of a line element will make these aspects clearer.

Let us consider the bar element with the local coordinate x and a natural coordinate system $r \in [-1, 1]$ as shown in Figure 1.20. The relationship between x and r is derived in Eq. (1.57) as the central point plus the half bar length multiplied by r, rewritten in compact form in Eq. (1.58). In turn, the bar local displacements are expressed by Eq. (1.59). As it is possible to observe, both x and u share the same interpolation/shape functions in terms of r.

$$x = \frac{x_1 + x_2}{2} + \frac{x_2 - x_1}{2} r = \frac{1}{2}(1-r)x_1 + \frac{1}{2}(1+r)x_2 \tag{1.57}$$

$$x = \sum_{i=1}^{2} N_i x_i \tag{1.58}$$

$$N_1 = \frac{1}{2}(1-r), \quad N_2 = \frac{1}{2}(1+r)$$

$$u \approx \frac{1}{2}(1-r)u_1 + \frac{1}{2}(1+r)u_2$$

$$u \approx \sum_{i=1}^{2} N_i u_i \tag{1.59}$$

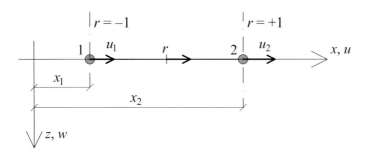

Figure 1.20 Bar element in local and natural coordinate systems.

To calculate the stiffness matrix, the strain ε is computed according to the chain rule as shown in Eq. (1.60). Then, in line with Section 1.2.2 and Eqs. (1.22) and (1.24), the strain-displacement transformation matrix **B** and the stiffness matrix **K** can be calculated according to Eqs. (1.61) and (1.62). In particular, the integral over the volume becomes an integral over the length of the element in both global and natural coordinate systems. It must be noted that the *Jacobian J* has been introduced in the variable transformation. This expresses the derivative of the local coordinate x with respect to the natural coordinate r, see Eq. (1.63). The reader should notice that Eq. (1.62) is the same stiffness matrix as in Section 1.2.2.

Following Eq. (1.12), the same reasoning can be applied in the calculation of the nodal forces. In particular, limiting the study to body forces b, the calculation is shown in Eq. (1.64). As expected, the resultant of the body force is equally split between the two nodes.

$$\varepsilon = \frac{du}{dx} = \frac{du}{dr}\frac{dr}{dx} = \frac{u_2 - u_1}{2}\frac{2}{x_2 - x_1} = \frac{u_2 - u_1}{l_e} \tag{1.60}$$

$$\varepsilon = \frac{1}{l_e}\begin{bmatrix} -1 & 1 \end{bmatrix}\begin{bmatrix} u_1 \\ u_2 \end{bmatrix} = \mathbf{Bd}, \qquad \mathbf{B} = \frac{1}{l_e}\begin{bmatrix} -1 & 1 \end{bmatrix} \tag{1.61}$$

$$\mathbf{K} = \int_V \mathbf{B}^T \mathbf{DB}\, dV = \frac{AE}{l_e^2}\int_{x_1}^{x_2}\begin{bmatrix} -1 \\ 1 \end{bmatrix}\begin{bmatrix} -1 & 1 \end{bmatrix} dx$$

$$= \frac{AE}{l_e^2}\int_{-1}^{1}\begin{bmatrix} -1 \\ 1 \end{bmatrix}\begin{bmatrix} -1 & 1 \end{bmatrix} J\, dr = \begin{bmatrix} \dfrac{EA}{l_e} & -\dfrac{EA}{l_e} \\ -\dfrac{EA}{l_e} & \dfrac{EA}{l_e} \end{bmatrix} \tag{1.62}$$

$$J = \frac{dx}{dr} = \frac{l_e}{2} \tag{1.63}$$

$$\mathbf{f} = \int_V \mathbf{N}^T b\, dV = A\int_{x_1}^{x_2}\mathbf{N}^T b\, dx =$$

$$= Ab\int_{-1}^{1}\frac{1}{2}\begin{bmatrix} 1-r \\ 1+r \end{bmatrix} J\, dr = \frac{Ab}{2}\int_{-1}^{1}\begin{bmatrix} 1-r \\ 1+r \end{bmatrix}\frac{l_e}{2}\, dr = \frac{Abl_e}{4}\begin{bmatrix} 2 \\ 2 \end{bmatrix} = \begin{bmatrix} \dfrac{Abl_e}{2} \\ \dfrac{Abl_e}{2} \end{bmatrix} \tag{1.64}$$

A similar approach can be pursued when plane elements are considered, such as plane stress, plane strain or axial symmetry, where strains involve the first derivatives of displacement (i.e. C^0 continuity). Eq. (1.65) shows the general relationships between internal coordinates and displacements with respect to the nodal ones. In particular, q is the number of element nodes and N_i are the shape functions. It is worth noticing that it is always possible to express the local coordinates x and y in terms of r and s but the inverse mapping is generally not available.

Let us consider the four-node quadrilateral element of Figure 1.21, where the local $(x - y)$ and natural coordinates $(r - s)$ are highlighted. The mutual relationship between displacements within the element and nodal displacements is now written in matrix form in Eq. (1.66). Also in this case, the interpolation functions lead to a unit displacement at the generic node i and zero at all other nodes. Being the natural variables always within $[-1, 1]$, N_i can be formulated in a systematic way, even including additional internal nodes. For the present case, the shape functions are calculated in Eq. (1.67) and depicted in Figure 1.22 (compare Figures 1.5 and 1.13).

$$x = \sum_{i=1}^{q} N_i x_i; \qquad\qquad y = \sum_{i=1}^{q} N_i y_i;$$

$$\tag{1.65}$$

$$u \approx \sum_{i=1}^{q} N_i u_i; \qquad\qquad v \approx \sum_{i=1}^{q} N_i v_i;$$

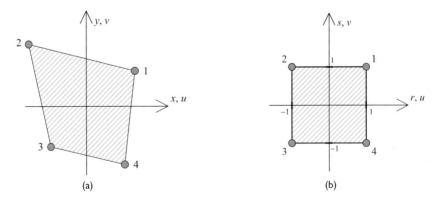

Figure 1.21 Isoparametric element described in (a) local and (b) natural coordinate systems.

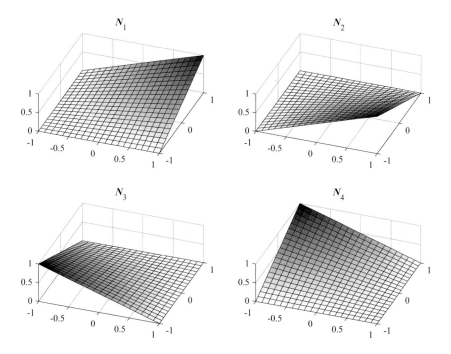

Figure 1.22 Shape functions for a four-node quadrilateral element (natural coordinate system).

$$\mathbf{u} = \begin{bmatrix} u \\ v \end{bmatrix} \approx \begin{bmatrix} N_1 & 0 & N_2 & 0 & N_3 & 0 & N_4 & 0 \\ 0 & N_1 & 0 & N_2 & 0 & N_3 & 0 & N_4 \end{bmatrix} \begin{bmatrix} u_1 \\ v_1 \\ u_2 \\ v_2 \\ u_3 \\ v_3 \\ u_4 \\ v_4 \end{bmatrix} = \mathbf{Nd}$$

$$\tag{1.66}$$

$$N_1 = \frac{1}{4}(1+r)(1+s) \qquad N_2 = \frac{1}{4}(1-r)(1+s)$$

$$\tag{1.67}$$

$$N_3 = \frac{1}{4}(1-r)(1-s) \qquad N_4 = \frac{1}{4}(1+r)(1-s)$$

When constructing the finite element matrices, the derivatives of u and v with respect to x and y are needed. However, as seen earlier, the spatial variation of the displacements is expressed through the shape functions, which depend on the natural coordinates r and s. To obtain the desired derivatives, the chain rule can be used as follows:

$$
\begin{aligned}
\frac{\partial}{\partial r} &= \frac{\partial x}{\partial r}\frac{\partial}{\partial x} + \frac{\partial y}{\partial r}\frac{\partial}{\partial y} \\
\frac{\partial}{\partial s} &= \frac{\partial x}{\partial s}\frac{\partial}{\partial x} + \frac{\partial y}{\partial s}\frac{\partial}{\partial y}
\end{aligned}
\rightarrow
\begin{bmatrix} \dfrac{\partial}{\partial r} \\ \dfrac{\partial}{\partial s} \end{bmatrix}
=
\begin{bmatrix} \dfrac{\partial x}{\partial r} & \dfrac{\partial y}{\partial r} \\ \dfrac{\partial x}{\partial s} & \dfrac{\partial y}{\partial s} \end{bmatrix}
\begin{bmatrix} \dfrac{\partial}{\partial x} \\ \dfrac{\partial}{\partial y} \end{bmatrix}
\rightarrow \frac{\partial}{\partial r} = \mathbf{J}\frac{\partial}{\partial \mathbf{x}}
$$

$$(1.68)$$

In this case, $\mathbf{J}$ is the *Jacobian operator*, i.e. a matrix with the derivatives of local coordinates with respect to the natural ones. Since the derivatives $\partial/\partial\mathbf{x}$ are needed to calculate the strain vector, it is possible to write the relation given in Eq. (1.69) provided that the inverse $\mathbf{J}^{-1}$ exists. It must be highlighted that the inverse of the Jacobian operator does not exist if the correspondence between natural and local coordinates is not unique, i.e. if the element is much distorted (e.g. convex shape with interior angles larger than 180°) or it folds back upon itself.

The calculation of the strain is derived in Eqs. (1.70) and (1.71) while the stiffness matrix in the local reference system is given in Eq. (1.72). In particular, since the strain-displacement transformation matrix $\mathbf{B}$ is a function of the natural coordinate system (due to $\mathbf{J}^{-1}$), the integral in Eq. (1.72) must be over the same coordinate system. With this aim, let us assume t as the thickness of the element (assumed constant) and det $\mathbf{J}$ the determinant of the Jacobian operator (compare Eq. (1.62)). In general, it is not efficient (or even possible) to explicitly evaluate the integral in Eq. (1.72) and numerical integration is employed instead (see Section 1.4.4). The following example clarifies the procedure by providing information regarding the body force and traction along one side of the element.

$$\frac{\partial}{\partial \mathbf{x}} = \mathbf{J}^{-1}\frac{\partial}{\partial \mathbf{r}}$$

$$(1.69)$$

$$
\begin{bmatrix} \dfrac{\partial}{\partial x} \\ \dfrac{\partial}{\partial y} \end{bmatrix}
=
\begin{bmatrix} J^{-1}{}_{11} & J^{-1}{}_{12} \\ J^{-1}{}_{21} & J^{-1}{}_{22} \end{bmatrix}
\begin{bmatrix} \dfrac{\partial}{\partial r} \\ \dfrac{\partial}{\partial s} \end{bmatrix}
\rightarrow \text{e.g.} \frac{\partial}{\partial x} = J^{-1}{}_{11}\frac{\partial}{\partial r} + J^{-1}{}_{12}\frac{\partial}{\partial s}
$$

$$
\boldsymbol{\varepsilon} =
\begin{bmatrix} \varepsilon_x \\ \varepsilon_y \\ \gamma_x \end{bmatrix}
=
\begin{bmatrix} \dfrac{\partial}{\partial x} & 0 \\ 0 & \dfrac{\partial}{\partial y} \\ \dfrac{\partial}{\partial y} & \dfrac{\partial}{\partial x} \end{bmatrix}
\begin{bmatrix} u \\ v \end{bmatrix}
\approx
\begin{bmatrix} \dfrac{\partial}{\partial x} & 0 \\ 0 & \dfrac{\partial}{\partial y} \\ \dfrac{\partial}{\partial y} & \dfrac{\partial}{\partial x} \end{bmatrix}
\mathbf{Nd} = \mathbf{Bd} \quad (1.70)
$$

$$\mathbf{B} = \begin{bmatrix} \dfrac{\partial}{\partial x} & 0 \\[2mm] 0 & \dfrac{\partial}{\partial y} \\[2mm] \dfrac{\partial}{\partial y} & \dfrac{\partial}{\partial x} \end{bmatrix} \mathbf{N} =$$

$$= \begin{bmatrix} \dfrac{\partial}{\partial x} & 0 \\[2mm] 0 & \dfrac{\partial}{\partial y} \\[2mm] \dfrac{\partial}{\partial y} & \dfrac{\partial}{\partial x} \end{bmatrix} \begin{bmatrix} N_1 & 0 & N_2 & 0 & N_3 & 0 & N_4 & 0 \\ 0 & N_1 & 0 & N_2 & 0 & N_3 & 0 & N_4 \end{bmatrix}$$

e.g. $\quad \mathbf{B}_{11} = \dfrac{\partial N_1}{\partial x} = \mathbf{J}^{-1}{}_{11} \dfrac{\partial N_1}{\partial r} + \mathbf{J}^{-1}{}_{12} \dfrac{\partial N_1}{\partial s}$ $\hfill (1.71)$

$$\mathbf{K} = \int_V \mathbf{B}^T \mathbf{D} \mathbf{B} \, dV = t \iint \mathbf{B}^T \mathbf{D} \mathbf{B} \, dx \, dy = t \int_{-1}^{1} \int_{-1}^{1} \mathbf{B}^T \mathbf{D} \mathbf{B} \det \mathbf{J} \, dr \, ds \qquad (1.72)$$

Let us consider a quadrilateral element subjected to a body force and traction on side 1–2, as depicted in Figure 1.23. According to Eq. (1.65), the relationship between the local and natural coordinates is derived in Eq. (1.73); the Jacobian operator $\mathbf{J}$ and its inverse are shown in Eq. (1.74).

$$\mathbf{x} = \begin{bmatrix} x \\ y \end{bmatrix} \approx \begin{bmatrix} N_1 & 0 & N_2 & 0 & N_3 & 0 & N_4 & 0 \\ 0 & N_1 & 0 & N_2 & 0 & N_3 & 0 & N_4 \end{bmatrix} \begin{bmatrix} 3 \\ 2 \\ 1 \\ 1 \\ 1 \\ 0 \\ 3 \\ 0 \end{bmatrix}$$

$$= \begin{bmatrix} r+2 \\[2mm] \dfrac{1}{4}(r+3)(s+1) \end{bmatrix} \qquad (1.73)$$

$$
\mathbf{J} = \begin{bmatrix} 1 & \dfrac{1}{4}(s+1) \\[2ex] 0 & \dfrac{1}{4}(r+3) \end{bmatrix}
\qquad
\mathbf{J}^{-1} = \begin{bmatrix} 1 & -\dfrac{s+1}{r+3} \\[2ex] 0 & \dfrac{4}{r+3} \end{bmatrix}
\tag{1.74}
$$

According to the previous calculated quantities, the strain-displacement transformation matrix $\mathbf{B}$ is derived in Eq. (1.75), where its first element is also detailed. Considering the linear elastic constitutive relations $\mathbf{D}$ in plane stress, it is possible to derive the stiffness matrix by numerical integration of Eq. (1.72), as shown below in text.

$$
\mathbf{B} =
\begin{bmatrix}
\dfrac{s+1}{2(r+3)} & 0 & -\dfrac{s+1}{r+3} & 0 & -\dfrac{r-2s+1}{2(r+3)} & 0 & \dfrac{r-s+2}{2(r+3)} & 0 \\[2ex]
0 & \dfrac{r+1}{r+3} & 0 & -\dfrac{r-1}{r+3} & 0 & \dfrac{r-1}{r+3} & 0 & -\dfrac{r+1}{r+3} \\[2ex]
\dfrac{r+1}{r+3} & \dfrac{s+1}{2(r+3)} & -\dfrac{r-1}{r+3} & \dfrac{s+1}{r+3} & \dfrac{r-1}{r+3} & -\dfrac{r-2s+1}{2(r+3)} & -\dfrac{r+1}{r+3} & \dfrac{r-s+2}{2(r+3)}
\end{bmatrix}
$$

$$
\mathbf{B}_{11} = \frac{\partial N_1}{\partial x} = \mathbf{J}^{-1}_{11}\frac{\partial N_1}{\partial r} + \mathbf{J}^{-1}_{12}\frac{\partial N_1}{\partial s} =
$$

$$
= 1\cdot\frac{1}{4}(1+s) - \frac{s+1}{r+3}\cdot\frac{1}{4}(1+r) = \frac{s+1}{2(r+3)}
\tag{1.75}
$$

Consistently with what was shown earlier for the beam element, the nodal forces associated to the body force can be calculated according to Eq. (1.76). In turn, for surface traction, integration must be performed along the loaded element boundary, i.e. along the coordinate c of Figure 1.23b (line where $s = +1$ in the natural system). The differential length along this coordinate dc and the consequent nodal forces can be calculated according to Eqs. (1.77) and (1.78), respectively. In particular, the shape function matrix should be evaluated with $s = +1$ and the traction $\mathbf{t}$ as a function of r (by means of interpolation functions). Consequently, the equilibrium equations are derived in Eq. (1.79). Applying the boundary conditions (i.e. nodes 3 and 4 are fully constrained), the solution is shown in Eq. (1.80).

$$
\mathbf{f}_b = \int_V \mathbf{N}^T \mathbf{b}\, dV = t \iint \mathbf{N}^T \mathbf{b}\, dx\, dy = t \int_{-1}^{1}\int_{-1}^{1} \mathbf{N}^T \mathbf{b}\, \det \mathbf{J}\, dr\, ds
\tag{1.76}
$$

$$
= \begin{bmatrix} 0 & -5{,}500 & 0 & -4{,}400 & 0 & -4{,}400 & 0 & -5{,}500 \end{bmatrix}^T
$$

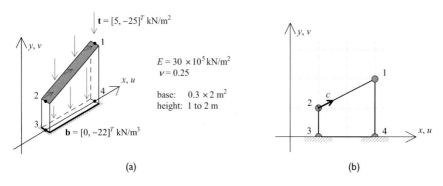

Figure 1.23 Quadrilateral isoparametric four-node element with body force and traction on one side (local coordinate system). (a) Three-dimensional representation and (b) FEM model.

$$dc = \det \mathbf{J}_c dr \qquad \mathbf{J}_c = \sqrt{\left(\frac{\partial x}{\partial r}\right)^2 + \left(\frac{\partial y}{\partial r}\right)^2} \qquad \frac{\partial x}{\partial r} = \frac{x_1 - x_2}{2}, \qquad \frac{\partial y}{\partial r} = \frac{y_1 - y_2}{2}$$

(1.77)

$$\mathbf{f}_t = \int_{-1}^{1} \mathbf{N}\Big|_{s=1}^{T} \mathbf{t}\, t\, \det \mathbf{J}_c\, dr$$

$$= \int_{-1}^{1}
\begin{bmatrix}
\dfrac{1+r}{2} & 0 & \dfrac{1-r}{2} & 0 & 0 & 0 & 0 & 0 \\[2mm]
0 & \dfrac{1+r}{2} & 0 & \dfrac{1-r}{2} & 0 & 0 & 0 & 0
\end{bmatrix}^{T}
\begin{bmatrix} 5,000 \\ -25,000 \end{bmatrix} 0.3 \frac{\sqrt{5}}{2}\, dr$$

$$= \begin{bmatrix} 1,677 & -8,385 & 1,677 & -8,385 & 0 & 0 & 0 & 0 \end{bmatrix}^{T}$$

(1.78)

$$10^7
\left[
\begin{array}{cccc:cccc}
25 & 9 & -13 & -8 & -10 & -11 & -2 & 10 \\
9 & 41 & 1 & 13 & -11 & -22 & 1 & -31 \\
-13 & 1 & 64 & -22 & -17 & 3 & -34 & 17 \\
-8 & 13 & -22 & 68 & 12 & -50 & 17 & -31 \\
\hdashline
-10 & -11 & -17 & 12 & 40 & 6 & -13 & -8 \\
-11 & -22 & 3 & -50 & 6 & 59 & 1 & 13 \\
-2 & 1 & -34 & 17 & -13 & 1 & 49 & -19 \\
10 & -31 & 17 & -31 & -8 & 13 & -19 & 50
\end{array}
\right]
\begin{bmatrix} u_1 \\ v_1 \\ u_2 \\ v_2 \\ \hline u_3 \\ v_3 \\ u_4 \\ v_4 \end{bmatrix}
=
\begin{bmatrix} 1,677 \\ -13,885 \\ 1,677 \\ -12,785 \\ \hline 0 \\ -4,400 \\ 0 \\ -5,500 \end{bmatrix}$$

(1.79)

$$\mathbf{d} = \begin{bmatrix} u_1 & v_1 & u_2 & v_2 \end{bmatrix}^T = \begin{bmatrix} 0.018 & -0.035 & -0.004 & -0.009 \end{bmatrix}^T \text{mm}$$

(1.80)

The reader should be aware that the solution encompasses the assumption that strains and stresses are constant within the element. In the present case, disregarding the stress/strain gradients close to corners and sides of the elements, the model does not approximate correctly the real behaviour of the wall and it is not adequate to represent the physical phenomenon. The error would be acceptable only if a larger number of elements would be adopted. Consistently with Section 1.3.1, mesh refinement is dealt with in Chapter 5.

In turn, it is worth noticing that the procedure adopted here for the quadrilateral element can be easily extended to any line, plane and solid elements. Additional interpolation nodes can be used to enrich the effectiveness of the shape functions (by means of higher-order polynomials, as shown in the next section), thus the approximation of the solution. Moreover, given the versatility of the shape functions, the same procedure can be adopted for elements with curved boundaries too.

As a final remark, two aspects need to be stressed. First, it is worth noticing that isoparametric elements satisfy the requirements of completeness and compatibility addressed in Section 1.3.1 for any geometrical (convex) shape. However, much distorted elements may affect the approximation of the numerical solution. In this regard, mesh regularity must be checked, as illustrated in Chapter 5. For instance, indicators such as *aspect ratio* and *skewness* are used to measure how much the given element is stretched and distorted when compared to regular shapes, such as equilateral triangle, square, isosceles tetrahedron and cube. Second, the finite element matrices have been described according to a local reference system, but they can be easily referred to the global reference system. Looking at Figure 1.23, should the $x - y$ system be changed with $x_g - y_g$, the calculations would have been the same. However, if more advantageous, it is always possible to first adopt the local coordinate system and then to transform the matrices by means of the rotation matrix $\mathbf{T}$ (given at the end of Section 1.4.1).

1.4.3 Completeness of shape functions for quadrilateral elements

In the previous examples, typical finite elements and related shape functions were addressed. As a matter of fact, once the element and the corresponding shape functions (complete and compatible) are determined, subsequent operations follow a standard and well-defined path, which belongs to algebra more than to engineering. However, the importance of the shape functions in the approximation of the displacement field has been stated, a matter the analyst should ponder carefully. In general, the more are the nodes for a single element, the more are the DOFs and the higher is the polynomial order (thus the complexity and computational costs); therefore, the higher is the accuracy and the more rapid is the convergence to the true solution. Ideally, optimal shape functions should be provided with the highest complete polynomial for a minimum of DOFs.

Regarding completeness, limiting the discussion to plane elements, it is possible to recall the Pascal triangle (Figure 1.24a). The complete polynomial of highest order in the Pascal triangle determines the order of convergence, see Eq. (1.31). The terms occurring in a polynomial in two variables x and y can be readily ascertained from Figure 1.24a. The shaded area highlights that, for instance, 10 terms are needed for a complete third-order polynomial.

Dealing with elements provided with multiple nodes, two main families are usually considered, namely, *Lagrange* and *serendipity*. The former is depicted in Figure 1.24b and is characterized by the presence of internal nodes and nodes on the edge. In this case, looking at the relative Pascal triangle, it is possible to demonstrate that a large number of polynomial terms are present above those needed for a complete expansion. For instance, if a third-order polynomial (i.e. cubic element) is considered, six surplus terms arise, which make richer the description of the displacement field and the consequent strain evaluation. Basically, if n is the order of the polynomial, the diamond shape illustrated in Figure 1.24b includes terms x^n, y^n and $x^n y^n$. It is evident that, moving to strains, the surplus terms provide a wider variability within the element. For instance, a nine-node quadrilateral isoparametric plane element (the first of Figure 1.24b) is based on quadratic interpolation. The polynomial for the displacements along x and y is shown

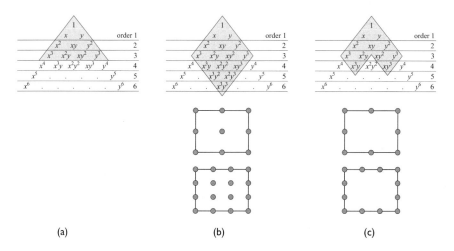

(a)   (b)   (c)

Figure 1.24 Pascal triangle and higher order finite elements. (a) Minimum number of terms for a complete third-order polynomial; (b) Lagrange and (c) serendipity families.

in Eq. (1.81). Typically, this polynomial yields a strain ε_x that varies linearly in x direction and quadratically in y direction. The strain ε_y varies linearly in y direction and quadratically in x direction. The shear strain γ_{xy} varies quadratically in both directions.

$$u(r,s) = a_0 + a_1 r + a_2 s + a_3 rs + a_4 r^2 + a_5 s^2 + a_6 r^2 s + a_7 rs^2 + a_8 r^2 s^2 \qquad (1.81)$$

Serendipity family, instead, considers nodes located only on the element boundary. In this case too, a routine can generate suitable shape functions according to the scheme of Figure 1.24c. Compared to Lagrange family, a cubic element is provided with only two surplus terms. If n is the order of the polynomial, the shaded shape illustrated in Figure 1.24c includes terms xy, x^n, y^n, $x^n y$ and xy^n. Basically, the shape is composed of two symmetric stripes. It is evident that complete polynomials beyond cubic order are not possible and additional internal nodes are usually employed to mitigate this aspect.

With illustrative purpose, an eight-node quadrilateral isoparametric serendipity plane element (the first of Figure 1.24c) is based on quadratic interpolation. The polynomial for the displacements along x and y can be expressed as shown in Eq. (1.82). If compared with Eq. (1.81), only one term is missing, and the interpolation function leads to the same strain distribution seen earlier (although with one node less). However, according to literature, in the case of angular distortion, the mentioned nine-node Lagrange element maintains the completeness of the second order, while the eight-node serendipity element loses one order. This aspect necessarily affects the convergence rate. In this regard, curved edges are also a feature that can influence the polynomial completeness, thus the convergence rate. For the mentioned elements, no differences are expected.

$$u(r,s) = a_0 + a_1 r + a_2 s + a_3 rs + a_4 r^2 + a_5 s^2 + a_6 r^2 s + a_7 rs^2 \qquad (1.82)$$

For both families, it is possible to extend the same reasoning to solid elements. Additionally, the shape functions can be determined according to a different number of nodes placed on each side. This aspect may be desirable if a transition between elements of different order is needed in the adopted discretization, enabling a different order of accuracy for different portions of the structure to be studied. From the computational point of view, it is easy to see that the shaded shapes in Figure 1.24b and c become nonsymmetric. However, to avoid compatibility issues, mixing elements with different order of polynomials is not recommended.

1.4.4 Numerical integration, integration points and integration schemes

As shown earlier, the calculation of the stiffness matrix and the nodal force vectors involves the numerical integration over the element volume or contour.

In general, the argument of the integration (i.e. the integrand) is a matrix and each element of the matrix must be integrated individually. In this regard, with the aim of describing the fundamentals of numerical integration, let us consider a generic function $F_{(r)}$. If the given function has been evaluated at $n + 1$ points, it is always possible to write a polynomial $p_{(r)}$ of degree at most n that passes through the points of the dataset. In numerical analysis, this calculation goes under the name of *polynomial interpolation* and it is possible to demonstrate that the polynomial $p_{(r)}$ exists and is unique.

Let us consider the function $F_{(r)}$ given in Eq. (1.83) and shown in Figure 1.25a evaluated at $n + 1 = 3$ equally spaced sampling points in the interval $[-1, 1]$, i.e. abscissas $r_0 = -1$, $r_1 = 0$ and $r_2 = 1$, such as $F_0 = 0$, $F_1 = \pi$ and $F_2 = 0$. Consequently, p is expected to be of second order (at the most) of the form $p_{(r)} = a_0 + a_1 r + a_2 r^2$. The given parameters are substituted in Eq. (1.84) in matrix form where the unknown coefficients are calculated inverting the coefficient matrix. Finally, the interpolating polynomial $p_{(r)}$ is reported. The comparison between $F_{(r)}$ and $p_{(r)}$ is shown also in Figure 1.25a. As clearly noticeable, $p_{(r)}$ provides the "true" values only in correspondence of the abscissa r_0, r_1 and r_2, missing the change of curvature of the original function $F_{(r)}$ (at about $r \approx 0.66$) as the order of the polynomial is not sufficient. Additionally, the interpolating curve is valid only in the given interval $[-1, 1]$, being erroneous outside (Figure 1.25b).

$$F_{(r)} = \frac{\sin \pi r}{r} \tag{1.83}$$

$$\begin{bmatrix} F_0 \\ F_1 \\ F_2 \end{bmatrix} = \begin{bmatrix} 0 \\ \pi \\ 0 \end{bmatrix} = \begin{bmatrix} 1 & -1 & 1 \\ 1 & 0 & 0 \\ 1 & 1 & 1 \end{bmatrix} \begin{bmatrix} a_0 \\ a_1 \\ a_2 \end{bmatrix} \rightarrow \begin{bmatrix} a_0 \\ a_1 \\ a_2 \end{bmatrix} = \begin{bmatrix} \pi \\ 0 \\ -\pi \end{bmatrix}$$

$$p_{(r)} = \pi - \pi r^2 = \pi \left(1 - r^2 \right) \tag{1.84}$$

Once the interpolating function is known, it is possible to assess the value of the integral of $F_{(r)}$ in a certain interval by exactly evaluating the integral of $p_{(r)}$. In particular, if the abscissas where $F_{(r)}$ is evaluated are equally spaced, it is possible to use the *Newton–Cotes formula* shown in Eq. (1.85), where C_i^n are the *Newton–Cotes constants* associated to the number n of intervals. These constants follow from the integral calculus of a polynomial with a given order and represent the *weight* of each F_i. In turn, F_i are the values of $F_{(r)}$ calculated at the $n + 1$ abscissas r_i. It is worth noticing that the points where the function is evaluated provide the so-called *integration points* (also named optimal or sampling points). The constants are summarized in Table 1.2 up to $n = 4$. For the previous example, being $n = 2$, the calculation shown in Eq. (1.86) leads to 4.1888, whereas the exact solution is 3.7039 (error of +13% with respect to the exact solution).

(a)

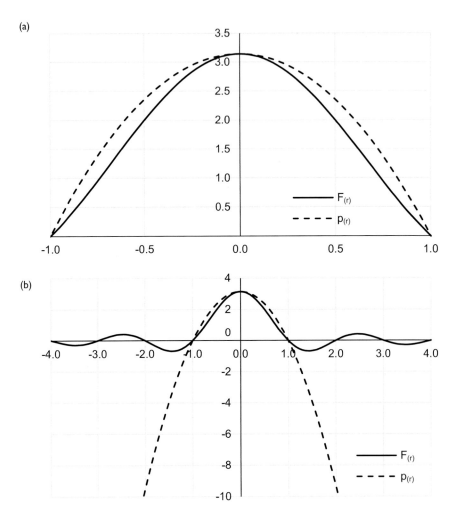

(b)

Figure 1.25 Comparison between function $F_{(r)}$ of Eq. (1.83) and the interpolating polynomial $p_{(r)}$ of Eq. (1.84) (a) within the range $[-1, 1]$ and (b) in a wider interval.

Table 1.2 Newton–Cotes constants according to the number of sampling points n

n	C_0^n	C_1^n	C_2^n	C_3^n	C_4^n
1	1/2	1/2	-	-	-
2	1/6	4/6	1/6	-	-
3	1/8	3/8	3/8	1/8	-
4	7/90	32/90	12/90	32/90	7/90

Two strategies are available to get better accuracy: (1) increase the number of points in which the function is evaluated, thus, the order of the interpolating polynomial and (2) break the interval [a, b] in several parts and apply the Newton–Cotes formula in each part separately. The latter option is based on the idea that a low-order polynomial approximates better $F_{(r)}$ in small portions of the interval than a high-order polynomial over the entire interval. For the same example as in Eq. (1.86), increasing the order of the polynomial to $n = 4$ leads to 3.6822 (−0.6% error), while dividing the interval in two parts and keeping the order of the polynomial to $n = 2$ leads to 3.7139 (+0.3% error). Upon further refinement, the "exact" solution can be found.

$$\int_a^b F_{(r)} \, dr \approx (b-a) \sum_{i=0}^n C_i^n F_i \tag{1.85}$$

$$\int_{-1}^1 \frac{\sin \pi r}{r} \, dr \approx (1+1)\left(\frac{1}{6}F_0 + \frac{4}{6}F_1 + \frac{1}{6}F_2\right) = \frac{1}{3}(0 + 4\pi + 0) = 4.1888 \tag{1.86}$$

Another approach to improve the accuracy of the numerical solution is by optimizing the position of the sampling points in which the function $F_{(r)}$ is evaluated. In this regard, the *Gauss quadrature* is one of the most important numerical integration procedures. Eq. (1.87) illustrates the relative approximation, where n is the number of the sampling points and F_i are the values of the function at the integration points r_i. Considering the integration interval [−1, 1], the abscissas r_i and the weights w_i are reported in Table 1.3. As it is possible to notice, the integration points do not include the ends of the interval. Again, for the given example, let us consider the cases with $n = 2$ and $n = 3$ sampling points illustrated in Eqs. (1.88) and (1.89), respectively. Comparing the results with the exact solution (equal to 3.7039), the error is equal to −9.2% and +0.6%, respectively, which is a superior performance when compared to Newton–Cotes formula.

$$\int_a^b F_{(r)} \, dr \approx \sum_{i=1}^n w_i F_i \tag{1.87}$$

$$r_1 = -0.57735 \qquad \rightarrow \qquad F_1 = 1.6812$$

$$r_2 = +0.57735 \qquad \rightarrow \qquad F_2 = 1.6812$$

$$\tag{1.88}$$

$$\int_{-1}^1 \frac{\sin \pi r}{r} \, dr \approx 1 \cdot 1.6812 + 1 \cdot 1.6812 = 3.3624$$

Table *1.3* Abscissas and weights in Gauss numerical integration procedure for $r \in [-1, 1]$ according to the number of sampling points n

n	r_i	w_i
1	0	2
2	$\pm\dfrac{1}{\sqrt{3}} = \pm 0.5774$	1
3	$\pm\sqrt{\dfrac{3}{5}} = \pm 0.7746$	$\dfrac{5}{9} = 0.5556$
	0	$\dfrac{8}{9} = 0.8889$
4	$\pm\sqrt{\dfrac{3+2\sqrt{6/5}}{7}} = \pm 0.8611$	$\dfrac{18-\sqrt{30}}{36} = 0.3479$
	$\pm\sqrt{\dfrac{3-2\sqrt{6/5}}{7}} = \pm 0.3400$	$\dfrac{18+\sqrt{30}}{36} = 0.6521$

$$r_1 = -0.7746 \quad \rightarrow \quad F_1 = 0.8397$$

$$r_2 = 0.0000 \quad \rightarrow \quad F_2 = 3.1416$$

$$r_3 = +0.7746 \quad \rightarrow \quad F_3 = 0.8397$$

$$\int_{-1}^{1} \frac{\sin \pi r}{r}\, dr \approx 0.5556 \cdot 0.8397 + 0.8889 \cdot 3.1416 + 0.5556 \cdot 0.8397 = 3.7255$$

$$(1.89)$$

As it is possible to grasp, shape functions and integration points are two connected features of finite elements. Basically, the integrations points inform about the accuracy of the integral of the shape functions. In this regard, as finite element matrices often deal with polynomials, it is important to summarize the accuracy of the two procedures presented earlier. Newton–Cotes formula employs n intervals (thus $n + 1$ sampling points) and integrates exactly polynomials of orders up to at most n (i.e. one unit less than the number of integration points); Gauss quadrature employs n sampling points and integrates exactly polynomials of order up to at most $2n - 1$. This is the common integration method adopted for isoparametric elements.

For the sake of completeness, *Lobatto quadrature* and *Simpson rule* must be also mentioned. Lobatto quadrature is similar to Gauss quadrature, but it includes the end points. In this case, by evaluating the function at n points, Lobatto method exactly integrates polynomials of degree

$2n - 3$. The advantage of Lobatto quadrature is clear if the value of the function to be integrated is already known at the end points. Accordingly, m new interior evaluations plus the known values at the end points gives $n = m + 2$ integration points, which can integrate polynomials of degree at most $2(m + 2) - 3 = 2m + 1$, two degrees higher than Gauss quadrature (if only m interior points were considered). Simpson rule considers equally spaced integration points, end points included. It approximates the integrand function by means of parabola arcs (i.e. quadratic polynomials) and, as expected, a second-order function can be integrated exactly. It provides a good approximation in the case of curvilinear integrand functions, and it is much used through the finite element thickness to obtain the values in top and bottom surfaces.

The extension to plane (or solid) elements is straightforward and based on the subsequent integration of two (or three) one-dimensional integration formulas successively in each direction. Figure 1.26 shows the position and weights of the integration points for usual finite elements. For quadrilateral elements, the analogies with the line elements (bar-type) are evident. In particular, the weight of each integration point is equal to the product of the

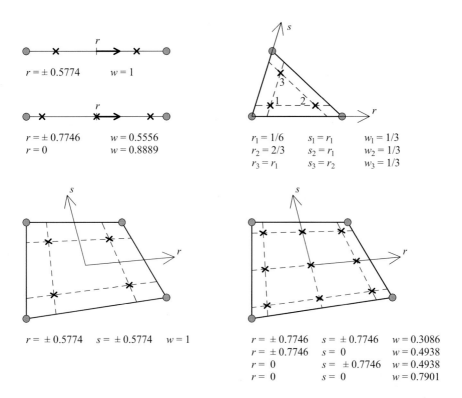

Figure 1.26 Position of the integration points and weights w in Gauss quadrature for line (i.e. bar-type), triangle and quadrilateral elements with $r, s \in [-1, 1]$.

one-dimensional integration weights considering both r and s coordinates (according to Table 1.3).

After introducing the integration points, the question is about the number to be used for the problem at hand. A minimum number of integration points is required by the numerical integration method and depends on the order of the shape function polynomial. The *full integration scheme* refers to the number of integration points required to integrate the polynomial terms in an element's stiffness matrix exactly when the element has a regular shape (i.e. geometrically undistorted). For hexahedral and quadrilateral elements, a "regular shape" means that the edges are straight and meet at right angles, and that any edge nodes are at the midpoint of the edge. As expected, the numerical integration errors are acceptably small assuming reasonable geometric distortions. Practically, fully integrated linear elements use two integration points in each direction; quadratic elements use three integration points in each direction.

However, especially in linear elastic problems, some higher order cross terms of the polynomial can be ignored safely, which reduces the required number of integration points, leading to the *reduced integration scheme*. For instance, for quadrilateral elements, reduced-integration elements may use one fewer integration point in each direction than the fully integrated elements. It is worth highlighting that a significant reduction of the integration points may lead to bad condition of the total stiffness matrix (see hourglassing in Chapter 5), with consequent less accurate or fully innacurate solution.

Going back to the line elements of Figure 1.26, the main integration is done along the longitudinal axis. In the case of linear elastic assumptions, this is sufficient to evaluate the behaviour of the element because the cross-sectional response is pre-integrated. This means that, as given in Eq. (1.52), the distribution of normal strains and stresses is already integrated in the stiffness matrix, and generalized quantities are used in their place. In other words, moving from nodal displacements, it is possible to calculate the beam response without passing through the internal strain (thus stress) distribution. However, this is not the case when nonlinear analysis is performed. In this case, the integration regards not only the longitudinal axis, but also the area of the cross-section. For three-dimensional analysis, considering the three directions, software codes may introduce hundreds of integration points per single beam element. The reason of such a large number of through-depth and through-thickness integration points lays on the need of accurately estimating the gradual stiffness reduction due to cracking and crushing. The two- and three-dimensional cases for beam elements are illustrated in Figure 1.27. For the sake of intelligibility, three integration points along the longitudinal axis (full squares) and five integrations points along the cross-sectional axes (crosses) are shown. Obviously, two integration points per cross-sectional axis would be sufficient to integrate exactly stresses and strains through the volumes with linear elastic analysis

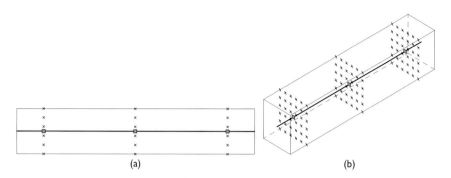

Figure 1.27 Example of integration schemes for beam elements. (a) Two- and (b) three-dimensional analysis.

and typical assumptions for beams. In turn, 7–11 points per cross-sectional axis is a suitable option to integrate the nonlinear constitutive relations in the cross-section. It is noted that, when nonlinear aspects are considered, stresses and strains are computed at the integration points, where the constitutive relation is enforced. By means of the rules described earlier, it is then possible to perform the integration over the volume of the different quantities used in the weak form of the governing equations.

The same reasoning used for beams can be extended to plane elements such as plate bending or shell elements. Compared to Figure 1.26, only through-thickness integration points are needed. Obviously, for solid elements, the integration points within the volume appear naturally.

1.4.5 Recovery process

The numerical integration presented in the previous section regards each element individually. If the shape functions do not guarantee C^1 continuity, strains and stresses will be discontinuous across the finite elements. This was already mentioned in Section 1.3.1 and is notable in the results of the axially loaded column shown in Figure 1.6. For the sake of clarity, two graphs of the mentioned examples are shown in Figure 1.28. As it is possible to notice, a linear approximation of the displacement field leads to piecewise constant strains. Additionally, two features must be pointed out: (1) displacements (the quantity of primary approximation in finite elements) attain the best accuracy at nodes and (2) strains (thus, stresses), involving gradients of the primary approximation (i.e. displacements), attain the best accuracy at points within elements, being poor at the nodes. Although intuitively, these results define some points in which the accuracy is larger than elsewhere. Such points have a quality known as *superconvergence* and they are often defined as *superconvergent points*.

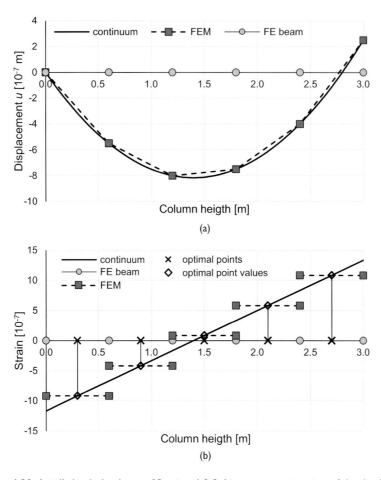

Figure 1.28 Axially loaded column of Section 1.2.2. Linear approximation of the displacement field: optimal sampling points for (a) the function (i.e. displacements)
and (b) its gradient (i.e. strains).

Another example was given in Section 1.3.3 where the displacement
field was approximated by means of a cubic function and the rotation
(being the first derivative of the deflection) by a quadratic function. For
the sake of clarity, the results are reported in Figure 1.29. As it is possible
to notice, the curvature (i.e. the second derivative of the deflection) is a
piecewise combination of straight lines. The deflection perfectly matches
the closed-form solution in terms of both values and slope (i.e. its first
derivative), whereas the curvature accurately interpolates the exact solution only in correspondence of the internal superconvergent points (two
per element).

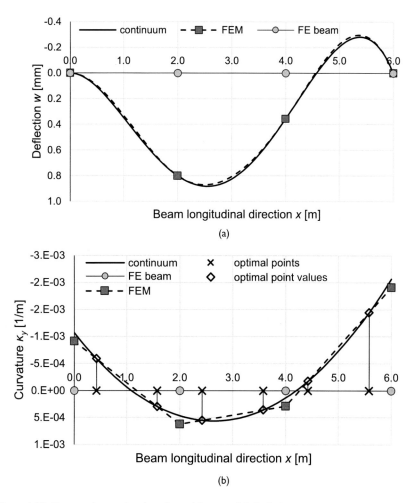

Figure 1.29 Beam subjected to bending of Section 1.3.3. Cubic approximation of the displacement field: optimal sampling points for (a) the function (i.e. deflection) and (b) its second derivative (i.e. curvature).

   The superconvergent points, in which the evaluation of gradients is optimal, coincide with the Gauss integration points in number according to the function to be integrated. At such points, the convergence is one order higher than anticipated from the polynomial approximation. The user should be aware of this feature and refer to these points to check the results of the analysis. However, although Gauss integration points still provide superior results, full superconvergence is not achieved with triangular elements and much distorted elements. Additionally, it must be stressed that Gauss integration points provide superconvergence only in the case of linear elasticity problems.

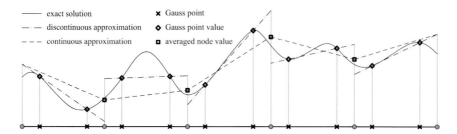

Figure 1.30 Schematic illustration of strain (or stress) recovery procedure: extrapolation of nodal values from Gauss integration points providing a discontinuous approximation (dash-dot line) and subsequent continuous approximation by averaging the nodal values (dashed line).

As superconvergent points provide a higher order of accuracy, it is possible to improve the overall accuracy of the finite element solution *after the analysis*. This process is referred to as *recovery of strains and stresses*. Various recovery methods are described in literature. Here, an example is given considering the recovered stresses obtained by interpolating nodal values the same way done for displacements, see Eq. (1.90). Nodal values of strains ε_{nodal} and stresses σ_{nodal} can be calculated by a two-step procedure: (1) extrapolating them from the values at the integration points and (2) averaging these values in the case the node is shared between two or more elements. Figure 1.30 schematically illustrates the method providing a good estimation of the nodal value. It is worth noticing that two Gauss integration points can be interpolated only with a straight line. Note also that averaging in Figure 1.28b provides an excellent estimation.

$$\varepsilon \approx N\varepsilon_{nodal} \qquad\qquad \sigma \approx N\sigma_{nodal} \qquad\qquad (1.90)$$

In Figure 1.31 the results of an example related to a cantilever beam discussed in Chapter 5 are shown. The beam is clamped in the left end and free in the right end, being subjected to a downward point load at the free extremity. As it is possible to notice, the values at the integration points (Figure 1.31a) are much lower than the maximum values found at the upper and lower sides and lateral edge of the beam because they indicate stress values internal to the beam. Moreover, the extrapolated values are not continuous at the edges of two neighbouring finite elements (Figure 1.31b), while smooth results are obtained after averaging (Figure 1.31c). Regarding integration points, some software codes may propose contour plots to visualize purely fixed location values. Practically, the values at the integration points are assigned to the closest nodes and then interpolated through the entire element. Although incompatible with the exact location of the integration points, the analyst should be aware of the postprocessing adopted and be sure to consider the needed information.

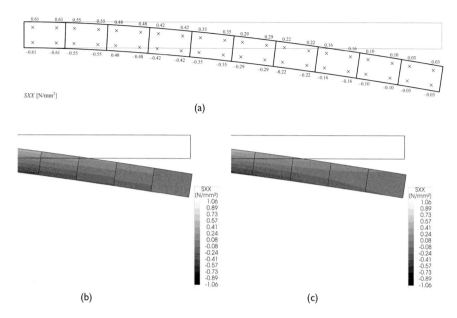

SXX [N/mm²]

(a)

(b)   (c)

Figure 1.31 Cantilever structure modelled with a single linear quadrilateral element through the height, 10 elements along the length and 2×2 integration points for each element. (a) Normal stress values at the integration points; contour plot with distribution (b) extrapolated to the nodes and (c) averaged.

Finally, it must be highlighted that stress or strain recovery (also called smoothing) should be used with caution in proximity of *internal boundaries* (i.e. interfaces between different materials) where the stress and strain fields are discontinuous in reality, *outer boundaries* (where extrapolation from optimum sampling points must be used instead) and locations where *stress* or *strain singularities* are expected (e.g. concentrated forces, point supports, sharp corners or crack tips). In the latter case, the smoothing procedure may overlook these singularities.

1.5 COMMONLY USED FINITE ELEMENTS

After the above discussion about the procedures that belong to the FEM, it is possible to describe the most common types of elements and their application fields. Although some consensus in terms of symbols, sign convention, etc., has been achieved by software developers and scientific community, the analyst should refer to the manuals of the adopted software code for further description. Here, the discussion is kept simple, e.g. avoiding curved geometries and changes of mechanical properties throughout the elements.

Table 1.4 Truss element

Displacement $u = a_0 + a_1 r$	Strain $\varepsilon_x = u'$ (constant)	Stress σ_x (constant)	Force N_x (constant)

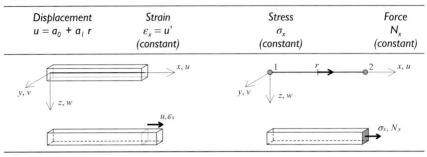

Truss elements (Table 1.4) are bars that model structural members in which the cross-sectional dimensions are small in relation to the element's length. The longitudinal displacement u is the only DOF, producing axial strain and stress, thus longitudinal force N_x, with no bending or shear deformation. The shape functions are usually linear, with consequent constant strain, stress and force throughout the element. Truss elements may be used to analyse bar structures with hinged connections, such as ties, stiffeners in frames or roof trusses.

Beam elements are geometrically similar to truss elements, but they are provided with flexural stiffness, with two additional DOFs, namely, deflection w and rotation θ_y for plane stress problems. Besides, since bending and axial effects are decoupled, Table 1.5 illustrates only bending features complementary to what is shown in Table 1.4. The shape functions, according to Hermite interpolation, are usually cubic, resulting in linear curvature and bending moment. Moreover, according to Bernoulli–Euler beam theory, cross-sections remain plane and perpendicular to the slope of the beam axis. Accordingly, the shear force is calculated satisfying equilibrium conditions (being equal to the first derivative of the bending moment).

Table 1.5 Beam element in plane stress state (Bernoulli–Euler theory)

Displacements $w = a_0 + a_1 r + a_2 r^2 + a_3 r^3$ $\theta_y = -w'$	Generalized strain $\kappa_y = -w''$ (linear variation)	Generalized stress M_y (linear variation)	Shear force Q_z (constant)

Table 1.6 Additional quantities for three-dimensional analysis of beam elements (Bernoulli–Euler theory)

Displacements	Generalized strain	Generalized stress	Shear force
$v = a_0 + a_1 r + a_2 r^2 + a_3 r^3$	$\kappa_z = -v''$	M_z	Q_y
$\theta_z = -v'$	*(linear variation)*	*(linear variation)*	*(constant)*
$\theta_x = b_0 + b_1 r$	$\Delta\theta$ *(constant)*	M_x *(constant)*	

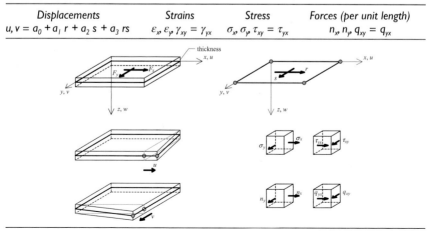

In turn, beam elements in three-dimensional analysis are supplemented with three additional DOFs (and relative generalized strains and stresses) as described in Table 1.6. It is worth reminding that a beam element, when simply pinned at the extremities (i.e. only the three translations are constrained) can rotate about its axis, in a torsional movement, leading to a rigid body motion and not-positive definite stiffness matrix. Beam elements may be used to analyse e.g. two- and three-dimensional frames or floor joists.

Plane stress elements (Table 1.7), sometimes called *membrane* elements, must: (1) have their coordinates in a common plane (*xy*, in this case); (2) have the thickness smaller than the dimensions in the plane of the element and (3) be subjected to loading acting in the plane of the element

Table 1.7 Plane stress elements

Displacements	Strains	Stress	Forces (per unit length)
$u, v = a_0 + a_1 r + a_2 s + a_3 rs$	$\varepsilon_x, \varepsilon_y, \gamma_{xy} = \gamma_{yx}$	$\sigma_x, \sigma_y, \tau_{xy} = \tau_{yx}$	$n_x, n_y, q_{xy} = q_{yx}$

(i.e. F_x and F_y). Similar to truss elements, the only components of displacements are longitudinal and transversal elongations of each node (u and v), leading to three components of strain, namely, ε_x, ε_y, $\gamma_{xy} = \gamma_{yx}$. According to the shape functions in Table 1.7, ε_x is constant in x direction and varies linearly in y direction, ε_y is constant in y direction and varies linearly in x direction and γ_{xy} varies linearly in both directions. The relative components of stresses and forces have the same behaviour. Additional nodes on the edges or within the element increase the polynomial order of the shape functions, thus of strains, stresses and forces approximation. It is worth reminding that plane stress is related to problems in which the stress component perpendicular to the element plane is null ($\sigma_z = 0$). Accordingly, this element may be used for modelling the in-plane behaviour of e.g. masonry walls or reinforced concrete deep beams.

Conceptually similar to plane stress, **plane strain elements** are used to describe long structures for which the longitudinal deformation ε_z can be considered null, e.g. tunnels, retaining walls or dams. Accordingly, it is possible to study the behaviour of a cross-sectional element with a unit thickness and loading on its plane. This element is similar to what was shown for the plane stress problems but, from the constitutive point of view, the imposition of $\varepsilon_z = 0$ represents a constraint that introduces a normal stress σ_z. This constraint may make the material stiffer and limit failure in the case of inelastic analysis due to volumetric locking. However, in terms of elastic analysis, the differences tend to be small and are controlled by the Poisson's ratio of the material. Given the mentioned similarities, plane strain elements are not discussed here.

Another type of element within plane stress problems is the **plate bending element**. If the (flat) plane stress element is the two-dimensional extension of the truss element, the plate bending element is the analogous extension of the beam element. Accordingly, besides the small thickness compared to the in-plane dimensions, force loading is perpendicular to the element face and moments act around the in-plane axes. Similar to Bernoulli–Euler theory, Kirchoff theory assumes that normals of the element plane remain straight and perpendicular to the element plane after the deformation. This means the slopes around the axes are coincident with the first derivatives of deflection. This element may be used for modelling e.g. building slabs.

Commercial codes often use the Mindlin–Reissner theory (normals are kept straight but not perpendicular to the element plane) and Table 1.8 illustrates the physical quantities of this type of element. Each node has three DOFs, namely deflection w and two rotations around the in-plane axes. The shape functions shown lead to curvature κ_x, moment m_x and the shear force q_{xz} constant in x direction and linear in y direction; curvature κ_y, moment m_y and shear force q_{yz} are constant in y direction and vary linearly in x direction. Additionally, as the moments imply a linear distribution of in-plane stresses within the thickness, it is possible to derive these

Table 1.8 Flat plate bending elements (Mindlin–Reissner theory)

Displacements	Generalized strains	Generalized stresses
$w = a_0 + a_1 r + a_2 s + a_3 rs$ $\theta_x, \theta_y = b_0 + b_1 r + b_2 s + b_3 rs$	$\kappa_x, \kappa_y, \kappa_{xy}, \psi_{xz}, \psi_{yz*}$	$m_x, m_y, m_{xy} = m_{yx}, q_{xz}, q_{yz}$

stresses for the upper, mid and lower planes ($+ t/2$, 0 and $- t/2$, respectively, being t the thickness of the plate).

Flat shell elements basically are a combination of plane stress elements and plate bending elements and are not discussed here. This element may be used for modelling e.g. building nucleus or walls subjected to a combination of vertical (gravitational) and horizontal (such as wind or earthquake) loading.

Solid elements are general purpose elements and are usually applied when other elements are unsuitable or would produce inaccurate analysis results. From the geometrical point of view, all element dimensions are of the same order of magnitude (and so does the structural part under consideration). The DOFs are three translations for each node and the consequent stress state is fully three-dimensional. Table 1.9 shows typical elements with a brief description of the shape functions and the relative strain/stress behaviour throughout the element. Typical applications of solid elements are voluminous structures like thick walls, concrete foundations and soil masses.

A different type of element is the **interface element**. This may be used to model e.g. discrete cracking, joints in masonry, no-tension bedding, bond-slip along reinforcement, friction between surfaces, soil–structure interaction or interaction between buildings. According to their shape and connectivity, interface elements can be nodal, linear (between edges of two-dimensional elements) or plane (between faces of three-dimensional elements). Table 1.10 shows the case of a three-dimensional

Table 1.9 Solid elements

Displacements: $u_i = a_0 + a_1 r + a_2 s + a_3 t$ Constant strain and stress distribution	
Displacements: $u_1 = 1/8 (1 - r) (1 - s) (1 - t)$ $u_2 = 1/8 (1 + r) (1 - s) (1 - t)$ $u_2 = 1/8 (1 + r) (1 + s) (1 - t)$ … Strain ε_x and stress σ_x are constant in x direction and vary linearly in y and z direction. A rotation of axes provides similar results for the other components.	
Displacements: $u_i = a_0 + a_1 r + a_2 s + a_3 t + a_4 rt + a_5 st$ Strain ε_x and stress σ_x are constant in x and y and vary linearly in z direction. The same holds for ε_y and stress σ_y.	
Displacements: $u_i = a_0 + a_1 r + a_2 s + a_3 t + a_4 rs + a_5 rt + a_6 st + a_7 rst$ Strain ε_x and stress σ_x are constant in x direction and vary linearly in y and z direction. A rotation of axes provides similar results for the other components.	

Table 1.10 Three-dimensional interface surface

Nodal displacements u, v, w	Relative displacements $\Delta x, \Delta y, \Delta z$	Tractions $\sigma_z, \tau_{zx}, \tau_{zy}$

interface surface. It is worth reminding that this element has no thickness and the distance between the two faces has an illustrative purpose only. According to the drawing, interface elements describe the relation between tractions (normal and shear) and relative displacements

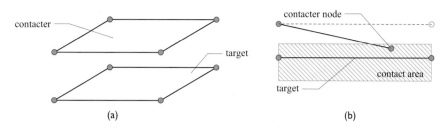

Figure 1.32 Contact element. (a) Definition of contacter and target elements and (b) contact area.

(normal and shear) across the interface. This represents a sort of soft contact, as interpenetration between the two faces occurs in the case of compressive normal action.

Conversely, **contact elements** are meant to detect and evaluate the interaction between elements that may get in contact in subsequent phases of the analysis. Figure 1.32a shows the two elements that, belonging to different parts of the structure, may interact as a consequence of the body deformation. To model this phenomenon, software codes usually adopt two types of contact elements, namely, *contacter* and *target* (Figure 1.32a), or master and slave. Considering line elements (the extension to plane elements is straightforward), their interaction is schematized in Figure 1.32b. If a contacter node falls within an assigned area in the neighbourhood of the target elements, then contact occurs. In this case, the contacter may stick or slide according to the frictional resistance, calculated according to e.g. a Mohr–Coulomb criterion. No interpenetration is allowed (i.e. there is no normal stiffness) and, compared to interface elements, contact elements lead to rigid contacts. For the given purpose, contact elements have no geometry and loading data are not applicable.

Spring elements can be used to model the mutual interaction between two points of the finite element model or between the model and the ground. Besides the specific cases of internal springs, Winkler foundation model represents one of the typical applications of nonrigid constraints. Spring elements request only the definition of the constitutive laws usually about forces and relative displacements. In the simplest cases, spring elements have only one DOF, namely, axial or rotational, but they can be used to model up to six DOFs (three displacements and three rotations) with possible coupling between them. This means that the activation of one DOF affects the response of the spring element regarding the others, e.g. rotational stiffness may be dependent on the axial force. Dual to spring elements, **rigid elements** are used to define a rigid connection (no mutual displacements or rotations) to or within the model.

Finally, it is noted that **tyings** may be applied between DOFs in a finite element model, such that linear relations between specific nodal displacements are ensured (e.g. to define symmetric response, a rigid rotation plate, a fixed boundary or other responses) so that there is a *master* node and one (or more) *slave* nodes, with displacements linearly dependent of the master node.

BIBLIOGRAPHY AND FURTHER READING

Bathe, K.-J. (2014) *Finite Element Procedures*. 2nd edn. Watertown, MA: Klaus-Jurgen Bathe.

Cook, R. D., Malkus, D. S. and Plesha, M. E. (1989) *Concepts and Applications of Finite Element Analysis*. Canada: John Wiley & Sons, Inc.

Zienkiewicz, O. C., Taylor, R. L. and Zhu, J. Z. (2013) *The Finite Element Method: Its Basis and Fundamentals* (vol.1). 7th edn. Oxford: Butterworth-Heinemann.

Chapter 2

Nonlinear structural analysis

INTRODUCTION

In Chapter 1, the fundamentals of finite element method have been discussed within the context of linear elastic problems. From the mathematical point of view, all the governing equations are linear with respect to the unknown variables. Besides the innumerable cases in which this theory provides acceptable results in engineering, deviance from the linear assumptions requests richer *nonlinear models*. In a simple approach to civil engineering practice, a nonlinear analysis should be performed when: (1) large displacements occur, and the configuration upon loading is substantially different from the initial one (even in the presence of small strains) and (2) nonlinear stress–strain relationships for the material behaviour (e.g. yielding steel or cracking concrete) have relevant consequences in the performance of the structures. Nonlinear material effects regard also the case of elastic nonlinear laws (as found in some soils, rubber or alloys) or in presence of important time-dependent effects (e.g. shrinkage or creep). Other sources of nonlinearities are large strains (generally not relevant in civil engineering practice), loads that change direction while the structure deforms, and the occurrence of new contacts. The latter may need to be considered due to changes in boundary conditions (e.g. modern bridge construction with temporary supports or ancient monumental structures with construction phases, alterations and additions). These additional cases fall outside the scope of this text and will not be considered. The first part of the present chapter deals with large displacements, which belong to the family of *geometrical nonlinear problems*, whereas *material nonlinearities* are addressed in Chapter 3. The second part of this chapter is devoted to nonlinear computation strategies and advanced solution procedures. For the sake of comparison, these techniques are employed to discuss the same one-dimensional example. The concluding example of the analysis of a steel frame structure allows discussing the impact of using different advanced solution procedures in an engineering problem.

DOI: 10.1201/9780429341564-2

2.1 POTENTIAL SOURCES OF NONLINEARITIES

Recalling the basic scheme of structural problems given in Chapter 1, Figure 2.1 shows the main sources of nonlinearities and how these influence the governing equations. Besides material nonlinearities, discussed in the following chapter, geometrical nonlinearities are here related to large, (or finite, contrarily to small or infinitesimal) displacements and rotations, but not large strains. Consequently, in evaluating strains, higher order derivatives of displacements may not be neglected. In turn, as the difference between undeformed and deformed configurations is of importance, the equilibrium equations need to be computed considering the final (deformed) configuration. In general, this configuration is unknown being, de facto, the primary goal of the analysis. Considering Figure 2.1, boundary conditions may also be a source of nonlinearity, such as unilateral constraints (e.g. the supports of an overturning block) or possible new contacts that may develop in the deformed configuration.

With the aim of introducing geometrical nonlinearities, let us consider the following examples. Figure 2.2 shows a rigid block with null mass and aspect ratio (i.e. height over width) equal to two, subjected to an overturning force on its top. Looking at Figure 2.2a, simple considerations of statics (namely, moment about the hinge on the bottom left corner) allow calculating the linear spring reaction equal to $2F$ and its elongation ε. The equilibrium condition is given in Eq. (2.1) where E and A are Young's modulus and the cross-section area of the spring, respectively. If equilibrium is computed in the deformed configuration, considering the overall lateral translation of the spring and the uplift of the application point of the external force, the new relation is presented in Eq. (2.2), together with the necessary manipulation to evaluate the nonlinear elongation ε^{nl}. The two approaches,

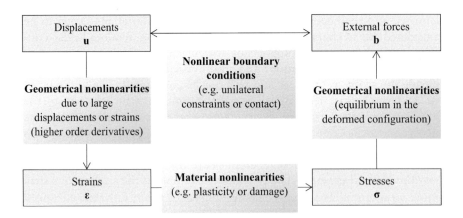

Figure 2.1 Schematic representation of the main sources of nonlinearities within the governing equations of structural problems.

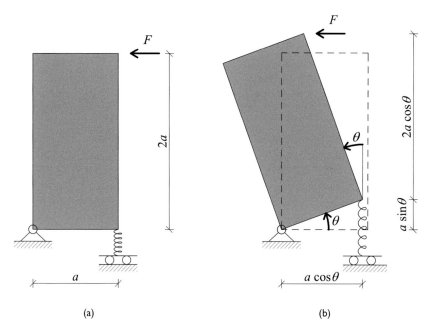

Figure 2.2 Rigid block subjected to an overturning force. (a) Linear and (b) nonlinear kinematics.

apparently different, provide coincident results for small values of θ (as $\theta \ll 1$ and $\tan \theta \to 0$), but differ for larger values of θ. For instance, when $\theta = 10°$ the difference between linear and nonlinear approaches is around 10%, achieving 50% when $\theta = 45°$. It is worth noticing that the possibility for the spring to laterally slide is a notable simplification: a fixed spring would have led the reaction to be inclined and more complicated calculations.

$$EA\varepsilon = 2F \qquad \rightarrow \qquad \varepsilon = \frac{2F}{EA} \tag{2.1}$$

$$Fa(2\cos\theta + \sin\theta) = \varepsilon^{nl} EA \, a\cos\theta$$

$$\varepsilon^{nl} = \frac{F(2\cos\theta + \sin\theta)}{EA\cos\theta} = \frac{F}{EA}(2 + \tan\theta) = \frac{2F}{EA}\left(1 + \frac{\tan\theta}{2}\right) \tag{2.2}$$

$$\varepsilon^{nl} = \varepsilon\left(1 + \frac{\tan\theta}{2}\right)$$

Let us now consider a rigid block with the same geometry, but with a certain mass, laying on a rigid support (Figure 2.3). If the only force acting is the self-weight W, the base reacts with a uniform stress distribution equal to the total

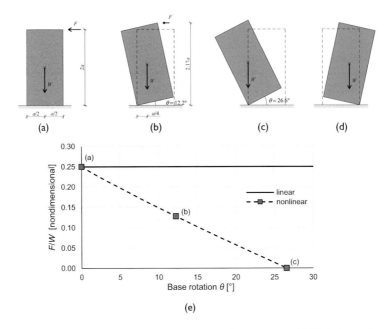

Figure 2.3 Rigid block subjected to an overturning force on top. (a) Undeformed configuration; (b) intermediate and (c) ultimate rotation; (d) possible rebound and
hinging on the other side; (e) force-rotation diagram.

weight divided by the contact area. However, if an overturning force F acts
on its top, and sufficient friction is present, the block starts rotating about its
edges (basically, the bottom corners in the elevation view). As a matter of fact,
the contact area becomes (theoretically) a line, i.e. a cylindrical hinge, and no
reactions are possible in the uplifted portion of the base. The correct relationship between the magnitude of the overturning force and the rotation at the
base can be estimated only if equilibrium is considered in the deformed configuration. Figure 2.3a–c describes the geometry of three significant steps of
the analysis, namely, undeformed, intermediate and ultimate configurations,
respectively. Calculating F and the rotation angle at the base θ in the three
cases is straightforward from basic geometry (see Figure 2.2b). Figure 2.3e
provides the full diagram F/W vs. θ where the markers are representative
of the three configurations discussed. It is worth noticing that, in the case
the force suddenly ceases before reaching the ultimate rotation, a dynamic
problem occurs as the block goes back to the rest position and rebounds on
the other side (Figure 2.3d). Clearly, the contact of the block with the support dramatically changes throughout the analysis. As shown, the difference
between linear analysis (constant force capacity given by $F/W = 0.25$) and
nonlinear analysis is impressive (varying force capacity F/W between 0.25 at
rest and the null value at a maximum rotation angle of 26.6°).

Another example about the differences between linear and nonlinear evaluation of the equilibrium conditions is shown in Figure 2.4. The cantilever

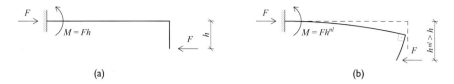

Figure 2.4 Cantilever beam subjected to an eccentric force. (a) Linear and (b) nonlinear kinematics.

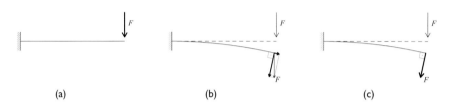

Figure 2.5 Cantilever subjected to a transversal force at its extremity. (a) Linear kinematics. Nonlinear kinematics with (b) force maintaining the vertical direction and (c) force maintaining the normality to the beam axis.

beam is subjected to an eccentric axial force F. By deforming the structure, the force lever arm passes from h in the undeformed configuration (Figure 2.4a) to h^{nl} in the deformed one (Figure 2.4b). Neglecting this change when the beam undergoes large displacements leads to a significant error in evaluating the bending moment at the fixed support.

Besides the force magnitude, which may follow a given law (either quasi-statically or dynamically), the examples in Figures 2.2 and 2.4 showed cases in which the external force keeps the direction in the deformed configuration. But this may not be the case and Figure 2.5 illustrates the case of a cantilever beam subjected to a transversal force at its extremity. If the external force maintains the vertical direction, e.g. due to gravitational effects (Figure 2.5b), a longitudinal component arises, producing tension in the beam in addition to the stress arising from bending. The case in which the force is always perpendicular to the beam axis, e.g. due to wind effects and air pressure normal to the surface, is shown in Figure 2.5c. As expected, the three cases reported in Figure 2.5 are only coincident if the deflection of the cantilever is small with respect to the dimension of the structural element.

2.2 GEOMETRICAL NONLINEARITIES AND STABILITY

The present section discusses selected paradigmatic structural problems to introduce the reader to geometrical nonlinearities, negative structural tangential stiffness and stability problems.

2.2.1 Introduction to finite displacements: the von Mises truss structure

The von Mises truss, named after the studies of Richard von Mises (1883–1953) in the early 1920s, is a simple structure consisting of two bars connected to each other by a common top joint. For the present discussion, the symmetric structure of Figure 2.6a is considered, where a vertical load is applied downward in correspondence of the top joint. Due to symmetry, the structure has a single degree of freedom and the vertical displacement w is chosen (positive downward). Figure 2.6b shows the deformed configuration when linear kinematics is considered. For the sake of clarity, linear kinematics assumes that an arc of a circumference can be approximated by the respective tangent. Accordingly, the two dashed lines of Figure 2.6b are assumed of the same length, as the bar elongation is assumed small. Certainly, this assumption can be accepted only for small values of w.

Due to symmetry, the strain is equal in the two bars and can be calculated according to Eq. (2.3), where the minus indicates contraction. The calculation of the stress σ and the resultant force N is straightforward knowing the Young's modulus E and the cross-section area A, as reported in Eq. (2.4). As the only kinematic variable is the vertical displacement w, the equilibrium equation regards the vertical components of the forces, as shown in Eq. (2.5). In the framework of small displacements and rotations, it is possible to write the equilibrium equation in the undeformed configuration

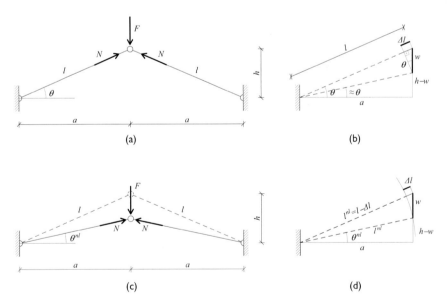

Figure 2.6 Symmetric von Mises truss subjected to a central vertical load. Linear analysis: (a) equilibrium and (b) kinematics. Nonlinear analysis: (c) equilibrium and (d) kinematics.

(Figure 2.6a). Accordingly, the vertical component of the force in each bar is $N\sin\theta$ upward (see Figure 2.6a) and the result is a linear relation between the displacement w and the external force F. The relative diagram is shown in Figure 2.7a, where $EA = 1,000\,\text{N}$ and $a/h = 5$ are assumed.

Despite the lack of apparent mistakes, the present solution is not acceptable when w becomes sufficiently large. For instance, let us consider the deformed configuration when $w = h$ (i.e. $w/h = 1$), as shown in Figure 2.7. According to the relation found in Eq. (2.5), the external force should be $15.1\,\text{N}$ but, upon basic statics considerations of equilibrium, null F is the only admissible value for this configuration. As expected, the assumption about linear kinematics leads to evident errors when large displacements are considered.

$$\varepsilon = \frac{\Delta l}{l} = -\frac{w\sin\theta}{l} = -\frac{wh}{l^2} \tag{2.3}$$

$$\sigma = E\varepsilon$$
$$N = EA\varepsilon \tag{2.4}$$

$$-2N\sin\theta = F$$
$$\frac{2EA\sin^2\theta}{l}w = F \tag{2.5}$$

Let us now consider the nonlinear approach. With the symbols shown in Figure 2.6d, strain (thanks to Pythagoras' theorem) and $\sin\theta^{nl}$ are calculated according to Eqs. (2.6) and (2.7), respectively. In Eq. (2.6) a minus is again introduced meaning contraction. Note that both calculations consider the correct geometry for the displacement (i.e. θ^{nl}) and length of the bar ($l^{nl} = l - \Delta l$), while the strain is still assumed small (i.e. $\Delta l/l$). Accordingly, the original cross-section A is assumed to remain constant throughout the

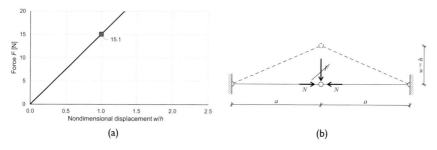

(a)   (b)

Figure 2.7 Symmetric von Mises truss subjected to a central vertical load ($EA = 1,000\,\text{N}$ and $a/h = 5$). (a) Linear response; (b) geometrical representation of the case $w = h$.

analysis, assumption that would be incorrect in the case of large strains. In line with previous discussion, the equilibrium condition is now calculated in Eq. (2.8), see Figure 2.6c. The new relationship found between w and F is nonlinear and the respective diagram is shown in Figure 2.8, together with the linear relation already found. The figure shows also the deformed configurations of significant points of the graph.

As expected, the linear approach approximates well the nonlinear diagram only for small values of w. It is possible to demonstrate that, by assuming $w/h \ll 1$ and with due simplifications, the set of nonlinear equations are coincident with the linear ones. In particular, Eq. (2.6) is rewritten in Eq. (2.9) by expanding the second power and indicating the ratio within the square root as η. Considering the undeformed configuration, $w = 0$, thus $\eta = 0$, the value of the strain in this neighbourhood can be calculated thanks to the Taylor's expansion. By neglecting the contribution of the order higher than two, the square root in Eq. (2.9) is evaluated in the brackets of Eq. (2.10). With subsequent substitutions and assuming $w/l \ll 1$, the consequent result is equal to what was calculated in Eq. (2.3). Analogously, Eq. (2.7) becomes Eq. (2.11). By combining Eqs. (2.8), (2.10) and (2.11), it is possible to obtain the same result of Eq. (2.5), i.e. the *linear theory can be obtained as an approximation of the nonlinear theory*.

In turn, for larger values of w, for instance, $w/h = 0.1$, the error rises to about 16%, becoming almost 100% when $w/h = 0.4$ (i.e. the linear approach overestimates the capacity of the structure providing a force F that is the double of the nonlinear one). The nonlinear approach shows interesting results in the terms of external load, being null in A, C and E, as required by equilibrium and symmetry. The force–displacement diagram has a local maximum in B and a local minimum in D (the word "local" is here referred to the function in Eq. (2.8) whose domain embraces the entire range of real numbers).

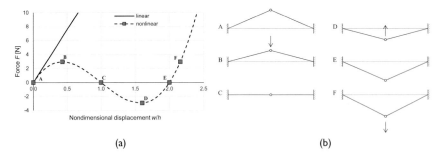

(a)   (b)

Figure 2.8 Symmetric von Mises truss subjected to a central vertical load ($EA = 1,000\,\text{N}$ and $a/h = 5$). (a) Comparison between linear and nonlinear response and (b) deformed configurations at key points in the graph.

$$\varepsilon = \frac{\Delta l}{l} = \frac{l^{nl} - l}{l} = \frac{\sqrt{(h-w)^2 + a^2} - \sqrt{h^2 + a^2}}{\sqrt{h^2 + a^2}} = \frac{\sqrt{(h-w)^2 + a^2}}{\sqrt{h^2 + a^2}} - 1 \quad (2.6)$$

$$\sin \theta^{nl} = \frac{h - w}{l^{nl}} = \frac{h - w}{\sqrt{(h-w)^2 + a^2}} \quad (2.7)$$

$$-2EA \underbrace{\left(\frac{\sqrt{(h-w)^2 + a^2}}{\sqrt{h^2 + a^2}} - 1 \right)}_{\varepsilon} \underbrace{\left(\frac{h - w}{\sqrt{(h-w)^2 + a^2}} \right)}_{\sin \theta^{nl}} = F \quad (2.8)$$

$$\varepsilon = \frac{\sqrt{(h-w)^2 + a^2}}{\sqrt{h^2 + a^2}} - 1 = \sqrt{\frac{h^2 + w^2 - 2wh + a^2}{h^2 + a^2}} - 1$$

$$= \sqrt{1 + \underbrace{\frac{w^2 - 2wh}{h^2 + a^2}}_{\eta}} - 1 = \sqrt{1 + \eta} - 1 \quad (2.9)$$

$$\varepsilon \approx \left(1 + \frac{1}{2} \eta \right) - 1 = \frac{1}{2} \frac{w^2 - 2wh}{h^2 + a^2} = \frac{1}{2} \frac{w^2 - 2wh}{l^2}$$

$$= \frac{1}{2} \left(\frac{w^2}{l^2} - \frac{2wh}{l^2} \right) \approx -\frac{wh}{l^2} \quad (2.10)$$

$$\sin \theta^{nl} = \frac{h - w}{l^{nl}} = \frac{h - w}{\sqrt{(h - w)^2 + a^2}} \approx \frac{h}{l} \quad (2.11)$$

A closer inspection of Figure 2.8a reveals that the relation between force F and displacement w is no longer univocal, e.g. $F = 2$ kN is associated to three different values of displacement w. At this point, it seems interesting to change the perspective and consider the problem from the displacement point of view. In the first part of the diagram, a positive relation between w and F is evident, i.e. if the displacement w increases, also the external force F increases, as usually expected in structural systems due to the reaction of the system to an external deformation. After point B, an increase of displacement is associated to a reduction of the force; after point C, the external force in the system becomes negative, meaning that the structure must be pulled to maintain the equilibrium. This is due to instability, as the system naturally buckles at point B. At this point the external force must be lowered and even pull back the structure to keep the system under control. Note that, after point E, the structure regains a typical response with increasing pushing force for increasing displacement.

Dealing with instability and nonlinear problems in general, it is worth considering the role of the structural stiffness, which is the relation between force and displacement. Looking at Figure 2.8a, this relation is nonlinear, and stiffness varies continuously. Accordingly, it is convenient to introduce the concept of (variable) *tangent stiffness*, i.e. the derivative of the force–displacement diagram, in opposition to the (constant) *linear elastic stiffness* of Figure 2.7a. Based on the definition of tangent stiffness, the diagram of Figure 2.8a can be divided into three branches: positive stiffness in A–B (force and displacement are both increasing), negative stiffness in B–D (force decreases while displacement increases) and positive stiffness again from D onward, being null in B and D (where the force is stationary, i.e. it maintains a constant value in a small neighbourhood of the correspondent displacement).

The sign of the stiffness is shown in Figure 2.9a. Without entering the merit of the discussion, it is possible to state that the equilibrium is stable, unstable and indifferent (or neutral) when the stiffness is positive, negative or null, respectively. As a general definition, considering an arbitrarily small displacement that introduces a perturbation of the original equilibrated state, the equilibrium is said (see Figure 2.9b):

- *stable* when the structure tends to restore the original equilibrium state (in the present case, the removal of load brings back the structure to the previous configuration);
- *unstable* when the structure moves even further away (in the present case, the load must be reduced and even change sign to avoid uncontrolled failure of the system or a dynamic response);

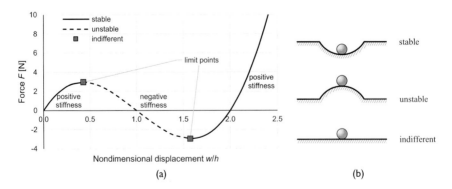

Figure 2.9 Stability definition. (a) Response of the symmetric von Mises truss subjected to a central vertical load; (b) illustrative examples.

- *indifferent* (or neutral) when the structure remains in the new equilibrium state (in the present case, this means that the structure has no opposition to movement).

Let us consider the truss problem from an experimental perspective. In the case a traditional experimental test setup is considered, the truss would be loaded by a jack applied at the top joint. Being the jack only able to increase the external force F by applying small increments in the oil pressure (this is often referred to as *force control*), the force cannot be reduced in the case of buckling, with a consequent collapse or a sudden uncontrolled jump in displacement (see Figure 2.10). Modern laboratories are equipped with displacement-controlled actuators with a control loop mechanism employing feedback given by a displacement transducer; as a consequence, it is possible to reduce the load when needed and impose a certain displacement rate to the system rather than a certain force rate (this is often referred to as *displacement control*). As small displacement rates are usually adopted, these tests are still considered quasi-static, and the inertia effects can be neglected.

If force control is enforced in the structural analysis of the von Mises truss (see Section 2.3.2 for further details), its response should be studied considering finite increments of the external forces (essentially *load steps* or *load increments*), which allows calculating the corresponding equilibrated displacements. In other words, the diagram drawn in Figure 2.8a may be idealized by a discrete number of points in which the magnitude of the external force is known, and the displacement must be found such that the internal and external forces are in equilibrium. Looking at Figure 2.10, this strategy inevitably leads to a jump in the force–displacement diagram according to the arrow (conceptually similar to a force-controlled test in

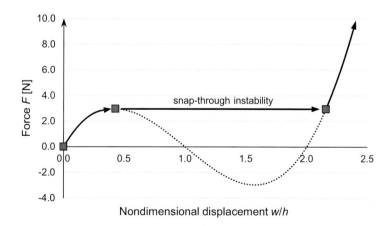

Figure 2.10 Symmetric von Mises truss subjected to a central vertical load. Snap-through instability.

the laboratory). This phenomenon is called *snap-through instability* and, as it is possible to observe, a substantial part of the structural behaviour would be missed in the computation (in the laboratory, a dynamic response would be faced with possible uncontrolled failure). It is worth mentioning that the point with zero stiffness prior to instability is called a *limit point*.

In general, large jumps (following either geometrical or material nonlinearities) are badly handled by numerical solution methods as they entail convergence issues and large errors in the integration of the nonlinear equations. Accordingly, adequate integration schemes of nonlinear quantities must be enforced, together with adequate solution procedures for nonlinear problems (as shown ahead in this and following chapters). These strategies can be gathered under the name of *load path–following methods*, whose name is self-explaining. The equilibrium states are "followed" throughout the entire diagram (i.e. load path, in which load is understood as any action, namely, forces and displacements), even in the unstable portions of it, rather than consider assigned increments of external force or displacement.

2.2.2 Geometric stiffness and linear buckling analysis

To highlight the role of geometry on equilibrium, let us focus on a similar truss under tensile action (the choice of tension in lieu of compression is meant to avoid instability, as explained in Section 2.2.3). This is shown in Figure 2.11a and represents a pretensioned cable truss (such as the one used to support a glass façade). As it is evident, the structure is not statically determined, as it is composed of three aligned hinges. According to Chapter 1, the stiffness matrix of a rigid mechanism is singular, and equilibrium cannot be satisfied (there are no vertical components of the reactions to counteract the external force F). This holds if equilibrium equations are written in the undeformed configuration. However, as done in the previous section, equilibrium is sought in the deformed configuration of Figure 2.11b where, due to symmetry, the central node undergoes only a vertical displacement w. For the sake of simplicity, the deformed configuration is addressed here considering small displacements. Accordingly, $w \ll l$ and the increase of deformation of the two trusses can be neglected.

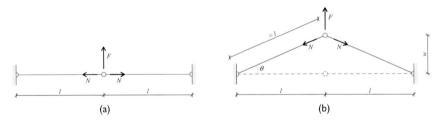

Figure 2.11 Pretensioned truss structure undergoing a subsequent transversal action. (a) Undeformed and (b) deformed configuration.

As a consequence, the pretension force is assumed as constant and the equilibrium equation regarding the vertical components of forces is derived in Eq. (2.12).

This is an important result as equilibrium (for small displacements) does not rely on elasticity but on the existing stress state of the structure and its geometry. Willing to extend the concept of stiffness, i.e. how a certain body deforms in response to an applied force, $2N/l$ is the *geometric stiffness* of the problem and the dependence from N is evident. It is rather intuitive that, N being dependent on the axial strain of the trusses, the larger is the vertical displacement, the larger is the value of the internal forces in the bars. In other words, the change of geometry produces an implicit increase of stiffness that goes under the name of *stress stiffening*. The next example shows the simultaneous presence of elastic and geometric stiffnesses.

$$2N \sin \theta = F$$

$$2N \frac{w}{l} = F$$

(2.12)

Let us consider the case of a constrained rigid column with height l and no weight, subjected to a tensile axial force F_v, as shown in Figure 2.12a. The axial stiffness of the spring on top is synthetically indicated with k. If a horizontal force F_h is applied (Figure 2.12b), equilibrium is guaranteed only in the deformed configuration, so that the spring can develop an elastic reaction. In Figure 2.12b, the deformed configuration is characterized by

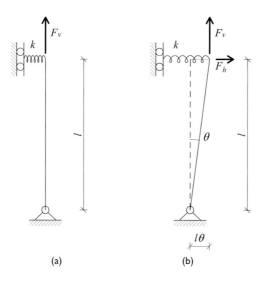

(a)   (b)

Figure 2.12 Rigid column under vertical force F_v. (a) Undeformed configuration and (b) subsequent application of a horizontal force F_h.

a clockwise rotation θ at the base. The consequent equilibrium equation (considering small displacements) regarding the moments about the hinge at the base is written in Eq. (2.13). In this case, besides the elastic stiffness of the spring, the presence of the vertical force F_v is beneficial as it provides a stabilizing effect. Intuitively, for a given horizontal force F_h, the larger is F_v, the smaller is the rotation θ.

$$F_h l - F_v l \theta - k \theta l^2 = 0$$

$$F_v \theta + k \theta l = F_h \qquad \rightarrow \qquad \underbrace{\left(F_v + kl \right) \theta = F_h}_{\text{stiffness}} \tag{2.13}$$

The previous unidimensional examples were straightforward and intuitive, and so were the respective equilibrium equations. However, a similar approach can be followed for multidimensional problems where the equilibrium equations are usually written in matrix form. Here unidimensional and multidimensional are referred to the number of unknown variables (i.e. DOFs), not to the geometrical dimension of the problem. Without loss of generality, based on Eq. (2.13), the generic equilibrium equation is shown in Eq. (2.14), where $\mathbf{K}_l$ and $\mathbf{K}_{nl}$ are the linear and nonlinear stiffness matrices, respectively. Whereas the $\mathbf{K}_l$ was introduced in Chapter 1, $\mathbf{K}_{nl}$ collects the contributions due to the existing stresses and loads in the deformed configuration, as for the two previous examples. In turn, $\mathbf{f}$ and $\mathbf{d}$ represent the external load and displacements to ensure equilibrium, respectively. In the case of nonlinear problems, both $\mathbf{K}_l$ and $\mathbf{K}_{nl}$ may change throughout the analysis, due to material (e.g. nonlinear constitutive law) and geometrical nonlinearities (e.g. finite displacements). Accordingly, $\mathbf{f}$ and $\mathbf{d}$ should be considered in their incremental form as usually done with nonlinear analysis (see Section 2.3).

$$\left(\mathbf{K}_l + \mathbf{K}_{nl} \right) \mathbf{d} = \mathbf{f} \tag{2.14}$$

The definition of the geometric stiffness matrix in continuum mechanics and its implementation in finite element method (FEM) falls outside the scope of the present book. However, the role of $\mathbf{K}_{nl}$ is discussed here keeping the focus on the structural stability of rigid bodies and elastic supports. Looking at Eq. (2.14), it is possible to consider $\mathbf{K}_{nl}$ as a nonlinear "correction" of the linear stiffness matrix $\mathbf{K}_l$. It is worth reminding the effect of $\mathbf{K}_{nl}$ was fundamental to solve the example of Figure 2.11 (where $\mathbf{K}_l$ was singular). In the example of Figure 2.12, the correction was beneficial but not essential.

From the mathematical point of view, the study of Eq. (2.14) deserves additional considerations. In Chapter 1, the stiffness matrix was assumed nonsingular (i.e. the matrix has an inverse). This means that, once rigid movements are prevented, the solution of any linear problem is unique and there is a one-to-one relationship between loads and displacements. A direct consequence of this statement is the fact that, in the absence of external forces, the only admissible solution of the problem $\mathbf{K}\mathbf{d} = 0$ is $\mathbf{d} = 0$, the so-called trivial solution. Considering the nonlinear correction, the nature of the total stiffness matrix $\mathbf{K}_l + \mathbf{K}_{nl}$ is not known. It may happen that, for particular values of existing stresses and loads (so-called *critical values*), this total stiffness matrix becomes singular and, besides the trivial solution, the problem may admit one or more solutions. The consequent solutions are of the form $\mathbf{d}_{cv} \neq 0$. If the solution is not unique, the force–displacement diagram, i.e. the graphical relationship between the external force and the kinematic variables, shows a *bifurcation*. These concepts will be clearer in Section 2.2.3.

Going back to Eq. (2.14), it is of interest to discuss the critical values of existing loads that lead to bifurcation (even before the application of subsequent increments of load). Mathematically, the problem is rewritten in Eq. (2.15) where the multiplier λ is introduced and the sign minus is adopted for conventional purposes. Assuming that $\mathbf{K}_{nl}$ is linearly dependent on the existing forces, and that all the existing forces are increased proportionally to the same scalar λ, the goal of the analysis is to find the critical values λ_{cv} that make the full stiffness matrix singular. In linear algebra, this is an *eigenvalue problem*, whose *characteristic equation* can be obtained from Eq. (2.16). To each eigenvalue λ_{cv}, a correspondent eigenvector $\mathbf{d}_{cv}$ is calculated from Eq. (2.15). In structural engineering, this is named *linear buckling analysis* and provides a multiplier of the existing forces for which the displacements are not bounded (i.e. for which uncontrolled collapse occurs).

$$\left(\mathbf{K}_l - \lambda \mathbf{K}_{nl} \right) \mathbf{d} = 0 \tag{2.15}$$

$$\det \left(\mathbf{K}_l - \lambda \mathbf{K}_{nl} \right) = 0 \tag{2.16}$$

It is noted that, by expanding Eq. (2.15), the equality $\mathbf{K}_l \, \mathbf{d} = \lambda \, \mathbf{K}_{nl} \, \mathbf{d}$ is evident, which can be synthesized as the equality between elastic and geometric forces. In other words, the eigenvalue problem is aimed at investigating the displacement field $\mathbf{d}$ for which the elastic response of the structure ($\mathbf{K}_l \, \mathbf{d}$) coincides with the geometric response induced by the existing loads ($\lambda \, \mathbf{K}_{nl} \, \mathbf{d}$). Additionally, $\mathbf{d}$ can be increased by an arbitrary scalar without affecting the equality of Eq. (2.15). Accordingly, the eigenvectors $\mathbf{d}_{cv}$ are named *modes*: they provide information about the shape of the deformed configurations, but they do not provide any information on the magnitude of the displacements. This is one of the drawbacks of linear buckling analysis, which provides reliable results only in a configuration close to the original one (i.e. under the assumption of small displacements).

2.2.3 Postbuckling analysis and geometrical imperfection

Let us now consider again the structure of Figure 2.12, this time undergo-ing an axial compressive force. The column is shown in Figure 2.13a, where the height of the column is $l = 5\,\text{m}$. The axial stiffness of the spring on top is $k = 10\,\text{N/m}$. If the equilibrium condition is written in the undeformed con-figuration, no matter the magnitude of F, the equilibrium is always satisfied thanks to the equal and opposite reaction provided by the hinge at the bot-tom. Consequently, the column does not undergo any rotation.

It is quite intuitive to understand that this result is not exhaustive: a small perturbation may lead the column to buckle to another configuration. In this regard, let us consider a small rotation as shown in Figure 2.13b. In the framework of linear analysis ($\sin\theta \to \theta$ and $\cos\theta \to 1$), the equilibrium of moments about the hinge leads to Eq. (2.17). Proceeding with a linear buckling analysis, it is possible to calculate the critical value F_{cv} and the related mode. Given the simplicity of the problem, the same result can be obtained by direct observation of Eq. (2.17). As the product must be zero, either $\theta = 0$ or $F = kl$, the former confirming the result already found for the undeformed column. The latter, instead, represents the critical value $F_{cv} = kl = 50\,\text{N}$. Once this value is reached, the equilibrium conditions are satisfied no matter the magnitude of rotation. In other words, the mode shape is known but not its magnitude: with $F = F_{cv}$, the structure may either stand or rotate indefinitely.

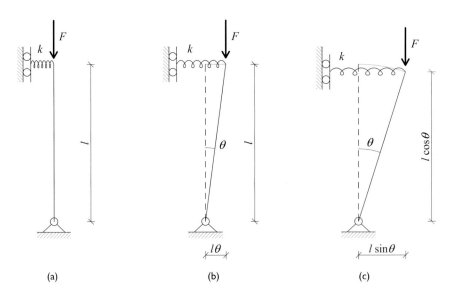

(a)   (b)   (c)

Figure 2.13 Rigid truss under compressive axial force. (a) Undeformed configuration; (b) linear (small rotation) and (c) nonlinear kinematics (large rotation).

As the obtained result of an arbitrary rotation violates the hypothesis of a small angle and of the intuitive insight of the problem, it is interesting to analyse the structure from the nonlinear point of view. In particular, considering Figure 2.13c, the new equilibrium condition is reported in Eq. (2.18). Again, the possibility for the spring to vertically slide is a nontrivial simplification. The nonlinear equation is similar to the previous one (they are equal in the case $\theta \ll 1$) and two possible solutions are evident: either $\sin \theta = 0$ (thus $\theta = 0$, as already discussed) or $F = kl \cos \theta$. The latter gives the nonlinear relation between the rotation θ at the base of the column and the magnitude of the external force. In the case $\theta \to 0$, the former critical value $F_{cv} = kl$ is obtained.

$$kl^2\theta = Fl\theta \qquad \rightarrow \qquad \left(kl - F\right)\theta = 0 \qquad\qquad (2.17)$$

$$k\left(l \sin \theta\right)\left(l \cos \theta\right) = Fl \sin \theta \qquad \rightarrow \qquad \left(kl \cos \theta - F\right)\sin \theta = 0 \qquad (2.18)$$

The results derived in the present subsection are drawn in Figure 2.14, and the following comments are noteworthy:

1. If $F < F_{cv}$, the column undergoes no rotation (i.e. *fundamental path*) and the equilibrium is stable.
2. When $F = F_{cv}$, a *bifurcation point* is reached with two different solutions: the structure may either remain perfectly straight or suffer a rotation (denoted here as *bifurcated path*) in the anticlockwise (negative) or clockwise (positive) direction. *This aspect is fundamental as it contradicts the uniqueness of the solution in linear elastic problems.* In the case linear kinematics is considered, the bifurcated path

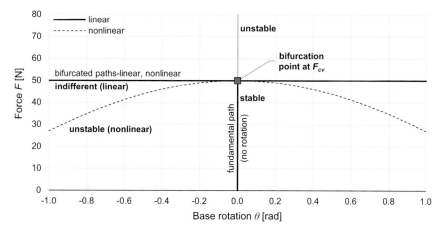

Figure 2.14 Rigid column under compressive axial force. Equilibrium paths and bifurcation point.

describes an indifferent equilibrium (incorrect for large rotations). In the case of nonlinear kinematics, the *bifurcated path* is unstable. As already mentioned, linear and nonlinear diagrams provide the same results only in the neighbourhood of $\theta = 0$.

3. $F > F_{cv}$ occurs only with null rotation. In this case, the equilibrium is unstable, and any perturbation may lead the structure to suddenly jump into the bifurcated path. Given the fact that $F \leq F_{cv}$ on this path, either F reduces in magnitude or the structure will collapse.

4. The bifurcation point occurs for a specific load F_{cv}. The critical force associated to this load level must be of a certain magnitude and direction. In the present case, bifurcation occurs only in presence of compressive (downward) force equal to 50 N.

This problem stresses one important difficulty related to the usage of FE in the nonlinear range, which is the appearance of multiple solutions. In mathematical terms, it is possible to state that the problem is not *well-posed* but it is *ill-posed*. This is demanding for the solution procedure, which, ideally, should be able to converge to the solution with the lowest structural capacity, and is demanding for the analyst, who may need to try different solution procedures in the case convergence cannot be found or the solution found represents the unloading path (and not the sought loading path). It is fair to state that different analysts with the same software code may obtain more or less complete results, depending on their skills and perseverance.

To mitigate this issue, imperfections can be considered in slender structures that are prone to instability. Several cases are shown in Figure 2.15, where a rotational spring with stiffness $k_r = 100$ Nm/rad is introduced, keeping the height of the column equal to $l = 5$ m. In all cases, imperfections are detrimental for clockwise rotation because they lead to an increase of the clockwise external moment. Figure 2.15d is further studied here (regarding an eccentric force), whose equilibrium equation of the moments about the hinge at the base is written in Eq. (2.19). In the case of small rotation and null eccentricity, the critical load is $F_{cv} = k_r/l = 20$ N.

$$k_r \theta = F(l \sin \theta + e \cos \theta) \quad \rightarrow \quad F = \frac{k_r \theta}{l \sin \theta + e \cos \theta} = \frac{k_r}{l} \frac{\theta}{\sin \theta + \frac{e}{l} \cos \theta} \quad (2.19)$$

Figure 2.16 shows the $F - \theta$ diagram regarding the cases with and without eccentricity (where e is assumed equal to $l/200$). The *fundamental path* of the column remaining straight is also reported, unstable when $F > F_{cv}$. More in detail, if eccentricity is null, the diagram is concave up (i.e. convex) and symmetric, i.e. the structure has the same response for either anti- or clockwise rotations. In turn, the eccentricity (to the right, in this case) pushes the structural response towards clockwise rotation and no bifurcation is present. Even small values of F lead the structure to move away from the rest

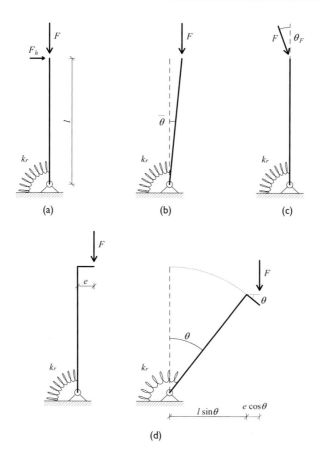

Figure 2.15 Rigid column under compressive axial force and geometrical imperfections. (a) Presence of a perturbative horizontal force; imperfect alignment of (b) column or (c) force; (d) eccentric force and related nonlinear kinematics.

position. In turn, in correspondence to $\theta = 90°$ (≈ 1.5 rad), the effect of the eccentricity is null, and the two diagrams meet.

The analysis after the bifurcation point is addressed as *postbuckling analysis*, either stable or unstable. As the exact postbuckling solution is hard to obtain due to bifurcation (looking at Figure 2.16, when $e = 0$, there are three solutions for a given $F > F_{cv}$), a continuous response, in some cases, can be triggered by introducing a geometric imperfection. In other words, it is possible to follow the solid line of Figure 2.16, avoiding, thus, bifurcation or unstable paths. This, ideally, will provide some response in the buckling configuration (i.e. related to the mode shape) before the critical load is reached. Usually, FE software codes allow to define an imperfection by adding to the original geometry some scaled superposition of the buckling configuration, displacements resulting from a static analysis or design code values directly. An example is discussed at the end of the following subsection.

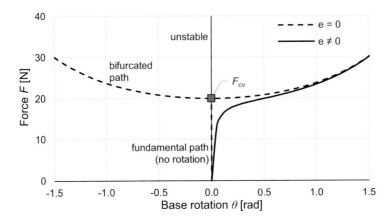

Figure 2.16 Rigid column under compressive axial force. Equilibrium paths with and without eccentricity.

2.2.4 Stability of compressed elements

For the sake of simplicity, the previous examples dealt with stability problems of rigid elements and springs. Let us now consider an elastic column with a cross-section inertia moment I and Young's modulus E. Similarly to Figure 2.13, a simply supported column subjected to an axial force is shown in Figure 2.17a. When F reaches a certain value, the column may start to bend due to instability (*buckling*), as shown in Figure 2.17b. Accordingly, the stability (thus, the equilibrium) of the column involves its bending capacity. The axial contribution is neglected next, and the discussion is limited to linear kinematics. The equilibrium relation is written in Eq. (2.20), where

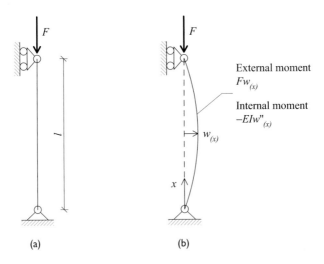

Figure 2.17 Elastic simply supported column under compressive axial force. (a) Undeformed configuration and (b) linear kinematics.

the external moment (Fw) is made equal to the internal bending moment ($-EIw''$), where w is the horizontal displacement of the column and the second derivative of w is the curvature w''. This relation represents the well-known Bernoulli–Euler beam equation.

Being Eq. (2.20) an ordinary second-order differential equation, w is obtained in Eq. (2.21), where C_1 and C_2 are constants of integration. These constants may be derived considering the boundary conditions (null displacement at the two extremities of the column), as shown in Eq. (2.22). The second of these is satisfied with either $C_1 = 0$ or by nullifying the sine function (i.e. the argument must be equal to $n\pi$, n being an integer positive number). The former leads to the trivial solution $w = 0$, i.e. the column keeps straight no matter the magnitude of F; the latter is discussed in Eq. (2.23). In particular, considering the inertia moment I (a symmetric cross-section is assumed for the column), the critical load that leads to buckling is indicated with F_{cv}. Neglecting the trivial case of $n = 0$, the minimum value of F_{cv} that leads to buckling is the so-called *Euler's critical load* calculated considering $n = 1$. The consequent displacement field is derived in Eq. (2.24). As the differential equation is satisfied for any value of C_1, Eq. (2.24) shows the configuration (i.e. the mode) of the deformed column rather than the exact value of deflection, which, as stated, cannot be found using linear buckling analysis.

To synthetize, similarly to the linear diagram of Figure 2.14, the column undergoes no horizontal displacements when $F < F_{cv}$, and the equilibrium bifurcates in correspondence of F_{cv}. The reader should be able to recognize the similarity with the eigenvalue problem in terms of both critical value and the consequent mode shape. Following from the initial assumption of linear kinematics, these results are acceptable only if displacements are kept small.

$$Fw = -EIw'' \qquad \rightarrow \qquad w'' + \frac{F}{EI}w = 0 \tag{2.20}$$

$$w = C_1 \sin\left(x\sqrt{\frac{F}{EI}} \right) + C_2 \cos\left(x\sqrt{\frac{F}{EI}} \right) \tag{2.21}$$

$$w_{(x=0)} = 0 \qquad \rightarrow \qquad C_2 = 0$$

$$w_{(x=l)} = 0 \qquad \rightarrow \qquad C_1 \sin\left(l\sqrt{\frac{F}{EI}} \right) = 0 \tag{2.22}$$

$$l\sqrt{\frac{F}{EI}} = n\pi \qquad \rightarrow \qquad F_{cv} = \frac{n^2\pi^2 EI}{l^2} \tag{2.23}$$

$$w = C_1 \sin\left(\frac{nx}{l}\pi \right) \tag{2.24}$$

Looking at Eqs. (2.23) and (2.24), it is interesting to discuss about values of n larger than 1. The consequent critical loads and modes are synthetically shown in Figure 2.18 for different values of n. This means that, as long as the critical load is smaller than the column axial capacity, multiple bifurcation points are possible. Looking at the figure, larger values of critical loads are associated to more complex shapes, whereas the case $n = 0$ is characteristic of the nonbuckled mode. From the physical point of view, although any buckling mode is possible, a quasi-static application of the external load is likely to yield the first modal shape (unless imperfections or additional constraints trigger the structure towards a different configuration). FE analysis may highlight the bifurcation points in correspondence of each critical load.

For the sake of completeness, the results obtained here can be easily extended to cases with different boundary conditions. The general formula is shown in Eq. (2.25) and is valid for any case, provided the buckling (or effective) length l_b is introduced (see also Figure 2.18). This can be associated to the length of a simply supported column, as shown in Figure 2.19, where the elements are geometrically equivalent. For the two last cases, the buckling length is associated to the position of inflection points in the deformed configuration, i.e. to a change of sign in the curvature. As it is possible to observe, taking the simply supported column as a base for comparison, the critical load of a column supported only at the base is four times as small, whereas it is four times as large for a fully clamped column, i.e. no rotation of the ends allowed but able to deform vertically.

$$F_{cv} = \frac{\pi^2 EI}{l_b^2} \tag{2.25}$$

Once the critical force is calculated, it is possible to evaluate the consequent compressive stress σ_{cv}. As shown in Eq. (2.26), this is basically F_{cv}/A, where A is the cross-section area (assumed constant). Additionally, the cross-section

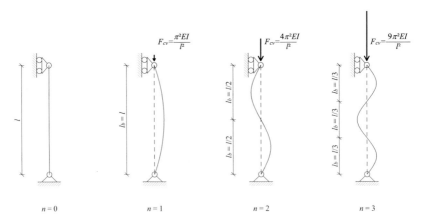

Figure 2.18 First four modes of buckling for a simply support column and relative critical loads ($n = 0$ is the nonbuckled mode).

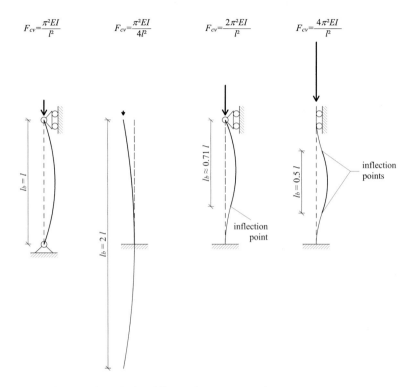

$$F_{cv} = \frac{\pi^2 EI}{l^2} \qquad F_{cv} = \frac{\pi^2 EI}{4l^2} \qquad F_{cv} = \frac{2\pi^2 EI}{l^2} \qquad F_{cv} = \frac{4\pi^2 EI}{l^2}$$

Figure 2.19 Buckling lengths for different boundary conditions on a compressed column.

inertia I is given by Ai^2, where i is the radius of inertia. Introducing the slenderness ratio λ as per Eq. (2.27), the compressive critical stress is derived by the hyperbola in Eq. (2.28). Although historically indicated with the same symbol λ, the reader should be able to recognize the profound difference between eigenvalues (see Section 2.2.2) and slenderness.

For the sake of clarity, let us consider the elastic and inelastic behaviour of a compressed element with the constitutive relationship σ–ε shown in Figure 2.20a. In particular, a generic material behaviour is assumed, linear elastic until $\sigma_y = 10\,\mathrm{MPa}$, with $E = 10{,}000\,\mathrm{MPa}$, followed by a parabolic branch and a plastic plateau in correspondence of $\sigma_u = 15\,\mathrm{MPa}$. The chosen relationship is similar to what is usually recommended for concrete in codes of practice; however, the elastic part is emphasized here (with a straight line) for illustrative purposes.

If the material behaviour is linear elastic, i.e. Young's modulus is kept constant, Eq. (2.28) provides the solid line of Figure 2.20b. As expected, the larger is the slenderness, the smaller is the critical stress. All the points below the diagram are representative of scenarios that would not cause failure. However, according to Figure 2.20a, this approach is correct for values of stress smaller than the limit elastic stress σ_y, where Young's modulus is constant. Following the nonlinear branches of the σ–ε diagram, the modulus must be updated to

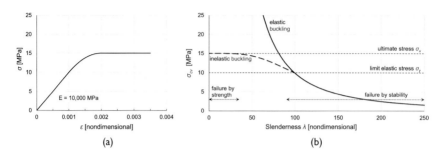

Figure 2.20 Elastic and inelastic buckling. (a) Elastoplastic constitutive relation σ–ε; (b) elastic and inelastic critical stress value according to the slenderness ratio λ.

the tangent value, reaching zero in correspondence of the plateau. The consequent simplified diagram is drawn with dashed line in Figure 2.20b (also the inertia I along the height of the column would need to be changed).

According to the diagram of Figure 2.20b, although not strictly defined, two regions can be recognized. Until about the limit of the elastic stress σ_y, failure is governed by stability rather than strength; close to the ultimate value σ_u, failure is governed by the material strength rather than stability. This means that, for large slenderness, failure is controlled by geometrical effects, and, for small slenderness, failure is controlled by material effects. In the intermediate region, both effects are of relevance.

$$\sigma_{cv} = \frac{F_{cv}}{A} = \frac{\pi^2 EI}{Al_b^2} \qquad \rightarrow \qquad \sigma_{cv} = \frac{F_{cv}}{A} = \frac{\pi^2 EAi^2}{Al_b^2} = \frac{\pi^2 Ei^2}{l_b^2} \tag{2.26}$$

$$\lambda = \frac{l_b}{i} \tag{2.27}$$

$$\sigma_{cv} = \frac{\pi^2 E}{\lambda^2} \tag{2.28}$$

This example illustrates an important physical phenomenon: geometrical and material nonlinearities are coupled and are both present in structural systems. In a FE analysis, it is the responsibility of the analyst to select the relevant inelastic phenomena. In civil engineering applications, material nonlinearities play the most relevant role and geometrical effects may be neglected in several cases. This choice is justified by the small error in evaluating the structural response (at least in the area of interest of the analysis with sufficiently small strains and displacements), and because the mathematical problem is simpler (thus, making solution procedures more robust and faster). Still, there are cases, such as a flat masonry arch, a slender modern pedestrian bridge and many steel structures, that may suffer global or local instability, in which neglecting buckling provides an erroneous response.

An example of nonlinear buckling analysis is shown in Figure 2.21. The photos of Figure 2.21a illustrate the typical failure of a silo made of an unstiffened steel plate due to buckling under vertical compression. With the aim of describing the phenomenon from the numerical point of view, nonlinear buckling analysis was performed on the simple cylindrical structure of Figure 2.21b, with a refined mesh using linear elements. The structure has a height of 14 m, with a diameter equal to 6.8 m, and constant thickness equal to 5 mm. As far as constraints are concerned, the cylinder is fixed in the bottom and pinned in the top (null horizontal displacements). The material is mild steel with $\sigma_y = 250$ MPa, $v = 0.3$ and $E = 210,000$ MPa. For the sake of simplicity, the material was considered with no mass and the silo is empty, with a unit force applied on top (downward) linearly distributed on the edge to lead to compression instability.

The imperfections for the nonlinear buckling analysis were made by superposing the scaled shapes of the first three modes to the original geometry (see Section 2.2.3). To obtain the modes configuration, a linear buckling analysis was considered (i.e. eigenvalue problem), where the first mode shape is shown in Figure 2.21c (buckling load = 19,030 kN). Considering the incremental force on top, significant steps of the nonlinear analysis are shown in Figure 2.21d. The nonlinear buckling load with imperfections equals 2,430 kN, almost eight times smaller than the one provided by the linear buckling analysis.

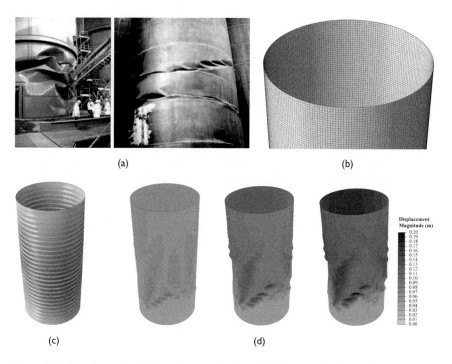

Figure 2.21 Nonlinear buckling of a steel silo. (a) Photos of observed behaviour; (b) detail of the FE model; (c) first eigen mode; (d) significant steps of the nonlinear analysis. (Pictures of the FE analysis courtesy of Pratik Gajjar.)

2.3 COMPUTATIONAL STRATEGIES

As shown in the previous section, when displacements are not infinitesimal, equilibrium equations show nonlinear relationships between external loads and displacements. Consequently, the *direct formulation* of the equilibrium equations in the deformed configuration cannot usually be written (even for simple geometries) and an analytical solution does not exist. With the aim of obtaining an approximated solution of the problem (according to the fundamental assumption of FEM), a different approach is preferred.

Without loss of generality, the objective of structural analysis is to determine a certain displacement field for which the external forces are in equilibrium with the internal ones (due to internal stresses), provided that compliance with the boundary conditions exists. In linear elasticity, the solution is unique, and the calculation is generally straightforward in a single step. When nonlinear aspects are considered (either geometrical or material), the superposition principle does not hold anymore, and external forces should be applied *incrementally* (i.e. in load steps or load increments). The accuracy of the consequent displacement field is affected by the level of nonlinearity of the structural system and an *iterative* solution algorithm is usually employed to ensure equilibrium.

As a matter of fact, the finite element procedure requires the discretization of the continuum (by means of finite elements) and the discretization of loading (by means of successive increments). The discretization in load increments can be associated to some sort of time discretization, such as the actual time needed for a generic load to achieve the final value. Although suggestive, this assumption does not hold in the present text because no dynamic effects are considered, and external loads are applied quasi-statically. Time should be understood only as a process of sorting successive events, passing from one load level to the next load level.

The usual strategy for the incremental approach is to apply a fraction of the external forces in sequence and follow step-by-step the response of the structure. In other words, the structural stiffness matrix is updated continuously throughout the analysis according to the previous-step solution. For the sake of clarity and conciseness, the example discussed in the present section addresses only material nonlinear behaviour, neglecting geometrical effects. Finally, for the explanation of common terms in nonlinear analysis, such as convergence and accuracy, the reader is referred to Section 2.3.3.

2.3.1 Incremental-iterative methods with material nonlinearities

The goal of structural analysis is to find a displacement field $\mathbf{d}$ for which the internal forces $\mathbf{f}_{int}$ (due to the internal stresses) are balanced with the external forces $\mathbf{f}_{ext}$. Let us consider a generic step n of the analysis, for which Eq. (2.29) holds, where the dependence of the internal forces on displacements at the step n (i.e. $\mathbf{d}^n$) is highlighted by the subscript. In FE analysis, according to what was stated in Chapter 1, the internal forces are computed with the product $\mathbf{Kd}$. However, the stiffness matrix $\mathbf{K}$ is usually not constant throughout the analysis, and must be updated.

To achieve the new equilibrium state, it is necessary to consider the structural response at the previous step of the analysis (step $n-1$) where all the data are known, and equilibrium holds, Eq. (2.30). As a load increment $\Delta \mathbf{f}_{ext}^n$ is imposed to the system by the FE analyst, all quantities at step n may be obtained as the summation of the known quantities (at the step $n-1$) plus an increment, as shown in Eq. (2.31). It is noted that the external force is updated exactly as defined by the analyst, whereas the displacement increment $\Delta \mathbf{d}^n$ must be calculated so that it ensures equilibrium, i.e. ensuring Eq. (2.32). For this purpose, Eq. (2.29) is expanded with Taylor's expansion in Eq. (2.33). The assumption of small increments of external forces leads to small increments of displacements, and the higher order terms can be neglected. Additionally, the underlined terms represent the equality of Eq. (2.29) and can be eliminated.

A closer inspection of Eq. (2.33) shows that the derivative of $\mathbf{f}_{int}$ replicates the tangent structural stiffness matrix $\mathbf{K}_{tan}$ evaluated at step $n-1$ (i.e. when $\mathbf{d} = \mathbf{d}^{n-1}$). The new incremental equilibrium equation is then updated in Eq. (2.34). According to the previous assumption, this is a linearized approach (namely, only the first derivate is considered in Eq. (2.33)), meaning that the solution obtained is not exact. It is important to highlight that, to obtain the increment of displacement $\Delta \mathbf{d}^n$, the inverse of the tangent stiffness matrix may be used, as in Eq. (2.35). However, as discussed in Chapter 1, this approach is computationally demanding for multidimensional problems, and different strategies are usually implemented to solve Eq. (2.34). Ultimately, the displacement at step n is updated according to the first part of Eq. (2.31).

$$\mathbf{f}_{int\left(\mathbf{d}^n\right)} = \mathbf{f}_{ext}^n \tag{2.29}$$

$$\mathbf{f}_{int\left(\mathbf{d}^{n-1}\right)} = \mathbf{f}_{ext}^{n-1} \tag{2.30}$$

$$\mathbf{d}^n = \mathbf{d}^{n-1} + \Delta \mathbf{d}^n$$
$$\mathbf{f}_{ext}^n = \mathbf{f}_{ext}^{n-1} + \Delta \mathbf{f}_{ext}^n \tag{2.31}$$

$$\mathbf{f}_{int\left(\mathbf{d}^{n-1} + \Delta \mathbf{d}^n\right)} = \mathbf{f}_{ext}^{n-1} + \Delta \mathbf{f}_{ext}^n \tag{2.32}$$

$$\underline{\mathbf{f}_{int\left(\mathbf{d} = \mathbf{d}^{n-1}\right)}} + \underbrace{\left.\frac{\partial \mathbf{f}_{int}}{\partial \mathbf{d}}\right|_{\mathbf{d} = \mathbf{d}^{n-1}}}_{\substack{\text{tangent structural} \\ \text{stiffness matrix}}} \Delta \mathbf{d}^n + \frac{1}{2} \left.\frac{\partial^2 \mathbf{f}_{int}}{\partial \mathbf{d}^2}\right|_{\mathbf{d} = \mathbf{d}^{n-1}} \cancel{\left(\Delta \mathbf{d}^n\right)^2} + \dots = \underline{\mathbf{f}_{ext}^{n-1}} + \Delta \mathbf{f}_{ext}^n \tag{2.33}$$

$$\mathbf{K}_{\text{tan}}^{n-1}\Delta\mathbf{d}^n = \Delta\mathbf{f}_{\text{ext}}^n \tag{2.34}$$

$$\Delta\mathbf{d}^n = \left(\mathbf{K}_{\text{tan}}^{n-1}\right)^{-1}\Delta\mathbf{f}_{\text{ext}}^n \tag{2.35}$$

Figure 2.22a shows a simple column with an axial compressive load. The cross-section is $0.3\times0.5\,\text{m}^2$ ($A = 0.15\,\text{m}^2$), the height is $l = 3\,\text{m}$ and buckling is prevented. The stress–strain relationship is parabolic with a maximum strain and stress equal to 0.003 (or 0.3% or 3‰) and 120 MPa, respectively, as shown in Figure 2.22b. The stress is calculated according to Eq. (2.36), whereas Young's modulus according to Eq. (2.37). The geometrical parameters and the constitutive relationship of the material are the only required data of the problem. The goal is to calculate the force–displacement diagram, i.e. the graphical relationship between the external force F and the longitudinal compression displacement of the column u, which is the only displacement variable (Figure 2.22a).

The analytical solution is derived in Eq. (2.38), where the dependence of F (external force applied to the structural system) and u (the unknown equilibrated displacement) is explicitly nonlinear, quadratic in this case. The equation is further expanded considering that the longitudinal strain is given by the shortening of the column divided by its height (u/h). Even if straightforward for simple structures, such as the column considered, the analytical solution of real engineering problems is impossible to obtain. In

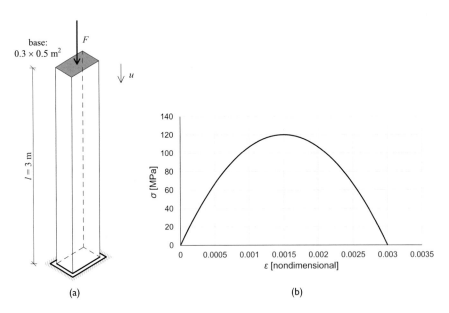

(a)   (b)

Figure 2.22 Column under axial compressive force. (a) Geometrical characteristics and (b) material stress–strain relationship adopted.

the present case, Eq. (2.38) will be used as term of comparison for the solution methods to be presented.

$$\sigma_{(\varepsilon)} = -\frac{16 \cdot 10^{13}}{3} \varepsilon^2 + 1.6 \cdot 10^{11} \varepsilon \qquad \text{[MPa]} \qquad (2.36)$$

$$E_{(\varepsilon)} = \frac{d\sigma}{d\varepsilon} = -\frac{32 \cdot 10^{13}}{3} \varepsilon + 1.6 \cdot 10^{11} \qquad \text{[MPa]} \qquad (2.37)$$

$$A\sigma_{(\varepsilon)} = F$$

$$A\left(-\frac{16 \cdot 10^{13}}{3} \varepsilon^2 + 1.6 \cdot 10^{11} \varepsilon \right) = F \qquad (2.38)$$

$$A\left(-\frac{16 \cdot 10^{13}}{3} \left(\frac{u}{h} \right)^2 + 1.6 \cdot 10^{11} \frac{u}{h} \right) = F$$

2.3.1.1 Forward Euler method

From Eqs. (2.33) and (2.34), the simplest solution method assumes the tangent structural stiffness equal to the one calculated at the previous load step, replacing the nonlinear problem by a first-order (linear) approximation. The term load will be dropped in the subsequent text for conciseness, whereas increment and step will be used indifferently. This is the strategy implemented in the so-called Forward Euler method, and shown in Figure 2.23. The calculations are made according to Eqs. (2.31) and (2.35), where the superscript $n - 1$ means the known info (of the previous step) and the superscript n indicates the result of the current step. In particular, Eq. (2.35) is detailed in Eq. (2.39) for the unidimensional case and applied in Eq. (2.40) for the case $n = 3$, with constant load steps $\Delta F = 5$ MN. As expected, $n = 0$ means the undeformed configuration where the structure is without any load (thus, displacement and force are null).

$$\Delta u^n = \frac{\Delta F_{\text{ext}}^n}{k^{n-1}} \qquad (2.39)$$

$$\Delta u^3 = \frac{\Delta F_{\text{ext}}^3}{k^2} = \frac{5}{5,599} \text{ m} = 0.89 \cdot 10^{-3} \text{m} \qquad (2.40)$$

Table 2.1 collects the results and it is split into two parts: on the left, the goal of the method, i.e. the evaluation of the displacement increment Δu (due to an external force increment ΔF_{ext}) and on the right, the opposite

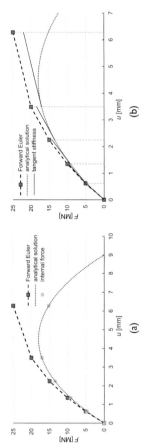

Figure 2.23 Forward Euler method with $\Delta F = 5$ MN. (a) Overall result and (b) description of the incremental approach.

Table 2.1 Step-by-step calculations for the Forward Euler method with constant $\Delta F = 5$ MN

n	ΔF^n [MN]	F^n_{ext} [MN]	u^{n-1} [10^{-3} m]	ε^{n-1} [10^{-3}]	E^{n-1} [MN/m²] Eq. (2.37)	k^{n-1} [MN/m] $E^{n-1}A/h$	Δu^n [10^{-3} m] Eq. (2.39)	u^n [10^{-3} m] $u^{n-1}+\Delta u^n$	ε^n [10^{-3}] u^n/h	σ^n [MN/m²] Eq. (2.36)	F^n_{int} [MN] $A\sigma^n$	Error [%]
1	5	5	0.00	0.00	160,000	8,000	0.63	0.63	0.21	31.02	4.65	7.5
2	5	10	0.63	0.21	137,778	6,889	0.73	1.35	0.45	61.23	9.18	8.9
3	5	15	1.35	0.45	111,971	5,599	0.89	2.24	0.75	89.84	13.48	11.3
4	5	20	2.24	0.75	80,217	4,011	1.25	3.49	1.16	113.96	17.09	17.0
5	5	25	3.49	1.16	35,893	1,795	2.79	6.28	2.09	101.30	15.19	64.5

reasoning, i.e. the evaluation of the internal response of the structure (F_{int}) once the displacement at the end of the current step (u^n) is known. As F_{int} comes from the displacement that the structure features at the current step according to the constitutive model (i.e. the stress–strain law), it generally differs from the applied external load. The last column shows the percentage in error between external and internal forces, where the latter is assumed as reference. The columns in grey provide the response of the structure in terms of load–displacement diagram shown in Figure 2.23a. It is worth noticing that the magnitude of the load step used here (and in the following) is only for didactic purpose and to provide a qualitative insight.

Schematically, the solution method is illustrated in Figure 2.23b. As discussed, and visible from the figure, the method is based on the tangent stiffness (i.e. linearization of the nonlinear problem) calculated at the previous step. Since all the data are known at this step, the tangent stiffness can be calculated directly. The dashed lines of the Forward Euler method are parallel to the tangent to the analytical diagram (thin solid lines). The intersection of the dashed lines with the target external force determines the result of the numerical method at the current step. Basically, it is possible to extrapolate the current solution from the previous solution.

When dealing with the analytical diagram, a few considerations apply. This diagram, in fact, should not mislead the reader as it is not known in real-case applications, and it is used here only as a term of comparison. Forward Euler method, and the other methods accordingly, provides only a numerical approximation of the real behaviour of the structural model. In other words, the dotted line of Figure 2.23 cannot be obtained numerically and the only available result is the dashed line, which is a multilinear response corresponding to the adopted load steps. The simplest way to refine the solution shown, which is explicit and noniterative, is to adopt smaller load increments. Following this approach, the graphs relative to load steps equal to $\Delta F = 1$ and $0.1\,\text{MN}$ are shown in Figure 2.24, where much better accuracy is evident (errors up to 1% are found for the latter). Again, the analyst is only aware of the dashed lines of

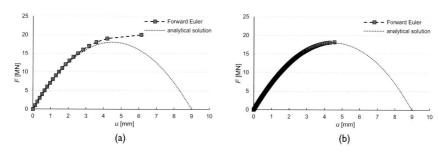

Figure 2.24 Forward Euler method with (a) $\Delta F = 1\,\text{MN}$ and (b) with $\Delta F = 0.1\,\text{MN}$.

Figures 2.23 and 2.24, as the dotted line is unknown. As a general observation, for noniterative procedures (and well-posed problems), the smaller is the increment, the higher is the accuracy (thus, the computational demand).

It must be stressed that the incremental solution method was stopped as soon as the tangential stiffness became negative and no positive increment of force could be further handled by the structure, i.e. the maximum capacity in terms of load was found. It is intuitive to imagine a negative increment of force and drive the numerical solution downward. If appealing in unidimensional problems, this is not evident in multidimensional ones and adequate methods are presented in Section 2.3.2.

A variant of the Forward Euler method may be introduced to limit the errors found. Basically, to calculate the incremental displacement at the current step, this variant of the method keeps track of the internal force calculated at the previous step, as shown in Eq. (2.41). Again, the formula is applied in Eq. (2.42) for $n = 3$, with constant load steps $\Delta F = 5\,\mathrm{MN}$. Looking at the results shown in Table 2.2, the errors in the last column are strongly reduced. Schematically, as shown Figure 2.25, the solution is sought at the intersection between the target external force and the tangent stiffness at the previous step starting from the value of the internal force, which is also available in FEM. Accordingly, the new numerical diagram is closer to the analytical one. Again, no iterations are implemented to refine the numerical solution and the errors can be reduced by decreasing the load increments, as already done in Figure 2.24 for the regular Forward Euler method.

$$\Delta u^n = \frac{F_{ext}^n - F_{int}^{n-1}}{k^{n-1}} \tag{2.41}$$

$$\Delta u^3 = \frac{F_{ext}^3 - F_{int}^2}{k^3} = \frac{15 - 9.46}{5,509}\,\mathrm{m} = 1.00 \cdot 10^{-3}\mathrm{m} \tag{2.42}$$

2.3.1.2 Newton–Raphson method (regular, modified and linear)

This family of methods includes iterations to ensure accuracy of the solution. With reference to Figure 2.26, the increment of displacement within each iteration is labelled as $\delta\mathbf{d}$, whereas $\Delta\mathbf{d}$ is used to compute the total increment of displacement since the beginning of the load step. The diagram is representative of the true response (unknown) of a unidimensional problem, but all the quantities are indicated by means of vectors and matrices, as per multidimensional problems. For the first iteration, all the incremental quantities needed are detailed in Eq. (2.43). In particular, the cardinal number after the comma in the superscript is the counter of the iterations.

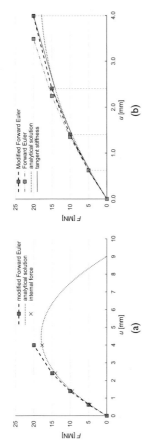

Figure 2.25 Modified Forward Euler method with $\Delta F = 5\,\text{MN}$. (a) Overall result and (b) description of the incremental approach.

Table 2.2 Step-by-step calculations for the modified Forward Euler method with constant $\Delta F = 5\,\text{MN}$ (step 5 provides a too distant solution and is not shown)

n	ΔF^n [MN]	F^n_{ext} [MN]	u^{n-1} [10^{-3} m]	ε^{n-1} [10^{-3}]	E^{n-1} [MN/m²] Eq.(2.37)	k^{n-1} [MN/m] $E^{n-1}A/h$	Δu^n [10^{-3} m] Eq.(2.41)	u^n [10^{-3} m] $u^{n-1}+\Delta u^n$	ε^n [10^{-3}] u^n/h	σ^n [MN/m²] Eq.(2.36)	F^n_{int} [MN] $A\sigma^n$	Error [%]
1	5	5	0.00	0.00	160,000	8,000	0.63	0.63	0.21	31.02	4.65	7.5
2	5	10	0.63	0.21	137,778	6,889	0.78	1.40	0.47	63.10	9.46	5.7
3	5	15	1.40	0.47	110,179	5,509	1.00	2.41	0.80	94.02	14.10	6.4
4	5	20	2.41	0.80	74,452	3,723	1.58	3.99	1.33	118.46	17.77	12.6

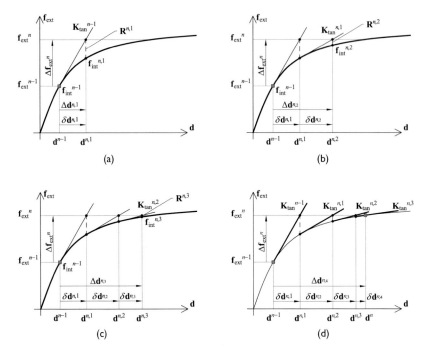

Figure 2.26 Graphical explanation of regular Newton–Raphson method for a unidimensional problem. (a) First, (b) second and (c) third iteration; (d) final iteration and overall evolution of the tangent stiffness matrix used in the iterative procedure.

$$\mathbf{d}^{n,1} = \mathbf{d}^{n-1} + \delta\mathbf{d}^{n,1} = \mathbf{d}^{n-1} + \Delta\mathbf{d}^{n,1}$$

$$\Delta\mathbf{d}^{n,1} = \delta\mathbf{d}^{n,1}$$

$$\mathbf{f}_{ext}^{n} = \mathbf{f}_{ext}^{n-1} + \Delta\mathbf{f}_{ext}^{n} \qquad (2.43)$$

$$\mathbf{f}_{int(d=d^{n,1})} = \mathbf{f}_{int}^{n,1}$$

Moving from the step $n-1$, for which all the data are known, by means of Taylor's expansion it is possible to linearize the equilibrium equation consequent to a small increment $\delta\mathbf{d}^{n,1}$, as shown in Eq. (2.44), and provide a first estimation of the final displacement increment. Considering that the equilibrium is satisfied at the previous step $n-1$, the underlined terms of the second of Eq. (2.44) can be ignored. The new equation is written in Eq. (2.45), where the first derivative shown in Eq. (2.44) is substituted with the structural tangent stiffness matrix at the previous step, and the vector $\delta\mathbf{d}^{n,1}$ is the only unknown of the problem.

However, as shown in Figure 2.26a, the linear approximation does not provide the true solution in nonlinear problems and an error is evident (see also Figures 2.23 and 2.24). In other words, the first line of Eq. (2.44), which indicates equilibrium, is not guaranteed. The difference between internal and external force is formally stated in Eq. (2.46), where the residual vector $\mathbf{R}^{n,1}$ is introduced. With no doubt, the closer $\mathbf{R}$ is to $\mathbf{0}$, the more accurate is the solution. To achieve this goal, the increment of displacement must be updated by a second iteration $\delta\mathbf{d}^{n,2}$.

$$\mathbf{f}_{\text{int}}^{n,1} = \mathbf{f}_{\text{ext}}^n$$

$$\mathbf{f}_{\text{int}}^{n-1} + \underbrace{\left.\frac{\partial\mathbf{f}_{\text{int}}}{\partial\mathbf{d}}\right|_{\mathbf{d}=\mathbf{d}^{n-1}}}_{\substack{\text{tangential structural}\\\text{stiffness matrix}}}\delta\mathbf{d}^{n,1} + \frac{1}{2}\left.\frac{\partial^2\mathbf{f}_{\text{int}}}{\partial\mathbf{d}^2}\right|_{\mathbf{d}=\mathbf{d}^{n-1}}\left(\delta\mathbf{d}^{n,1}\right)^2 + \ldots = \mathbf{f}_{\text{ext}}^{n-1} + \Delta\mathbf{f}_{\text{ext}}^{n,1} \quad (2.44)$$

$$\mathbf{K}_{\text{tan}}^{n-1}\delta\mathbf{d}^{n,1} = \Delta\mathbf{f}_{\text{ext}}^n \tag{2.45}$$

$$\mathbf{f}_{\text{int}}^{n,1} - \mathbf{f}_{\text{ext}}^n = \mathbf{R}^{n,1} \tag{2.46}$$

According to Figure 2.26b, all the incremental quantities at the second iteration are detailed in Eq. (2.47). Conversely to the displacement, it must be noted that the external force is updated only at the end of the load step, being kept constant within the iterative process. The new Taylor's expansion is written and developed in Eq. (2.48), where all the data at iteration n, 1 are known. Neglecting higher order terms, the underlined terms of the last line of Eq. (2.48) can be substituted with the equality of Eq. (2.46), yielding Eq. (2.49). It is worth noticing that $\mathbf{K}_{\text{tan}}^{n,1}$ is the matrix used in the iterative procedure, which is similar to the structural stiffness matrix, even if it is not calculated in an equilibrium configuration and, thus, entails errors in the case the load steps are not infinitesimal. For the sake of simplicity, this matrix will be denoted here as tangent stiffness.

By analogy, the new increment $\Delta\mathbf{d}^{n,2}$ can be used to calculate the new residual $\mathbf{R}^{n,2}$, and a new iteration can start. Figure 2.26b–d illustrates the second, third and final iteration of the method with a progressive reduction of the residuals (i.e. $\mathbf{R} \rightarrow \mathbf{0}$). In engineering applications, the number of iterations required to achieve suitable accuracy depends on the complexity of the problem and the degree of nonlinearity.

$$\mathbf{d}^{n,2} = \mathbf{d}^{n,1} + \delta\mathbf{d}^{n,2} = \mathbf{d}^{n-1} + \Delta\mathbf{d}^{n,2}$$

$$\Delta\mathbf{d}^{n,2} = \Delta\mathbf{d}^{n,1} + \delta\mathbf{d}^{n,2}$$

$$\mathbf{f}_{\text{ext}}^n = \mathbf{f}_{\text{ext}}^{n-1} + \Delta\mathbf{f}_{\text{ext}}^n \tag{2.47}$$

$$\mathbf{f}_{\text{int}\left(\mathbf{d}=\mathbf{d}^{n,2}\right)} = \mathbf{f}_{\text{int}}^{n,2}$$

$$\mathbf{f}_{\text{int}}^{n,2} = \mathbf{f}_{\text{ext}}^{n}$$

$$\mathbf{f}_{\text{int}}^{n,1} + \frac{\partial \mathbf{f}_{\text{int}}}{\partial \mathbf{d}}\bigg|_{\mathbf{d}=\mathbf{d}^{n,1}} \delta \mathbf{d}^{n,2} = \mathbf{f}_{\text{ext}}^{n-1} + \Delta \mathbf{f}_{\text{ext}}^{n} \qquad (2.48)$$

$$\underline{\mathbf{f}_{\text{int}}^{n,1}} + \mathbf{K}_{\text{tan}}^{n,1} \delta \mathbf{d}^{n,2} = \underline{\mathbf{f}_{\text{ext}}^{n}}$$

$$\mathbf{K}_{\text{tan}}^{n,1} \delta \mathbf{d}^{n,2} = -\mathbf{R}^{n,1} \qquad (2.49)$$

The method illustrated so far goes under the name of *Regular* Newton–Raphson method, whose iterative process is synthetically indicated in Eqs. (2.50) and (2.51), where $i = 1,2,3...$ is the counter of the iterations at the step n. As it is possible to see, $\delta \mathbf{d}^{n,i}$ depends on the iteration n, $i - 1$ for which all data are known. The displacements in Eq. (2.52) should be updated until a convergence criterion is met (see Section 2.3.3).

A closer inspection of Eq. (2.51) reveals that the tangent stiffness matrix is updated at each iteration (Figure 2.26d), a procedure that may result computationally very demanding and, in cases of ill-posed problems, make convergence impossible. Accordingly, variants of the method consider the possibility of keeping the tangent stiffness matrix constant throughout the load increment or throughout the entire analysis, leading to the *Modified* and *Linear* variants, respectively.

$$\mathbf{R}^{n,i} = \mathbf{f}_{\text{int}}^{n,i} - \mathbf{f}_{\text{ext}}^{n} \qquad (2.50)$$

$$\mathbf{K}_{\text{tan}}^{n,i-1} \delta \mathbf{d}^{n,i} = -\mathbf{R}^{n,i-1} \qquad (2.51)$$

$$\Delta \mathbf{d}^{n,i} = \Delta \mathbf{d}^{n,i-1} + \delta \mathbf{d}^{n,i}$$

$$\mathbf{d}^{n,i} = \mathbf{d}^{n,i-1} + \delta \mathbf{d}^{n,i} = \mathbf{d}^{n-1} + \Delta \mathbf{d}^{n,i} \qquad (2.52)$$

Going back to the practical example dealt with in this section, Table 2.3 collects the results for $\Delta F = 5\,\text{MN}$, and the calculation for $n = 3$ and $i = 2$ is given in Eq. (2.53). As it is possible to notice, the solution method converges to the true values (with an error lower than 1%) with only two iterations. Figure 2.27a shows the overall result for the mentioned example, where the solution method was stopped when an evident divergence arose. However, the results of the iterations for $n = 4$ are shown in Table 2.3 and illustrated in Figure 2.27b. As the solution is sought for $F_{\text{ext}} = 20\,\text{MN}$ and collapse is found at $F_{\text{ext}} = 18\,\text{MN}$, the solution method fails (Section 2.3.2 provides methods to address this problem). In this regard, it must be stressed that the collapse force and the analytical solution are not known in engineering problems, and the increasing error is an indicator of the lack of convergence of the method.

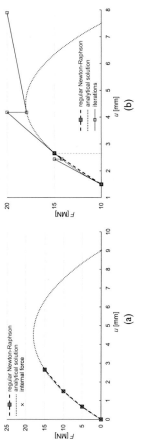

Figure 2.27 Regular Newton–Raphson method. (a) Overall result and (b) incremental-iterative approach for load levels $F = 15$ MN and $F = 20$ MN (note the different scales in the two graphs).

Table 2.3 Step-by-step calculations for the regular Newton–Raphson method with constant $\Delta F = 5$ MN

n	i	ΔF^n [MN]	F_{ext}^n [MN]	$u^{n,i-1}$ [10^{-3} m]	$\varepsilon^{n,i-1}$ [10^{-3}]	$E^{n,i-1}$ [MN/m²]	$k^{n,i-1}$ [MN/m]	$\delta u^{n,i}$ [10^{-3} m]	$u^{n,i}$ [10^{-3} m]	$\varepsilon^{n,i}$ [10^{-3}]	$\sigma^{n,i}$ [MN/m²]	$F_{int}^{n,i}$ [MN]	Error [%]
						Eq. (2.37)	$E^{n,i-1}A/h$	Eq. (2.51)	Eq. (2.52)	$u^{n,i}/h$	Eq. (2.36)	$A\sigma^{n,i}$	
1	1	5	5	0.00	0.00	160,000	8,000	0.63	0.63	0.21	31.02	4.65	7.5
	2			0.63	0.21	137,778	6,889	0.05	0.68	0.23	33.32	5.00	0.0
2	1	5	10	0.68	0.23	135,986	6,799	0.74	1.41	0.47	63.46	9.52	5.1
	2			1.41	0.47	109,827	5,491	0.09	1.50	0.50	66.62	9.99	0.1
3	1	5	15	1.50	0.50	106,712	5,336	0.94	2.44	0.81	94.78	14.22	5.5
	2			2.44	0.81	73,348	3,667	0.21	2.65	0.88	99.73	14.96	0.3
4	1	5	20	2.65	0.88	65,759	3,288	1.53	4.18	1.39	119.41	17.91	11.7
	2			4.18	1.39	11,252	563	3.71	7.90	2.63	51.63	7.74	158.3

$$\delta u^{3,2} = \frac{F_{\text{ext}}^3 - F_{\text{int}}^{3,1}}{k^{3,1}} = \frac{15 - 14.22}{3,667}\,\text{m} = 0.21 \times 10^{-3}\text{m} \qquad (2.53)$$

The modified Newton–Raphson method assumes a constant stiffness for each step (equal to the one at the first iteration). With reference to Eq. (2.51), this method assumes $\mathbf{K}^{n,i} = \mathbf{K}^{n,1}$ throughout the step n. For the given example, the results are collected in Table 2.4. As shown in Figure 2.28a, the lower computational effort of the stiffness matrix is counterbalanced by the increased number of iterations needed. In general, the modified Newton–Raphson method needs more iterations, but every iteration is faster than with the regular method. As already described, the last iteration $n = 4$, which is inherent to $F_{\text{ext}} = 20\,\text{MN}$, does not lead to convergence because there is no solution for the problem.

The second variant of Newton–Raphson method considers the linear stiffness along the entire analysis and it is never updated, i.e. $\mathbf{K}^{n,i} = \mathbf{K}^{1,1}$ (the matrix coincides with the one computed at the first iteration of the first step). In practical problems, as stiffness is computed once for all, the computational effort is drastically reduced, thus, with the least time per iteration. In turn, the number of iterations is larger than regular and modified Newton–Raphson, putting this method amid the ones with the slowest convergence. Table 2.5 collects the results for the mentioned example. Again, the last step does not converge and the result with $i = 12$ is shown at the end of table. Figure 2.28b illustrates the iterations needed to converge to the result for $F_{\text{ext}} = 15\,\text{MN}$ where the comparison with Figure 2.28a is self-explaining.

In general, assuming the linear stiffness throughout the analysis guarantees that the same is always symmetric and well-posed, making the method robust and to be a valuable option for complex problems. However, it is possible that this method follows unstable equilibrium paths in the case of bifurcation and large load increments, and the analyst must carefully analyse the results obtained, particularly when large tolerances are used. In a simple problem, as the one shown, all methods give the same result (in the case a tight tolerance is used).

2.3.1.3 Quasi-Newton methods

Quasi-Newton methods are also called Secant methods and they abandon the idea of the tangent stiffness in favour of the secant one. This stifness matrix is computed considering the two last iterations. Although straightforward for unidimensional problems, in multidimensional problems the secant stiffness matrix is not unique and different approaches and mathematical manipulations may be implemented to simplify or improve its calculation. The use of the secant stiffness is particularly helpful when the tangent stiffness matrix becomes nonsymmetric, when

Table 2.4 Step-by-step calculations for the modified Newton–Raphson method with constant $\Delta F = 5$ MN

n	i	ΔF^n [MN]	F_{ext}^n [MN]	$u^{n,i-1}$ [10^{-3} m]	$\varepsilon^{n,i-1}$ [10^{-3}]	$E^{n,i}$ [MN/m²] Eq. (2.37)	$k^{n,i}$ [MN/m] $E^{n,i}A/h$	$\delta u^{n,i}$ [10^{-3} m] Eq. (2.51)	$u^{n,i}$ [10^{-3} m] Eq. (2.52)	$\varepsilon^{n,i}$ [10^{-3}] $u^{n,i}/h$	$\sigma^{n,i}$ [MN/m²] Eq. (2.36)	$F_{int}^{n,i}$ [MN] $A\sigma^{n,i}$	Error [%]
1	1	5	5	0.00	0.00	160,000.00	8,000.00	0.63	0.63	0.00021	31.02	4.65	7.5
	2			0.63	0.21	160,000.00	8,000.00	0.04	0.67	0.00022	33.00	4.95	1.0
	3			0.67	0.22	160,000.00	8,000.00	0.01	0.67	0.00022	33.28	4.99	0.1
2	1	5	10	0.67	0.22	136,012.79	6,800.64	0.74	1.42	0.00047	63.68	9.55	4.7
	2			1.42	0.47	136,012.79	6,800.64	0.07	1.48	0.00049	66.06	9.91	0.9
	3			1.48	0.49	136,012.79	6,800.64	0.01	1.50	0.00050	66.54	9.98	0.2
3	1	5	15	1.50	0.50	106,795.50	5,339.78	0.95	2.45	0.00082	95.09	14.26	5.2
	2			2.45	0.82	106,795.50	5,339.78	0.14	2.59	0.00086	98.33	14.75	1.7
	3			2.59	0.86	106,795.50	5,339.78	0.05	2.63	0.00088	99.38	14.91	0.6
4	1	5	20	2.63	0.88	66,324.85	3,316.24	1.58	4.22	0.00141	119.53	17.93	11.5
	2			4.22	1.41	66,324.85	3,316.24	0.62	4.84	0.00161	119.31	17.90	11.8
	3			4.84	1.61	66,324.85	3,316.24	0.63	5.48	0.00183	114.35	17.15	16.6
	4			5.48	1.83	66,324.85	3,316.24	0.86	6.34	0.00211	100.03	15.00	33.3

Table 2.5 Step-by-step calculations for the linear Newton–Raphson method with constant $\Delta F = 5\,MN$

n	i	ΔF^n [MN]	F_{ext}^n [MN]	$u^{n,i-1}$ [10^{-3} m]	$\varepsilon^{n,i-1}$ [10^{-3}]	$E^{1,1}$ [MN/m^2] Eq. (2.37)	$k^{1,1}$ [MN/m] $E^{1,1}A/h$	$\delta u^{n,i}$ [10^{-3} m] Eq. (2.51)	$u^{n,i}$ [10^{-3} m] Eq. (2.52)	$\varepsilon^{n,i}$ [10^{-3}] $u^{n,i}/h$	$\sigma^{n,i}$ [MN/m^2] Eq. (2.36)	$F_{int}^{n,i}$ [MN] $A\sigma^{n,i}$	Error [%]
1	1	5	5	0.00	0.00	160,000	8,000	0.63	0.63	0.21	31.02	4.65	7.5
	2			0.63	0.21	160,000	8,000	0.04	0.67	0.22	33.00	4.95	1.0
	3			0.67	0.22	160,000	8,000	0.01	0.67	0.22	33.28	4.99	0.1
2	1	5	10	0.67	0.22	160,000	8,000	0.63	1.30	0.43	59.30	8.90	12.4
	2			1.30	0.43	160,000	8,000	0.14	1.44	0.48	64.43	9.66	3.5
	3			1.44	0.48	160,000	8,000	0.04	1.48	0.49	65.94	9.89	1.1
	4			1.48	0.49	160,000	8,000	0.01	1.49	0.50	66.43	9.96	0.4
3	1	5	15	1.48	0.50	160,000	8,000	0.64	2.12	0.71	86.38	12.96	15.8
	2			2.12	0.71	160,000	8,000	0.26	2.37	0.79	93.20	13.98	7.3
	3			2.37	0.79	160,000	8,000	0.13	2.50	0.83	96.32	14.45	3.8
	4			2.50	0.83	160,000	8,000	0.07	2.57	0.86	97.93	14.69	2.1
	5			2.57	0.86	160,000	8,000	0.04	2.61	0.87	98.81	14.82	1.2
	6			2.61	0.87	160,000	8,000	0.02	2.63	0.88	99.31	14.90	0.7
4	1	5	20	2.63	0.88	160,000	8,000	0.64	3.27	1.09	111.02	16.65	20.1
	2			3.27	1.09	160,000	8,000	0.42	3.69	1.23	116.09	17.41	14.9
	3			3.69	1.23	160,000	8,000	0.32	4.01	1.34	118.58	17.79	12.4
⋮	12			6.59	2.20	160,000	8,000	0.73	7.32	2.44	72.88	10.93	82.9

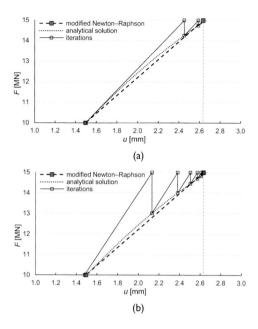

Figure 2.28 Incremental-iterative approach for load levels $F = 15\,MN$ with (a) modified and (b) linear Newton–Raphson methods.

the derivation of the consistent tangent stiffness matrix is excessively complex or when a fast approximation of the tangent stiffness is needed, such as time history analysis in earthquake engineering and several thousands of time steps. These aspects may make a Quasi-Newton method more stable, robust and faster than the methods described earlier, with a convergence rate usually between that of regular Newton–Raphson and modified Newton–Raphson schemes. This generally holds also for time per iteration.

As the rigorous dissertation of this family of methods is outside the scope of the book, let us focus on the simple unidimensional problem discussed so far. From the numerical point of view, it is possible to calculate the slope of the secant line (i.e. the stiffness, in structural terms) considering two subsequent iterations for which all the data are known. In other words, looking at Eq. (2.54), to compute the stiffness at iteration n,i, data at iterations $n,i - 1$ and $n,i - 2$ must be considered. Note that, as the secant stiffness cannot be calculated at beginning of the step $n = 1$, the method starts with the linear stiffness. In addition, the first iteration of a new step n may start from the last iteration of the previous step $n - 1$, which is available. These aspects are synthetically illustrated in Figure 2.29a, which may be compared with Figure 2.26d to illustrate the difference between tangent and secant stiffness.

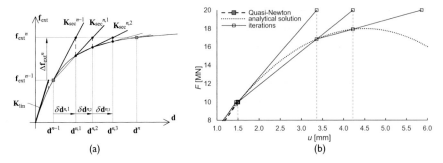

(a)   (b)

Figure 2.29 Quasi-Newton method with secant stiffness. (a) Description of the method and (b) example.

$$k^{n,i-1} = \frac{F_{int}^{n,i-1} - F_{int}^{n,i-2}}{\delta u^{n,i-1}} \qquad (2.54)$$

For the given example, the results of this method are shown in Table 2.6 and illustrated in Figure 2.29b, where ΔF was chosen equal to 10 MN for an illustrative purpose. Although the increment of force is rather large, since the initial behaviour is almost linear, only three iterations are needed to achieve an error of 0.1% at the first step. In turn, as already stated, the solution method does not converge for an external force equal to $F_{ext} = 20$ MN and the algorithm quits. This is also clear from Figure 2.29b.

2.3.2 Advanced solution procedures

The incremental-iterative strategies described in the previous section face convergence issues after the peak load. According to Section 2.2.1, the onset of negative stiffness defines a limit point, and a different strategy is needed to describe the structural behaviour after this point (i.e. the softening branch). Before introducing advanced solution procedures, let us consider a more practical approach to the solution of nonlinear problems.

Most of the strategies followed for nonlinear analyses use a *load* (or *force*) *control method*, as introduced in Section 2.2.1. The solution at each step is sought increasing the load by a finite amount that is kept fixed throughout the equilibrium iterations until convergence is reached. A schematic illustration of the method is depicted in Figure 2.30a. No matter the amount of the increment of force, once the structure reaches the maximum capacity in terms of load, the numerical model is not able to provide any solution (as it does not exist) and the incremental-iterative procedure diverges. A possible alternative is a *displacement control method*, as shown in Figure 2.30b. This works such that the external force is calculated starting from a displacement increment (of selected nodes). The method allows to obtain results in the postpeak behaviour, because, in this case, the displacement

Table 2.6 Step-by-step calculations for the Quasi-Newton method with constant $\Delta F = 10$MN

n	i	ΔF^n [MN]	F_{ext}^n [MN]	$u^{n,i-1}$ [10^{-3} m]	$\varepsilon^{n,i-1}$ [10^{-3}]	$E^{n,i-1}$ [MN/m²] Eq. (2.37)	$k^{n,i-1}$ [MN/m] Eq. (2.54)	$\delta u^{n,i}$ [10^{-3} m] Eq. (2.51)	$u^{n,i}$ [10^{-3} m] Eq. (2.52)	$\varepsilon^{n,i}$ [10^{-3}] $u^{n,i}/h$	$\sigma^{n,i}$ [MN/m²] Eq. (2.36)	$F_{int}^{n,i}$ [MN] $A\sigma^{n,i}$	Error [%]
1	1	10	10	0.00	0.00	160,000	8,000	1.25	1.25	0.42	57.41	8.61	16.1
	2			1.25	0.42	115,556	6,889	0.20	1.45	0.48	64.93	9.74	2.7
	3			1.45	0.48	108,387	5,599	0.05	1.50	0.50	66.60	9.99	0.1
2	1	10	20	1.50	0.50	106,735	5,378	1.86	3.36	1.12	112.29	16.84	18.7
	2			3.36	1.12	40,555	3,682	0.86	4.22	1.41	119.52	17.93	11.6
	3			4.22	1.41	10,077	1,266	1.64	5.85	1.95	109.15	16.37	22.2

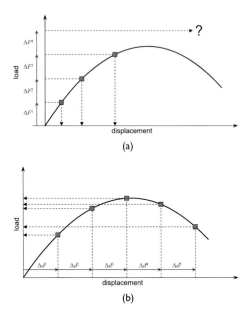

Figure 2.30 Comparison between (a) load and (b) displacement control methods. The input adopted by the analyst to carry out the analysis, either in terms of load increments or displacement increments is also shown.

monotonically increases in the structural response. The method is discussed ahead, together with the *indirect displacement control method.*

Both control methods (load and displacement) suffer convergence problems in proximity of *critical points.* When load control is implemented, Figure 2.31a shows the so-called *snap-through* instability. In correspondence of the *limit point,* a subsequent increment of load may make a displacement jump (provided the absence of convergence issues). In turn, Figure 2.31b shows the analogous case faced with the displacement control method, the so-called *snap-back* instability. In correspondence of the *turning point,* convergence may not occur due to the large jump of the load. In both cases, the analyses may provide incomplete or misleading information about the stability of the structure. In this regard, *arc-length control* provides a valid support to correctly handle problems of instability, as shown next.

2.3.2.1 Arc-length control

The main feature of this solution procedure is to abandon the idea of load or displacement control of Figure 2.31 and to provide a mutual relationship between increments of load and displacement. The solution is not sought for a target force or displacement, but at a certain "distance" from

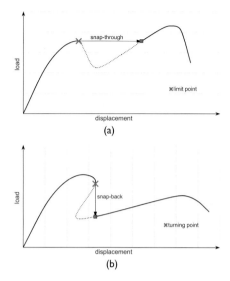

(a)

(b)

Figure 2.31 Difficulties related to load and displacement control methods. (a) Snap-through (load control) and (b) snap-back instability (displacement control).

the last converged step. Figure 2.32 exemplifies the method combining snap-back and snap-through instabilities, such as independently described in Figure 2.31. From the mathematical point of view, the formulation of the problem remains almost unchanged, but one additional equation is requested for the solution to lay on the circle of Figure 2.32. As it is possible to observe in the scheme, all the solutions are distant Δl from each other, load increments are no longer fixed and the algorithm follows the entire equilibrium path.

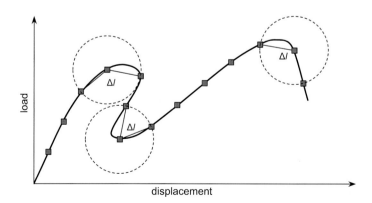

Figure 2.32 Arc-length control solution scheme.

In multidimensional problems, provided the parametrization of the external load, arc-length control follows an incremental-iterative approach. It must be highlighted that parametrization means that the external load is not increased by a subsequent summation of $\Delta\mathbf{f}_{\text{ext}}$, but by means of subsequent summation of a load multiplier increment $\Delta\mu$. Accordingly, the initial value of the external load $\mathbf{f}_{\text{ext}}$ is not determinant because it is tuned by the load multiplier μ (in a simplistic way, in engineering practice, $\mathbf{f}_{\text{ext}}$ could represent e.g. the self-weight and μ equal to 1.0 would represent the actual load in the structure, or $\mathbf{f}_{\text{ext}}$ could represent a set of characteristic loads per the applicable code of practice and μ could indicate some safety level of the structure). An advantage of parametrization is that the load multiplier can be positive or negative, avoiding the main drawback of the load control method, according to which the load is always increasing. In any case, it is assumed that the set of loads within $\mathbf{f}_{\text{ext}}$ are all changing proportionally to μ.

Adopting the same nomenclature and approach of the regular Newton–Raphson method, the novelty of arc-length method involves the increment of the external force. By analogy, the related quantities are described in Eq. (2.55), compare with Eq. (2.52) relative to the displacement increment. Figure 2.33 shows the scheme of the arc-length control strategy, detailed only for the first three iterations. Similarly to the Newton–Raphson method, the residual at the generic iteration $n, i-1$ is detailed in Eq. (2.56). As this is generally different from $\mathbf{0}$, to refine the solution, iterative corrections must be considered in terms of displacement and load multiplication factor, $\delta\mathbf{d}^{n,i}$ and $\delta\mu^{n,i}$, respectively.

$$\mathbf{f}_{\text{ext}}^{n,i} = \left(\mu^{n-1} + \Delta\mu^{n,i}\right)\mathbf{f}_{\text{ext}} \tag{2.55}$$

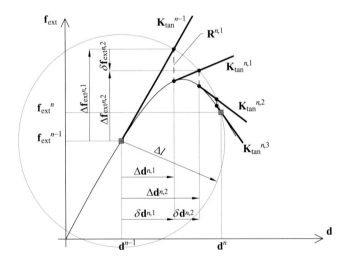

Figure 2.33 Scheme of the incremental-iterative approach with arc-length control.

$$\mathbf{R}^{n,i-1} = \mathbf{f}_{int}^{n,i-1} - \mathbf{f}_{ext}^{n,i-1}$$

$$\mathbf{R}^{n,i-1} = \mathbf{f}_{int\left(d=d^{n,i-1}\right)} - \left(\mu^{n-1} + \Delta\mu^{n,i-1}\right)\mathbf{f}_{ext}$$

(2.56)

To evaluate equilibrium at the step n,i, the Taylor's expansion is developed in Eq. (2.57) limited to the first-order term. In analogy with Eq. (2.52), Eq. (2.58) describes the load increment following from the correction $\delta\mu^{n,i}$. Accordingly, Eq. (2.57) becomes (2.59), where the underlined terms can be substituted with the equality of Eq. (2.56). Consequently, the equation to be adopted is given in Eq. (2.60). The possibility of a negative increment of the load multiplier $\delta\mu^{n,i}$ is illustrated in Figure 2.33 for $i = 3$ (it is noted that the stiffness matrix is evaluated at step $n, i-1$).

$$\mathbf{f}_{int}^{n,i} = \mathbf{f}_{ext}^{n,i}$$

$$\mathbf{f}_{int}^{n,i-1} + \left.\frac{\partial \mathbf{f}_{int}}{\partial \mathbf{d}}\right|_{\mathbf{d}=\mathbf{d}^{n,i-1}} \delta\mathbf{d}^{n,i} = \left(\mu^{n-1} + \Delta\mu^{n,i}\right)\mathbf{f}_{ext}$$

(2.57)

$$\Delta\mu^{n,i} = \Delta\mu^{n,i-1} + \delta\mu^{n,i}$$

(2.58)

$$\mathbf{f}_{int}^{n,i-1} + \mathbf{K}_{tan}^{n,i-1}\delta\mathbf{d}^{n,i} = \left(\mu^{n-1} + \Delta\mu^{n,i-1} + \delta\mu^{n,i}\right)\mathbf{f}_{ext}$$

$$\underline{\mathbf{f}_{int}^{n,i-1} + \mathbf{K}_{tan}^{n,i-1}\delta\mathbf{d}^{n,i}} = \left(\mu^{n-1} + \Delta\mu^{n,i-1}\right)\mathbf{f}_{ext} + \delta\mu^{n,i}\mathbf{f}_{ext}$$

(2.59)

$$\mathbf{K}_{tan}^{n,i-1}\delta\mathbf{d}^{n,i} - \delta\mu^{n,i}\mathbf{f}_{ext} = -\mathbf{R}^{n,i-1}$$

(2.60)

The similitude of Eq. (2.60) with Eq. (2.51) is evident, but the unknowns with the arc-length control method are now given by two groups, namely, $\delta\mathbf{d}^{n,i}$ and $\delta\mu^{n,i}$. Whereas the former is already contemplated by the set of equilibrium equations, the latter needs an additional equation for the new set to be solved. This is illustrated in Eq. (2.61) and can be understood as a geometrical constraint. This is evident if a unidimensional problem is considered, such as the one illustrated in Figure 2.33. Eq. (2.61) describes, in fact, a circle whose centre lays on the last converged step $(n - 1)$ and with a radius equal to Δl. In other words, at each iteration, the solution is sought on this circle and Δl essentially defines how far to search the solution of the current step. In this regard, it is worth noticing that Δl is not a physical distance and combines quantities with different units and magnitudes. For this reason, a scaling factor (or weight) c is introduced, in absence of which Eq. (2.61) would be unbalanced and dependent on the measurement units adopted by the analyst, thus misinterpreting the influence of the two terms (namely, displacement and force). For instance, in the example analysed in the previous subsection, an increment $\Delta F = 5\,\text{MN}$, i.e. $5 \times 10^6\,\text{N}$, led to an average $\Delta u = 10^{-3}$m. Computationally, Δl is calculated at the first iteration of each step (namely, at iteration $n,1$) as a function of the correspondent load step ($\delta\mu^{n,1}$) and kept constant for all the following iterations of the given step.

$$\left(\Delta\mathbf{d}^{n,i-1} + \delta\mathbf{d}^{n,i}\right)^{T}\left(\Delta\mathbf{d}^{n,i-1} + \delta\mathbf{d}^{n,i}\right) + c^{2}\left(\Delta\mu^{n,i-1} + \delta\mu^{n,i}\right)^{2}\mathbf{f}_{ext}{}^{T}\mathbf{f}_{ext} = \left(\Delta l^{n,1}\right)^{2} \quad (2.61)$$

As shown in Eq. (2.60), the method assumes that the tangential stiffness is calculated at each iteration. The evolution of the tangential stiffness is depicted in Figure 2.33, where the arc-length control is meant to "contain" the subsequent iterations and provide results even in the case of strong changes of the stiffness. The difference with Figure 2.27b is evident and mainly justifiable by the fixed target external load. It is rather intuitive that other iterative methods, such as modified Newton–Raphson, or Quasi-Newton methods, can be also adopted together with the arc-length method, with the due consequences in terms of number of iterations and computational effort.

For the sake of correctness, it must be highlighted that, although didactic and conceptually valid, the schemes of Figures 2.32 and 2.33 are not totally accurate. Basically, due to the presence of the scaling factor c and load multiplier μ, as presented in Eq. (2.61), the circular shape is more complex than how it is shown.

Going back to the example discussed so far, the force–displacement diagram of Figure 2.34 shows the solution with arc-length control activated, assuming $c = 0.0005$ and ΔF initially equal to 5.0 MN and to 2.5 MN in the postpeak branch in combination with the Newton–Raphson method. As it is possible to notice, the unstable branch of the equilibrium path is followed precisely. As expected, according to the main idea of arc-length, the converged points do not show constant increments of force or displacement.

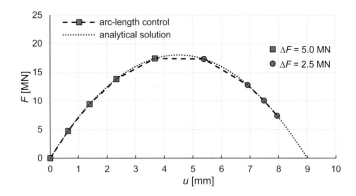

Figure 2.34 Arc-length control in combination with regular Newton–Raphson method.

Note that, in engineering applications, it is usual that around the peak, smaller load step sizes are needed, as the stiffness decreases.

Looking at Eqs. (2.60) and (2.61), arc-length requires the new constraining equation to be solved together with the set of equilibrium equations. In other words, despite the advantages of the method, the system of equations of the form $\mathbf{Kd = f}$ must be supplemented with an additional equation and a consequent new unknown μ. Whereas, on the one hand, a single new variable among a large number of displacement variables is not relevant in terms of dimension of the nonlinear problem, on the other hand, in the case of the tangential stiffness matrix, the augmented problem has a coefficient matrix neither symmetric nor banded. To avoid this drawback, it is possible to isolate $\delta \mathbf{d}^{n,i}$ from the second line of Eq. (2.59), as done in Eq. (2.62). The underlined terms of this equation are known quantities and can be gathered under the symbol $\delta \mathbf{d}_r$ and $\delta \mathbf{d}_\mu$, see Eq. (2.63). By substituting this into Eq. (2.61), a scalar quadratic equation is found (not reported here), which is independent from the rest of the equilibrium equations and can be solved with standard solution methods.

$$\delta \mathbf{d}^{n,i} = \underbrace{-\left(\mathbf{K}_{\tan}^{n,i-1}\right)^{-1}\left[\mathbf{f}_{\text{int}}^{n,i-1} - \left(\mu^{n-1} + \Delta\mu^{n,i-1}\right)\mathbf{f}_{\text{ext}}\right]}_{\delta \mathbf{d}_r^{n,i-1}} + \delta\mu^{n,i}\underbrace{\left(\mathbf{K}_{\tan}^{n,i-1}\right)^{-1}\mathbf{f}_{\text{ext}}}_{\delta \mathbf{d}_\mu^{n,i-1}} \quad (2.62)$$

$$\delta \mathbf{d}^{n,i} = \delta \mathbf{d}_r^{n,i-1} - \delta\mu^{n,i}\delta \mathbf{d}_\mu^{n,i-1} \quad (2.63)$$

This approach goes under the name of *spherical arc-length method* and is recommended in practical applications. The name is related to the circle of Figure 2.33 that becomes a sphere in the multidimensional space of the unknown variables ($\delta \mathbf{d}$ and $\delta\mu$). Although the formulation presented in Eq. (2.63) simplifies the augmented problem, the quadratic form of the equation inherent $\delta\mu^{n,i}$ leads to two distinct roots (even in apparent absence of bifurcation). From the mathematical point of view, in order to avoid this drawback, several authors have proposed different formulations about a *linearized* form of the arc-length method. Without entering the merit of these, the two main strategies are depicted in Figure 2.35. Riks–Wempner method is illustrated in Figure 2.35a, according to which the iterative changes are always orthogonal to the initial iteration. A further refinement is offered by Ramm's method, illustrated in Figure 2.35b. It assumes that the iterative change is always orthogonal to the "secant change". Although computational appealing due to the linearity of the equation for computing $\delta\mu^{n,i}$, the spherical arc-length method is more stable and can converge when the linearized form misses the equilibrium path. It is not difficult to see that, looking at Figure 2.35a and b, both methods would have faced convergence issues in the case of a steeper softening branch (compare with Figure 2.33). For both strategies, the variants of the Newton–Raphson method or Quasi-Newton methods can be adopted for the iterative process.

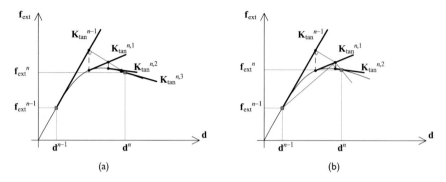

Figure 2.35 Linearized arc-length control. (a) Riks–Wempner method and (b) Ramm's method.

As introduced for the spherical method, the quadratic equation to find $\delta\mu^{n,i}$ leads to two roots. In a simplistic way, if the structural response is associated to the curved line in Figure 2.32, once a circle is established with its centre on the line, there are always two intersections with the equilibrium path. The main issue is thus the choice of the correct set of $\delta\mu^{n,i}$ and $\delta d^{n,i}$ to update the structural response. With this aim, robust algorithms are usually implemented so that the solution *evolves forward*. Mathematically, it is not easy to discuss about what is meant with the term forward. Looking at the previous examples, it is intuitive to associate forward with a positive increment of displacement. However, this definition fails if the diagram of Figure 2.32 is considered. Additionally, the displacement variables may display positive and negative increments within the same iteration, posing more issues about the correct solution to choose, especially close to limit points.

Besides the previously converged step, a typical case of "not-forward" evolution is represented by a solution on the unloading path. This is evident in Figure 2.36a and it is of scarce interest from the engineering point of view. The elastic unloading is easily identifiable from the overall deformation of the model, particularly by looking at the displacement increments of the last load step Δd (note that these cannot usually be observed in an experiment, which features only total displacements in a body). As better explained in Chapter 3, unloading means the external load is reduced while the structure recovers part of its deformation, in agreement with the constitutive material model. If the deformed configuration shows a recovery, the structure is in the unloading path and a different solution must be sought. For instance, a larger step may overcome the bifurcation issue and provide a correct result on the nonlinear branch (Figure 2.36b), forcing the solution to evolve forward. Moving from the last converged step on the nonlinear response, also smaller steps, different iteration methods or different control methods can be helpful in avoiding the issue of unwanted unloading.

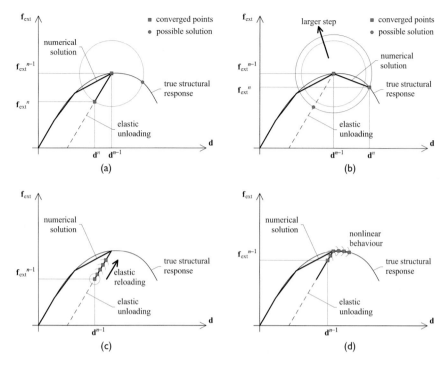

Figure 2.36 Multiple solutions with arc-length control. (a) Generic step with the solution on the elastic unloading path; (b) larger step to avoid the issue. Moving from (a), (c) small steps along the elastic reloading path and (d) subsequent loading until the nonlinear branch is reached.

It is important to note that a misinterpretation of the results of Figure 2.36a may yield the analyst to assume the structure as susceptible of snap-back instability (see Figure 2.31b). In this case, it is not uncommon that smaller steps are selected to deal with the supposed strong nonlinearity. The subsequent result is shown in Figure 2.36c where the results simply follow the elastic reloading path, possibly reaching the nonlinear branch again, such as in Figure 2.36d, provided the instability is no longer present. Despite the correctness of the results, the investigation of the elastic appendage is not worth the computational effort and should be avoided. Again, careful inspection of the results at these load steps, particularly in terms of displacement increments is helpful to understand the obtained response and make informed decisions on changes required in the solution process.

From the computational point of view, in the case of multiple solutions, a software code needs to adopt algorithms to evaluate these solutions and choose e.g. the one that lies closest to the old incremental direction $\delta \mathbf{d}^{n-1}$. In any case, the final check about the correctness of the results is left to the scrutiny of the analyst.

2.3.2.2 Direct and indirect displacement control

The displacement control was introduced at the beginning of the subsection and illustrated in Figure 2.31b. The strategy follows what was already stated in Section 1.3.2 of Chapter 1. For the sake of clearness, Eqs. (1.32) and (1.34) are rewritten in Eqs. (2.64) and (2.65), where the subscripts t and u indicate submatrices and subvectors related to unconstrained nodes and to nodes with constrained displacements, respectively. The assigned displacement $\mathbf{d}_u$ produces the external load $\mathbf{K}_{tu}\mathbf{d}_u$ to be added to the assigned $\mathbf{f}_t$. In turn, $\mathbf{d}_t$ collects the unknown displacements of the problem, together with $\mathbf{f}_u$ that collects the reactions of the constraints.

Willing to extend Eq. (2.64) to nonlinear analysis, in the case of direct displacement control, $\mathbf{d}_u$ must be incremented of a certain $\Delta\mathbf{d}_u$, Eq. (2.66). By retracing the mathematical passages used before, the first iteration of the generic step n is described in Eq. (2.67), where the equilibrium equation due to an increment of the assigned displacement $\Delta\mathbf{d}_u$ is balanced by an increment of displacement $\delta\mathbf{d}_t$. As usual, Taylor's expansion is used to evaluate the new equilibrium state, neglecting higher order terms. Looking at Eq. (2.67), the internal force is evaluated at the step $n-1$ and the external force is supplemented by the contribution due to the assigned displacement. In turn, when $i > 1$ the internal force already includes the increment of displacement $\Delta\mathbf{d}_u$ and Eq. (2.68) is formally the same of Newton–Raphson method, see Eqs. (2.50) and (2.51).

$$
\begin{bmatrix}
\mathbf{K}_{tt} & \mathbf{K}_{tu} \\
\mathbf{K}_{ut} & \mathbf{K}_{uu}
\end{bmatrix}
\begin{bmatrix}
\mathbf{d}_t \\
\mathbf{d}_u
\end{bmatrix}
=
\begin{bmatrix}
\mathbf{f}_t \\
\mathbf{f}_u
\end{bmatrix}
\tag{2.64}
$$

$$
\mathbf{K}_{tt}\mathbf{d}_t = \mathbf{f}_t - \mathbf{K}_{tu}\mathbf{d}_u \tag{2.65}
$$

$$
\mathbf{d}_u^n = \mathbf{d}_u^{n-1} + \Delta\mathbf{d}_u^n \tag{2.66}
$$

$$
\mathbf{f}_{\text{int}\left(\mathbf{d}^{n-1}\right)} + \mathbf{K}_{uu}^{n-1}\delta\mathbf{d}_t^{n,1} = \mathbf{f}_t - \mathbf{K}_{tu}^{n-1}\Delta\mathbf{d}_u^n
$$

$$
\mathbf{d}^{n-1} = \left[\mathbf{d}_t^{n-1}, \mathbf{d}_u^{n-1}\right]^T \qquad i = 1 \tag{2.67}
$$

$$
\mathbf{f}_{\text{int}\left(\mathbf{d}^{n,i-1}\right)} + \mathbf{K}_{uu}^{n,i-1}\delta\mathbf{d}_t^{n,i} = \mathbf{f}_t
$$

$$
\mathbf{d}^{n,i-1} = \left[\mathbf{d}_t^{n,i-1}, \mathbf{d}_u^n\right]^T \qquad i > 1 \tag{2.68}
$$

Besides the mathematical aspects, let us focus on the application of the direct displacement control. One typical example is the three-point bending flexural test on a reinforced concrete beam shown in Figure 2.37. Picturing a jack pushing downward, the structural behaviour can be described by means of subsequent increments of external force (Figure 2.37a). However, using a control loop mechanism, it is also possible to impose a displacement in the middle and evaluate the jack pressure as a reaction $\mathbf{f}_u$, as shown in Figure 2.37b. The method can, theoretically, be extended to a set of displacements, as long as their profiles are known (i.e. the entire vector $\Delta\mathbf{d}_u$), even if this is often nonpractical. Basically, all displacements of a given set evolve proportionally to the same scalar. In turn, Figure 2.38a shows a case when direct displacement control fails and an alternative procedure is needed, as discussed ahead.

Instead of direct control of a given displacement, it is possible to indirectly control the displacement according to the formulation of the spherical arc-length method. This can be done if $c = 0$ in Eq. (2.61): the equation becomes a constraint about the Euclidean norm of the incremental displacement. In this case, the method is usually referred to as *cylindrical* due to the relaxation of the constraint on the external force.

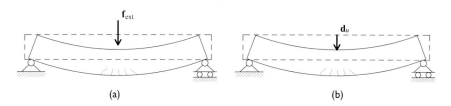

(a) (b)

Figure 2.37 Three-point bending flexural test. (a) Force control and (b) direct displacement control.

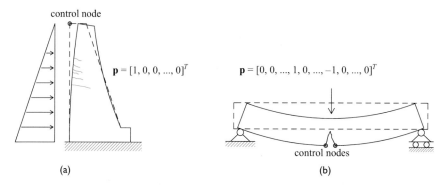

control node

$\mathbf{p} = [1, 0, 0, ..., 0]^T$

$\mathbf{p} = [0, 0, ..., 1, 0, ..., -1, 0, ..., 0]^T$

control nodes

(a) (b)

Figure 2.38 Examples of the arc-length variants. (a) Dam with indirect displacement control and (b) beam under a three-point bending flexural test with CMOD.

Another possible variant is to consider Eq. (2.69) in lieu of Eq. (2.61). A vector **p** of projection coefficients full of zeroes and only one nonzero component is now added in the additional arc-length equation. The goal is to select a single control node within Δ**d** for which the displacement is being monitored. More concisely, the vector **p** picks the displacement variable that controls the solution procedure without modifying the system of equations and the strategy normally adopted for force control methods with arc-length method. Accordingly, this type of control is called *Indirect Displacement Control*.

Let us consider the example in Figure 2.38a. The dam is subjected to hydrostatic pressure due to the reservoir water and the consequent load is applied along one boundary with a well-known triangular distribution. The use of load control is evident but, in the case of displacement control, the analyst needs to know the displacement profile of the same boundary. This is not possible because not only the shape of the initial displacement profile is not known in advance but also because, in the case of nonlinear analysis, the displacement profile changes as cracks initiate. The full nonlinear response, including the postpeak behaviour, can be obtained by means of the indirect displacement control of the horizontal displacement on its top, which is the largest. The consequent vector **p** is detailed in Figure 2.38a and, based on Eq. (2.69), it implies that the increment of displacement of the first node (i.e. the one on the top) is controlling the solution procedure. This provides some physical and geometrical meaning to Δl.

$$\mathbf{p}^T \Delta \mathbf{d}^{n,i} = \Delta l^{n,1} \tag{2.69}$$

Another variant of the arc-length control is shown in Figure 2.38b. In this case, the **p** vector considers two unitary values of opposite sign, meaning that the difference of displacement between the two corresponding nodes (actually, a given component of the displacement of these nodes, in this case the horizontal one) is used to control the structural response. In other words, the beam is studied according to the crack opening, similarly to laboratory tests in which a displacement transducer is placed at a notch to control the postpeak response of a specimen. In this case, the method takes the name of *Crack Mouth Opening Displacement* control, usually called CMOD. Note that any combination of node displacements may, in fact, be used for the **p** vector, according to the physical problem at hand.

2.3.2.3 Line search

As discussed in Section 2.3.1, the iterative methods presented are rooted in the linearization of the equilibrium equations. In other words, they are based on the Taylor's theorem that says that the first derivative (i.e. the tangent line for a single variable function) is a good approximation of the function itself *close to the point* it has been calculated. Accordingly, by assuming the sought displacement field $\mathbf{d}^{n,i}$ is in proximity of $\mathbf{d}^{n,i-1}$ (calculated at the previous iteration), the higher order terms of Taylor's expansion can be neglected, and $\mathbf{d}^{n,i}$ can be evaluated thanks to a linear approximation. It is evident that, in the case the sought displacement field is far from the previous one, Taylor's theorem does not apply anymore. In other words, in the case of strong discontinuities in the displacement field, iterative methods may not converge. The same applies in the case multiple solutions exist due to bifurcations.

From the practical point of view, the increments of displacement of each node are likely to be found along "directions" close to the previous increment of displacement. A logical and reasonable prediction of displacement for the nodes of the dam of Figure 2.38a is that they keep moving towards the right-hand side, unless the analyst defines an unloading stage (e.g. because earthquake action is being simulated). This occurs despite the fact that most structures tend to find alternative loading paths upon cracking, crushing or yielding, meaning that stress redistributions are normal. But these tend to be slow processed upon loading application.

Different is the case shown in Figure 2.38b: a sudden crack opening may alter the previous "direction" of displacement of some nodes, due to the sudden unloading of the crack sides. In the presence of such strong nonlinearities, which are not uncommon in civil engineering structures, it is possible that the iteration process will not converge. This may be due to the difficulties to identify the part of the structure where inversion occurs or because there are multiple solutions with several cracks competing. In such cases, possible solutions are to adopt smaller load steps (to contain the jump) or to use an optimization technique to tune the increment of displacement.

Regarding the latter, the *line search* algorithm is a widely used technique to increase both the convergence rate and the convergence likelihood. In mathematical terms, the method makes the solution procedure globally convergent, i.e. convergent to some solution independently of the starting point. The method is schematized in Eq. (2.70). Compared with the last of Eq. (2.52), the formula means that the method scales the increment of displacement through a factor α. The goal of this factor is to force the current iteration to be a better estimation than the previous iteration. To have a visual intuition of the method, let us consider the equilibrium from energetic point of view. Basically, equilibrium is satisfied when the work of residuals on displacement increments Ω reaches the null value, which is also the global minimum if Ω is considered with its absolute value. With reference to Eq. (2.70), as all data are known at the iteration $i - 1$, the work Ω^{i-1} is also known: the objective is to find a value of α for which the work in correspondence of $\mathbf{d}^{n,i}$ is lower than Ω^{i-1}. In this case, α works as a *convergence enhancer* (or *catalyst*, in the chemical lexicon). If the work of residuals decreases at each iteration, i.e. $\Omega^i < \Omega^{i-1}$, the solution method converges to its minimum.

$$\mathbf{d}^{n,i} = \mathbf{d}^{n,i-1} + \alpha \delta \mathbf{d}^{n,i} \qquad (2.70)$$

The illustrative example of Figure 2.39a sheds light on the procedure. Operatively, a generic iteration method (e.g. regular Newton–Raphson) is used to calculate $\delta \mathbf{d}^{n,i}$. Consequently, if $\alpha = 0$, $\mathbf{d}^{n,i} = \mathbf{d}^{n,i-1}$, i.e. no progress in the incremental analysis; if $\alpha = 1$, $\mathbf{d}^{n,i} = \mathbf{d}^{n,i-1} + \delta \mathbf{d}^{n,i}$, i.e. line search strategy is not active. However, the work of residuals can be calculated for both the mentioned cases and, as intuitively grasped, the latter can be larger or smaller than the previous one, i.e. ascendant and descendent behaviour, respectively. In the first case, the factor α should be found within an optimal range so that the increment of displacement at the given iteration (i.e. $\alpha \delta \mathbf{d}^{n,i}$) actually leads to a lower work and, iteratively, to its minimum. In the second case, instead, if the solution obtained is already better than the previous one, line search is not required.

Figure 2.39b shows a simple approach to evaluate α. Basically, the minimum of the work Ω^i in the line search direction requires that its derivative with respect to α must be zero. Knowing the value of the derivatives in correspondence of $\alpha = 0$ and $\alpha = 1$, several iterative approaches can be used to calculate a better estimation of α. In Figure 2.39b, as the two derivatives have opposite sign, the Regula–Falsi method (or the method of false position, which is basically an interpolation line) is used, whereas extrapolation would be needed in the case of same sign. According to the idea of *convergence enhancer*, line search iterative procedure usually quits when the absolute value of the first derivative of Ω^i lays within a certain pre-assigned interval (such as $\pm 80\%$ of the value of the derivative in correspondence of

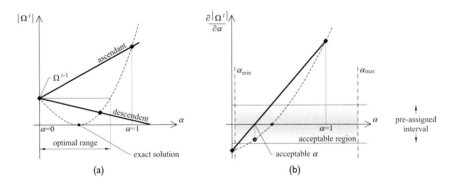

Figure 2.39 Line search strategy. (a) Equilibrium as minimum of work of residuals; (b) its derivative with respect to α and acceptable region for α. Note that only the three diamond markers can be calculated, and dashed curves are not known.

$\alpha = 0$), in any case within a minimum and maximum values (α_{min} and α_{max}, respectively). The latter, in particular, allows the analysis to handle larger increments of external load in one step (i.e. $\alpha_{max} > 1$), even if large steps typically entail errors in the linearization process. Note also that the involved computational effort is small as it is only necessary to calculate new residuals in the process, without a need to assemble a new stiffness matrix or calculate new displacement increments from the global nonlinear problem.

2.3.3 Convergence criteria

A numerical method is *convergent* when it is able to reach the exact solution. In structural analysis, this means the method converges towards an equilibrium state (a configuration in which external and internal forces are equal). The graphs of the Newton–Raphson method given earlier (Figure 2.26 *et seq.*) showed how the iterations came closer and closer to the true structural response. The speed at which a convergent sequence approaches its final value is called *rate of convergence*, an indicator about the efficiency of the method. From the computational point of view, only regular Newton–Raphson method is discussed here because, among the solution methods described, it provides the highest rate of convergence (i.e. *quadratic*) when it starts from a displacement field sufficiently *close* to the solution. This is an important aspect as a given method not necessarily converges to the exact solution from any initial guess (in structural analysis, very large load increments typically mean that the exact solution is rather far from the initial guess; additionally, multiple solutions are often present around peak load). As seen earlier, the efficiency of regular Newton–Raphson method is counterbalanced by the high computational costs in terms of resources required (time, elementary operations and memory allocated), in addition to being less robust when large load increments are adopted by the analyst.

For the sake of simplicity, let us consider a unidimensional function $f_{(x)} = 0$ for which a solution is sought by means of Newton–Raphson method. Retracing the previous discussion, at a generic step $n - 1$, the solution found is x^{n-1} with $f_{(x^{n-1})} \neq 0$. A certain increment δx^n is thus necessary to update the solution, Eq. (2.71), and Taylor's expansion can be adopted to evaluate the new value of the function. Neglecting the higher order terms, it is possible to calculate the increment δx^n as shown in Eq. (2.72). The combination of Eqs. (2.71) and (2.72) is at the base of the iterative procedure of the Newton–Raphson method. This process stops when $f_{(x^n)}$ is below a certain tolerance.

$$x^n = x^{n-1} + \delta x^n \tag{2.71}$$

$$f_{(x^{n-1})} + f'_{(x^{n-1})} \delta x^n = 0 \quad \rightarrow \quad \delta x^n = -\frac{f_{(x^{n-1})}}{f'_{(x^{n-1})}} \tag{2.72}$$

It can be demonstrated that the regular Newton–Raphson method converges *quadratically* (Figure 2.40), i.e. the residual or error in the solution decreases to the power of two (such as $10^{-1} \rightarrow 10^{-2} \rightarrow 10^{-4}$). This is synthetically illustrated in Eq. (2.73), where M is a generic positive constant. In general, a solution method converges *linearly* to the true solution if Eq. (2.74) holds, where μ is the ratio of the absolute values of two subsequent errors. When the number of iterations $n \rightarrow \infty$, if μ is closer to zero, the method is said to converge *superlinearly*; if μ is closer to one, the method converges *sublinearly*. The comparison between the rates of convergence is depicted in Figure 2.40. For the sake of readability, the linear convergence depicted in the figure represents the case of $\mu = 0.5$, i.e. the residual (and the absolute error) is halved at each iteration.

$$\left| r^n \right| = M \left(r^{n-1} \right)^2 \tag{2.73}$$

$$\left| r^n \right| = \mu \left| r^{n-1} \right| \qquad \text{with} \qquad \mu \in (0,1) \tag{2.74}$$

Dealing with convergence, the numerical solution will finally coincide with the true one (i.e. the residuals of Figure 2.40 tend to be zero). In this regard, it is possible to define a certain level of *accuracy* within which the solution is considered satisfactory. Accordingly, accuracy is defined as the degree of closeness of the numerical solution to the exact/true value. From the opposite viewpoint, there is no need to refine the numerical solution below a certain satisfactory level of accuracy. The condition adopted to evaluate the accuracy of the iterations is called *convergence criterion*. Once the criterion is met, the iteration process stops, and a new load step can be considered.

For engineering applications, software codes offer different measures to evaluate the accuracy of the result. In general, they consider the norm of out-of-balance forces, incremental displacements and energy. With

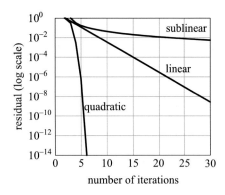

Figure 2.40 Schematic comparison of the rates of convergence.

reference to the nomenclature of Figure 2.41 (compare with Figure 2.26), these norms are calculated comparing the results of the current and first iteration (to scale the error and make it independent from the units of measurement adopted by the analyst), as detailed in Eq. (2.75). As it is possible to see, when $i = 1$, only the first norm can be calculated, as the other two become unitary. Typical tolerance values for the convergence criteria are 10^{-3}–10^{-2} for the force norm, 10^{-4}–10^{-3} for the displacement norm and 10^{-4}–10^{-3} for the energy norm. As expected, the choice of the criterion to be adopted depends on the type of analysis. In the case of doubt, more than one convergence criterion may be used in a given analysis and, if the criteria are performed independently, the most conservative one should be chosen.

$$\text{force norm ratio} = \frac{\sqrt{\mathbf{R}^{n,i^T}\mathbf{R}^{n,i}}}{\sqrt{\mathbf{f}_{\text{int}}^{n,1^T}\mathbf{f}_{\text{int}}^{n,1}}} \qquad i \geq 1$$

$$\text{displacement norm ratio} = \frac{\sqrt{\delta\mathbf{d}^{n,i^T}\delta\mathbf{d}^{n,i}}}{\sqrt{\Delta\mathbf{d}^{n,1^T}\Delta\mathbf{d}^{n,1}}} \qquad i > 1 \qquad (2.75)$$

$$\text{energy norm ratio} = \left|\frac{\delta\mathbf{d}^{n,i^T}\mathbf{R}^{n,i}}{\Delta\mathbf{d}^{n,1^T}\mathbf{R}^{n,1}}\right| \qquad i > 1$$

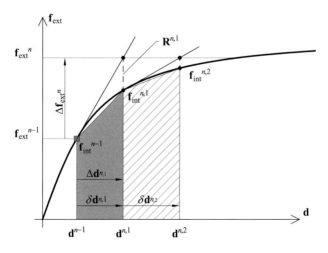

Figure 2.41 Quantities adopted in the convergence criteria.

As a general recommendation, the convergence criterion should investigate the most sensitive aspect of the structural response. With reference to Figure 2.41, when the diagram approaches the peak load ($K_{tan} \rightarrow 0$), a small variation of the force provides a large variation of displacement. The displacement norm is therefore more stringent. The opposite reasoning can be adopted in the first part of the diagram (closer to the elastic condition), where force norm may be used instead (a small variation of the displacement provides a large variation of the force). In general, the energy norm, being composed of internal forces and relative displacements, represents a valid compromise and it is often suggested as a standard criterion. Nonetheless, the analyst should be aware that the product of force and displacement gives an overall accuracy, but both quantities may result inaccurate. In a simplistic way, the energy being an area, no direct information is conceived about its sides. In the case an unexpected structural response is found, e.g. when a large force increment or decrement suddenly occurs (usually in combination with the arc-length method and a deficient control of the solution procedure) or a very large load increment with a less adequate stiffness matrix is chosen to overcome a bifurcation or convergence difficulties, it is highly recommended to check the force norm in order to ensure that equilibrium has been reached.

Finally, the analyst can always set a maximum number of iterations after which, in absence of convergence, the analysis stops or proceeds to the next load step. It usually depends on the algorithms adopted for the iterative process and the complexity of the model, in terms of number of displacement variables and nonlinearities. As a rule of thumb, if convergence is not reached after 25–50 iterations, it is advisable to check the model for input inaccuracies or change the iterative methods/incremental procedure. Alternatively, the analysis may be continued (as a particularly difficult convergence condition was momentarily found, e.g. due to competing cracks opening or local crushing) and convergence may occur in subsequent load steps. In this case, the analyst must ensure that, for the specific load nonconverged step, the convergence norm in the last iteration is not excessively large, so that a fully erroneous solution is not obtained, and the result is not compromised. Automatic procedures for load application are discussed in Chapter 5.

2.3.4 Example

With the aim of providing a general example of the mentioned methods, let us consider the steel frame structure in Figure 2.42a. This is a plane frame 6.0 m long and 3.0 high, where the column cross-section is $0.30 \times 0.30\,m^2$ and the beam cross-section is $0.30 \times 0.40\,m^2$ (these would be unexpected cross-sections in steel and are only used here for illustration purposes). Columns and beam are made of the same material, which is elastic-perfectly plastic, with yield stress equal to 200 MPa and Young's modulus equal to 200,000 MPa. The diagram in Figure 2.42b shows the relationship between

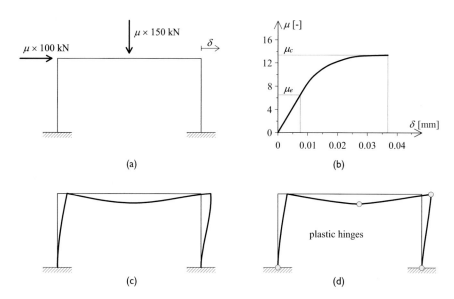

Figure 2.42 Plane frame under incremental vertical and horizontal loads. (a) Overview of the problem; (b) diagram of load multiplier vs. horizontal displacement; (c) and (d) deformed configurations at the elastic limit state and at collapse, respectively (not to scale).

load multiplier μ of the steel frame and the horizontal displacement of the beam δ, as calculated with FE analysis.

According to linear structural analysis, it is possible to calculate exactly the load multiplier $\mu_e = 6.618$ corresponding to the elastic limit of the structure. Figure 2.42c represents the deformed diagram for this load multiplier value. The collapse load multiplier $\mu_c = 13.6$ can be also calculated exactly using plastic hinge structural analysis. In this case, the deformed configuration is shown in Figure 2.42d, where the plastic hinges are highlighted. Although apparently similar, the deformed diagrams of Figure 2.42c and Figure 2.42d have different representation scales, being the latter almost four times larger than the former.

The FE analysis was performed considering a subdivision of each structural member into 20 beam elements. Additionally, as $\mu_e = 6.618$, a first load step equal to 6.5 was used. Then, six unitary load (multiplier) increments were applied, raising the load multiplier to $\mu = 12.5$. Three iterative methods were used, namely, regular, modified and linear Newton–Raphson, and tolerances equal to 10^{-2} and 10^{-3} for the force norm ratio. The results are shown in Table 2.7, according to which the following conclusions can be drawn:

- No substantial differences in terms of horizontal displacement are notable between the results with the two selected tolerances, and no substantial differences are notable between the results with different

Table 2.7 Comparison in terms of horizontal displacement δ [mm] for the problem in Figure 2.42 between regular, modified and linear Newton–Raphson methods (number of iterations in brackets)

μ	Step	Force norm tolerance = 10^{-2}			Force norm tolerance = 10^{-3}		
		Regular	Modified	Linear	Regular	Modified	Linear
7.5	2	8.681 (1)	8.677 (2)	8.677 (2)	8.681 (1)	8.681 (5)	8.681 (5)
8.5	3	9.931 (1)	9.920 (1)	9.917 (2)	9.931 (1)	9.931 (4)	9.929 (4)
9.5	4	11.41 (2)	11.38 (11)	11.38 (18)	11.41 (3)	11.41 (29)	11.41 (45)
10.5	5	13.60 (1)	13.44 (3)	13.48 (59)	13.60 (2)	13.59 (15)	13.59 (202)
11.5	6	17.03 (1)	16.96 (5)	16.90 (140)	17.05 (2)	17.05 (300)[a]	17.03 (256)
12.5	7	21.24 (1)	21.24 (150)[a]	21.03 (145)	21.26 (2)	21.25 (19)	21.23 (274)

[a] Maximum number of iterations reached without convergence.

solution methods, getting closer among themselves with tighter tolerance. Still, all results differ slightly given the tolerance defined by the convergence criteria. In this simple problem, all methods are able to converge to the same solution, which will be the *exact* solution, if small tolerances are adopted. *So, the structural response (in well-posed problems) is independent of solution procedure.*

- For both tolerances, and in a few load increments, modified Newton–Raphson method was not able to achieve convergence. This is due to the fact that the solution was oscillating between two possible configurations (in successive iterations the number of inelastic integration points was not constant and jumped repeatedly over these two configurations, because the stiffness matrix adopted is frozen and unsuitable for the specific structural configuration at the given load step). The results were still considered acceptable, and the analysis proceeded in finding the correct solution. *So, it is usually correct to continue with the analysis even in the case convergence is not found (and the error is small).*

- The quadratic convergence of the regular Newton–Raphson method is evident when comparing the number of iterations needed for the procedure to converge. As the problem is weakly nonlinear, a very low number of iterations (1–3) are needed, which is often not the case in real engineering problems. The linear method increases the number of required iterations as the nonlinear processes increase, given the deviation between the linear elastic stiffness and the current structural stiffness. The modified Newton–Raphson method provides a number of iterations between the other two methods, with the possibility of no convergence if the stiffness matrix becomes inadequate due to nonlinear processes occurring within a given load step.

The use of line search control can significantly improve the convergence of the iterative methods, as shown in Table 2.8. As it is possible to notice, line search is beneficial in all cases (reducing the number of iterations

Table 2.8 Comparison in terms of horizontal displacement δ [mm] for the problem in *Figure 2.42* between regular, modified and linear Newton–Raphson methods with line search control (number of iterations in the brackets)

μ	Step	Force norm tolerance = 10^{-2}			Force norm tolerance = 10^{-3}		
		Regular	*Modified*	*Linear*	*Regular*	*Modified*	*Linear*
7.5	2	8.681(1)	8.677(2)	8.677(2)	8.681(1)	8.681(5)	8.681(5)
8.5	3	9.931(1)	9.920(1)	9.917(2)	9.931(1)	9.931(4)	9.929(4)
9.5	4	11.41(2)	11.40(6)	11.38(3)	11.41(3)	11.41(9)	11.41(13)
10.5	5	13.60(1)	13.44(3)	13.49(18)	13.60(2)	13.60(11)	13.59(54)
11.5	6	17.03(1)	16.88(3)	16.81(26)	17.05(2)	17.03(28)	17.02(60)
12.5	7	21.24(1)	21.06(4)	21.04(37)	21.26(2)	21.25(11)	21.24(73)

significantly and eliminating the problem of lack of convergence), except for the regular Newton–Raphson method that provides the best possible solution in this simple structural example. Therefore, line search is not activated. In general, it is recommended to always adopt line search control because of the advantage in terms of convergence rate (a solution may be achieved with a lower computational effort when compared to the case of absence on this control) and also because of the increasing likelihood of finding a solution in the case of complex problems or larger load steps.

To proceed with the incremental analysis, as the structure approaches a null stiffness (see Figure 2.42b), it is convenient to abandon the load-control strategy because this would request a continuous (and strong) reduction of the load increment in order to be able to obtain convergence. As discussed above, arc-length control circumvents this problem. Five increments of the load multiplier $\Delta\mu = 0.25$ were applied using arc-length control to all nodes of the structure, obtaining the final load multiplier–horizontal displacement diagram shown in Figure 2.42b. It should be noted that the collapse load calculated with these additional five increments is $\mu_c = 13.4$ (error of 0.1% with respect to the exact value of 13.6), and not $12.5 + 5 \times 0.25 = 13.75$. This is evident because the arc-length method does not consider a fixed (target) load but assumes that the load multiplier is also a variable.

BIBLIOGRAPHY AND FURTHER READING

de Borst, R. et al. (2012) *Nonlinear Finite Element Analysis of Solids and Structures.* 2nd ed. Chichester: John Wiley & Sons, Ltd.

Crisfield, M. A. (1991) *Non-Linear Finite Element Analysis of Solids and Structures* (vol.1 Essentials). Chichester: John Wiley & Sons, Ltd.

Crisfield, M. A. (1997) *Non-Linear Finite Element Analysis of Solids and Structures,* Volume 2: Advanced Topics. Chichester: John Wiley & Sons, Ltd.

Zienkiewicz, O. C. and Taylor, R. L. (2013) *The Finite Element Method for Solid and Structural Mechanics* (vol.2). 7th ed. Oxford: Butterworth-Heinemann.

Chapter 3

Constitutive models

INTRODUCTION

Despite the great differences in terms of microscopic composition (such as crystals of atoms in metals, granular aggregates in concrete or fibres of cells in timber), structural materials show common features in their macroscopic behaviour. In this context, it is possible to deal with general *constitutive models* using different formulations such as elasticity, plasticity, damage and fracture. As these may describe rather different responses, the analyst is requested to develop some understanding of their complexity before adopting a specific nonlinear *constitutive relationship*. In this regard, it is worth mentioning that linear elasticity is hardly an acceptable representation of the behaviour of existing structures (given the impact of an overconservative performance assessment in terms of cost, user distress, effective use of resources or loss of cultural value) and nonlinear constitutive laws are normally used in their place, both for the design of new buildings and the assessment of existing buildings. This is aimed at balancing adequate performance and safety levels, on the one hand, with minimally invasive intervention, respect of cultural heritage values (if applicable) and possible aesthetic implications, on the other hand. Constitutive models able to replicate relevant physical phenomena are thus needed. In the first part of this chapter, the main material properties are introduced by means of simple experimental tests. Elastoplasticity, damage and cracking are described next. As softening leads to strain localization phenomena (and the need for a mesh size regularization technique), this aspect is also discussed. The distinctive features of the most used constitutive models in commercial software codes are presented at the end of the chapter, referring the reader to Chapter 4 for recommended structural material properties. It is noted that the influence of environmental effects (e.g. temperature or humidity) and of time (e.g. fatigue, creep or shrinkage), which may be relevant in specific applications, are outside the scope of the present text.

DOI: 10.1201/9780429341564-3

3.1 MATERIAL BEHAVIOUR

Dealing with the material behaviour, experimental evidence and test results are the fundamental starting point of the so-called *phenomenological approach*. From the numerical point of view, modelling the constituents of a structural element (such as units and mortar for masonry, or aggregates and cement matrix for concrete) is often unfeasible and the assumption of *homogeneous* materials is made in most engineering applications. Accordingly, stress and strain variations within the constituents are neglected and average stresses and strains are considered. It is noted that, in the text below, unless evident from the context or explicitly stated, tensile stresses are assumed positive and compressive stresses are assumed negative.

3.1.1 Main constitutive models

According to the experimental results of material samples (or specimens) under uniaxial loading, Table 3.1 introduces idealized material models, together with the essential behaviour description and possible cases of application. As expected, these material models are a simplification of the real response, which may be rather complex, also entailing different mechanical behaviour in compression and in tension (e.g. masonry and concrete, which are strong in compression and weak in tension) or along different directions (i.e. a material such as masonry and timber with anisotropy defined, respectively, by the horizontal joints and fibre orientation).

More in detail, *elasticity* guarantees a one-to-one (either linear or nonlinear) univocal relationship between stress and strain. Both can increase indefinitely (in the more traditional case of linear elasticity), and, upon load removal, the material goes back to the initial conditions following the (same) loading path with no residual strain. In this case, strain is said to be *reversible*. The slope of the $\sigma-\varepsilon$ curve is called *tangent modulus* and, for the most usual linear elastic materials, it is constant and called Young's modulus E. The simplest evolution of this model is by defining an *elastic threshold stress* σ_y (also called *yield stress*) to limit elasticity, thus reversible strain, and elastoplastic after reaching the threshold, when the material exhibits instantaneously increasing deformation under constant stress. This model is designated as *elastic perfectly plastic* or *ideal plasticity*. Possible unloading–reloading follows lines parallel to the elastic branch. The main feature of plasticity is represented by the *additive strain decomposition*, according to which the total strain $\varepsilon = \varepsilon_e + \varepsilon_p$, namely, the sum of a reversible and irreversible part (elastic and plastic, respectively). As it is possible to notice, the stress is kept constant along the plastic branch of the diagram (thus the term "perfectly"). If this condition is violated, the stress can increase or decrease with respect to strain increments, leading to models with *hardening* or *softening*, respectively. In these cases, additive strain decomposition still holds. A limit case of the mentioned models can be obtained considering infinitely

Table 3.1 Examples of idealized material models for uniaxial behaviour relating stress σ and strain ε

Definition	σ–ε curve	Description	Application
Elastic (linear and nonlinear)		Either linear or nonlinear; the (*elastic*) *deformation is fully reversible*. In the case of unloading–reloading, the response follows the same path.	Any material with stresses within the elastic limits (typical of most civil engineering materials at low stress levels). Usually, linear elasticity is used.
Elastic perfectly plastic		Elastic behaviour for $\sigma < \sigma_y$ (yield stress) and perfectly plastic for $\sigma = \sigma_y$ ($\sigma > \sigma_y$ is not possible). Upon unloading–reloading, the response follows a path parallel to the elastic branch, with an *irreversible (plastic)* strain.	Mild steels subjected to tension and compression (if buckling is ignored); concrete subjected to compression and shear.
Elastoplastic hardening/ softening		After reaching the yield stress σ_y, the stress can increase (*hardening*) or decrease (*softening*). Unloading–reloading paths remain parallel to the initial elastic branch.	Hard steels usually exhibit hardening and concrete subjected to crushing exhibits softening, both with plastic strains.

(*Continued*)

Table 3.1 (Continued) Examples of idealized material models for uniaxial behaviour relating stress σ and strain ε

Definition	σ–ε curve	Description	Application
Rigid perfectly plastic/with hardening		Similar to previous models but without the elastic branch, which is assumed negligible.	Often used for hand calculations at failure (e.g. plastic hinge structural analysis, soil limit analysis or masonry macroblock analysis).
Elastic – damage with softening		Reversible strain but lower *mechanical properties* upon unloading–reloading (secant stiffness).	Concrete and masonry subjected to cracking and cyclic loading, such as earthquakes.

large values of Young's modulus, which is often used in engineering for hand calculations using limit analysis. In this case, the structure remains undeformed or rigid, with the exception of localized regions defining a failure mechanism, such as hinges, yield lines or failure surfaces. Accordingly, the elastic branch can be ignored, and the material is said to be *rigid*, either perfectly plastic or with hardening or softening. As a last example, after the elastic threshold, the material properties may undergo a detrimental process that reduces both the material strength and the stiffness. This process is called *damage* and considers the *secant modulus* for the unloading–reloading process, with null residual strain upon unloading.

In the following, elastoplasticity (originally used for perfect plasticity and hardening of metals) is discussed in Section 3.2, whereas damage and cracking models (originally used for softening of concrete-like materials) are discussed in Sections 3.3 and 3.4.

The material models introduced in Table 3.1 do not provide any information about the ultimate behaviour, i.e. when the capacity ceases to be adequate for the requested performance. This is an issue that can be tackled at different scales, such as (1) the microscopic one (typical of fracture mechanics) with the occurrence and propagation of cracks, (2) the homogeneous material scale where, irrespective of the elementary components, failure occurs at a representative material volume and (3) the global structural scale of the building as a whole. The focus here is in the constitutive model at the homogenous material scale. In this case, considering a specimen subjected to uniaxial tension, three main failure categories (brittle, quasi-brittle and ductile) are shown in Figure 3.1, drawn with the same maximum stress for comparative purpose. It is worth mentioning that, dealing with homogenized properties, strain must be considered carefully as the notion of average response (thus, homogeneity) may be lost in the case of localized failure.

Looking at Figure 3.1a, in brittle materials, fracture in tension is characterized by one main crack. Microcracks develop soon after the elastic limit stress is reached making the main crack to spread swiftly with little or no further deformation. It is worth reminding that the area underneath the σ–ε curve represents the energy per unit volume stored by the material. As it is

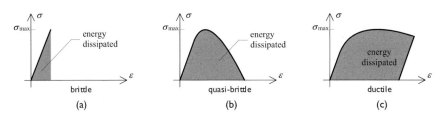

Figure 3.1 Main failure categories and respective shape of stress–strain diagrams. (a) Brittle, (b) quasi-brittle and (c) ductile.

observed, brittle materials, such as cast iron, glass or carbon fibres (used as strengthening in civil engineering applications) are able to store and dissipate (during failure) only a relatively small amount of energy that coincides with the elastic energy. Failure of quasi-brittle materials (Figure 3.1b), such as concrete, clay brick, mortar, soils and rocks, is characterized by *softening*. These materials fail due to a process of progressive internal damage and microcrack growth. At the level of the specimen (and structure), softening describes the gradual decrease of mechanical resistance under a continuous increase of deformation. For quasi-brittle materials, the softening branch is able to guarantee a larger dissipated energy when compared with brittle materials. Finally, ductile failure (typical of steel or polymers) is associated with large plastic strains due to progressive alterations of the material in proximity of the fracture zone (Figure 3.1c). The process propagates at a stable speed (fracture is not sudden) and significant energy is usually dissipated by the plastic branch. Once failure occurs, upon load removal, the elastic deformation is recovered, and so does the elastic energy; the energy dissipated is thus the one delimited by the elastic unloading.

Now, it is possible to discuss the typical behaviour of materials such as concrete and steel, as shown in Figure 3.2, where the stress–strain diagram for uniaxial standard characterization tests and the main features are highlighted. For both cases, an initial linear elastic branch is notable, but the subsequent behaviours are markedly different, being quasi-brittle for concrete and ductile for steel. Concrete exhibits hardening and a subsequent softening. Steel, once the yield limit is reached, shows a long elastoplastic behaviour with hardening, until fracture is reached.

As already mentioned, the constitutive models must be tuned to represent as much as possible the natural behaviour displayed in experimental tests. If this appears not too complex for uniaxial tests, the same does not hold for three-dimensional stress states and for the real internal force (thus stress) evolution conditions within buildings.

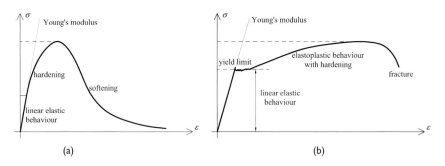

Figure 3.2 Stress–strain diagrams for the most used materials in modern civil engineering (different graphical scale). (a) Concrete in compression (note that compressive stresses are here assumed positive) and (b) steel in tension.

3.1.2 Failure of concrete-like materials

In civil engineering, quasi-brittle materials, such as concrete and masonry, are often designated as concrete-like materials, as they all share similar macroscopic features, and concrete is the most used material today in civil engineering structures. Concrete-like materials exhibit good compressive strength with hardening before reaching the maximum (or peak) stress, followed by softening after the peak stress; in turn they are characterized by a vanishing tensile strength with a dramatic softening. In the following, tensile and shear effects are introduced first, being crushing failure under compression presented at the end of the present subsection.

Failure by tension and shear can occur according to three elementary modes. They are defined in accordance with the relative displacement of the crack faces (Figure 3.3) and detailed as follows: Mode I: opening mode (displacement along y axis); Mode II: in-plane shear or sliding mode (displacement along x axis); Mode III: antiplane or out-of-plane shear, also called tearing mode (displacement along z axis). Of the above, only Mode I and II will be further discussed, as Mode III can be considered an extension of Mode II.

As far as Mode I is concerned, evidence from uniaxial tensile tests has shown that failure under tension is usually triggered by the longest most favourably oriented flaw or crack. As its location is not known a priori, to limit uncertainties of laboratory experiments, notches are often made to localize the crack occurrence and control its propagation. With reference to Figure 3.4a, where the stress σ is drawn according to the displacement u, the specimen behaves elastically up to the tensile strength f_t, exhibiting a strong softening until reaching the ultimate displacement u_{cr}, i.e. when the crack is fully developed and the crack edges are completely separated. It should be noted that the measured displacement u is preferable to strain, which becomes meaningless in the case of a discrete, localized crack. Additionally, u can be decomposed in the elastic part of the specimen outside the notch, and the crack opening w. Only the latter accounts for dissipated energy (Figure 3.4b), as the elastic part is fully recoverable. Basically, when the crack is fully developed, there is no stress at the crack faces and the adjacent

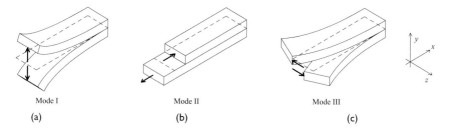

Mode I	Mode II	Mode III
(a)	(b)	(c)

Figure 3.3 Fracture modes (with reference to an orthogonal set of axes x-y-z). (a) Mode I, (b) Mode II and (c) Mode III.

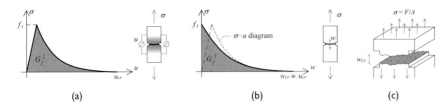

Figure 3.4 Typical Mode I behaviour of quasi-brittle materials. (a) Idealized stress–
total displacement and (b) stress–crack opening diagrams (with *u* the total
displacement and *w* the crack opening); (c) detail of the crack opening.

material (with shaded light grey in the specimen of Figure 3.4a) releases the internal stress by recovering the elastic displacement. This means that, at the ultimate stage, elastic displacement is null, and the total ultimate displacement u_{cr} coincides with the crack displacement w_{cr}.

It has been experimentally observed that the area under any of the curves of Figure 3.4a and b is a material constant and is independent of structure size or geometry. Its calculation is shown in Eq. (3.1) using both diagrams, which lead to the same result. As the stress is a force over the unit area, when integrated over a displacement, it provides work over the unit area (i.e. J/m² or N/m). Accordingly, G_f^I is defined as the *fracture energy* of Mode I and represents the amount of energy necessary to create a unit area of a fully developed crack. Figure 3.4c illustrates the phenomenon by magnifying the displacements (note the simplified representation, as no stress transfer is possible between the two faces of a fully opened crack). Conversely to brittle materials, where the elastic energy represents the only form of energy dissipated, fracture energy is much larger for quasi-brittle materials (compare Figure 3.1).

$$G_f^I = \int_0^{u_{cr}} \sigma_{(u)} du = \int_0^{w_{cr}} \sigma_{(w)} dw \qquad (3.1)$$

It is worth noticing that the σ–u diagram of Figure 3.4a is univocally determined once, in addition to the elastic properties, G_f^I, f_t and the shape of the softening curve are known. In structural modelling, these parameters are assumed to be fixed material properties. However, if there is a certain consensus on likely values for the first two parameters (confirmed by experimental evidence), the shape of tensile-softening curve receives far less consensus and depends on the material being considered. Experimental results show uniformity in the identification of a concave curve with a steep decline just after the peak. The most common softening diagrams adopted are linear, bilinear and exponential, and the reader is referred to Section 3.5 for more information.

The analysis of Mode II is shown in Figure 3.5a where the response of a masonry specimen, with two masonry units and one horizontal mortar joint, is analysed. As clearly noticeable, the behaviour of the specimen is strongly influenced by the compressive stress, as the resistance increases with normal confining stress σ, which is a typical behaviour of frictional materials. The example shows the slip between two bricks where the mortar follows the transversal displacement until a sliding crack is fully developed. After the tangential stress peak, the softening shear stress has a gradual decrease (i.e. softening) and stabilizes in a residual value, given by friction or mechanical interlocking at the joint. The softening behaviour in shear is assumed governed by the Mode II fracture energy G_f^{II} equal to the area of the shaded region in Figure 3.5a (in this case v is the shear displacement). The fracture energy has been found to be linearly dependent from the confining stress σ. However, in numerical simulation, a constant value of G_f^{II} is generally used, calculated in correspondence of $\sigma = 0$ (area shaded with diagonal lines in Figure 3.5a), without considering any evolution. It must be noted that, with null confining load, the maximum tangential stress coincides with the material cohesion c (see also Mohr–Coulomb yield function in Figure 3.13).

For the sake of completeness, compressive loading is described in Figure 3.5b. Under uniaxial compression tests, given the heterogeneity of concrete-like materials, many microcracks usually propagate with no favourable directions. The initial stresses and cracks present in the material, as well as the scatter of internal stiffness and resistance in the material, cause progressive crack growth with increasing deformation. Around the peak load, microcracks swiftly propagate leading to macrocracks with an overall softening of the σ–u diagram. Note that the postpeak is dependent on the boundary conditions of the specimen and requires careful testing. The maximum stress is the compressive strength f_c, whereas the area under the σ–u curve is the compressive fracture energy G_c.

On closer look, the total energy dissipated can be considered as the sum of the energy associated with microcrack initiation and propagation (until the peak stress), the energy dissipated by the coalescence of microfractures

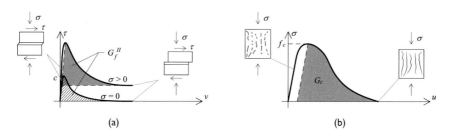

Figure 3.5 Typical (a) Mode II and (b) crushing behaviour of quasi-brittle materials (and relative fracture energy G_f^{II} and G_c, respectively).

(initiated before the peak stress) and the fracture energy consumed in the localized macrofracture zone (after the peak stress). To consider this localization and the consequent unloading of the undamaged continuum (see Section 3.5 for strain localization in finite element method [FEM]), the inelastic displacement is calculated by subtraction of the inelastic prepeak displacement to the total displacement. Accordingly, in the framework of FEM, the fracture energy G_c can be defined as the shaded area of Figure 3.5b.

As a final remark, it is important to underline that softening (either in tension or compression) is always indicative of an unstable process, i.e. the load must decrease to avoid an uncontrolled failure. According to Chapter 2, a deformation-controlled test is thus needed to describe the growth of cracks and their localization, while the rest of the specimen unloads.

3.1.3 Simple nonlinear problems

Let us now tackle two illustrative examples about the differences of adopting linear and nonlinear constitutive laws. Figure 3.6a shows a rigid block connected to three bars (modelled with springs) of the same length $l = 1.0\,\text{m}$ and the same cross-section area $A = 1,000\,\text{mm}^2$. According to Figure 3.6b, they have also the same Young's modulus $E = 200,000\,\text{MPa}$, but they exhibit different yield stresses equal to 100, 200 and 300 MPa for bars 1, 2 and 3, respectively. The rigid body is constrained such as the only kinematic variable is the horizontal displacement u. It is of interest to understand which is the relationship between the horizontal force F (pulling the bars) and the longitudinal elongation u.

With this aim, an incremental procedure is pursued. Since the problem deals with parallel springs, it is possible to assume that, for the three bars, the strain is the same and equal to $\varepsilon = u/l$. Accordingly, Table 3.2 describes the calculations considering a constant increment of strain equal to $\Delta\varepsilon = 0.1 \times 10^{-3}$. Once the strain is known, the related stress is derived for each bar from Figure 3.6b, thus the related force (the subscripts in Table 3.2 indicate which bar is considered). It must be noted that the shaded cells highlight the strains when each bar reaches the yield limit. According to the elastic perfectly plastic model, after this point, increments of strain do not lead to increments of stress. The two last columns collect the overall results shown in the diagram of Figure 3.6c about the entire structure. As it is possible to notice, once the yielding limit $F = 600\,\text{kN}$ is reached, displacement increases indefinitely and only an incremental approach (in displacement control) can track the plastic behaviour of the structure.

As shown in Table 3.2, the overall behaviour of the structure is a consequence of the behaviour of each spring (namely, the material level). To get more insight on the nonlinear behaviour of materials, Figure 3.7 illustrates the unloading behaviour after having reached its maximum force. To facilitate the comparison between material behaviour (Figure 3.7a) and

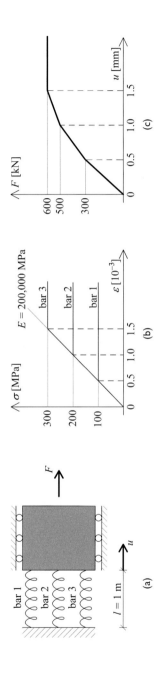

Figure 3.6 Three-bar structure subjected to uniaxial load. (a) Overall description of the structure, (b) constitutive law of the three bars and (c) force–displacement diagram of the structure.

Table 3.2 Step-by-step calculations for the problem of Figure 3.6 with constant $\Delta\varepsilon = 0.1 \times 10^{-3}$

ε [10^{-3}]	σ_1 [MPa]	$F_1 = A\sigma_1$ [kN]	σ_2 [MPa]	$F_2 = A\sigma_2$ [kN]	σ_3 [MPa]	$F_3 = A\sigma_3$ [kN]	$u = l\varepsilon$ [mm]	$F = F_1 + F_2 + F_3$ [kN]
0.0	0	0	0	0	0	0	0.0	0
0.1	20	20	20	20	20	20	0.1	60
0.2	40	40	40	40	40	40	0.2	120
0.3	60	60	60	60	60	60	0.3	180
0.4	80	80	80	80	80	80	0.4	240
0.5	100	100	100	100	100	100	0.5	300
0.6	100	100	120	120	120	120	0.6	340
0.7	100	100	140	140	140	140	0.7	380
0.8	100	100	160	160	160	160	0.8	420
0.9	100	100	180	180	180	180	0.9	460
1.0	100	100	200	200	200	200	1.0	500
1.1	100	100	200	200	220	220	1.1	520
1.2	100	100	200	200	240	240	1.2	540
1.3	100	100	200	200	260	260	1.3	560
1.4	100	100	200	200	280	280	1.4	580
1.5	100	100	200	200	300	300	1.5	600
1.6	100	100	200	200	300	300	1.6	600
1.7	100	100	200	200	300	300	1.7	600
1.8	100	100	200	200	300	300	1.8	600

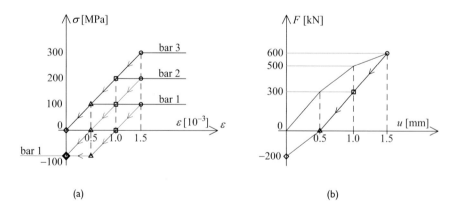

(a)   (b)

Figure 3.7 Three-bar structure subjected to uniaxial unloading after reaching the maximum force. Unloading history (a) at the material level (stress–strain relationship) and (b) at the structural level (force–displacement diagram).

structural behaviour (Figure 3.7b), the key conditions are highlighted with markers. Starting from the point of maximum force in the structural system (indicated by a circle), and assuming ideal elastoplasticity, the reduction of strain from 1.5×10^{-3} to 1.0×10^{-3} leads the three springs to the square markers. As expected, the first bar is totally unloaded, whereas the other two bars provide a total force of 300 kN (100 and 200 kN from bars 2 and 3, respectively). A subsequent reduction of strain to 0.5×10^{-3} (triangle marker) moves the first bar to its yielding stress in compression: this negative value counterbalances the positive stress of the bar 3, ending up with a null overall force. This result is somewhat unexpected: the structure is in equilibrium with null applied force, but compression and tension are evident at the material level (the so-called *residual stresses*). A final reduction to zero strain (diamond marker) means that bar 1 is on the plastic branch, bar 2 is still on the elastic branch (both with the same compressive stress equal to 100 MPa) and bar 3 is unloaded. As a result, the overall force on the rigid block is −200 kN (compression force). Basically, due to plasticity, a pushing force is now needed to bring the structure to the original kinematic "undeformed" configuration ($u = 0$ mm). Note that the possibility of buckling of the bars in compression was ignored.

Although it may seem complicate, Figure 3.7 is a good approximation of the behaviour of real structures and, also, of how a finite element model works. Accordingly, by simply assigning the constitutive law of the material (uniaxial stress–strain relationship and ideal plasticity in this case), it was possible to *track* the loading and unloading diagram at the structural level. Following the same reasoning of Figure 3.7, it is evident that any combination of force and displacement below the diagram of Figure 3.6c can be achieved by simply changing the unloading (and, possibly, reloading) starting point. Intuitively, starting from another point of the plastic branch of Figure 3.6c would have led the unloading path to simply move towards the positive u. This aspect shows that multiple forces can be related to the same displacement, and vice versa, meaning that the univocal correlation between stress (or force) and strain (or displacement), typical of linear elastic materials (and structures), is lost. The only way to describe the structural response is by keeping track of the load history, because every condition of materials (and structure) is path dependent. In this case, the incremental procedure could have been carried out also using an *event-driven approach*. This means that the structural system is updated every time a singular event occurs, which in this case is evident with the yielding of one of the bars. As shown in Figure 3.6c and marked with shaded cells in Table 3.2, the structural system evolves from the configuration of three bars initially active (in terms of stiffness), to two bars active, one bar active and a mechanism. The consequence is that the global stiffness of the system (or structural stiffness) reduces from the elastic value, successively to 2/3 and 1/3 of this, until a null value. Upon unloading from 1.5×10^{-3}, again the original structural stiffness is found, see Figure 3.6c.

The second example, given in Figure 3.8a, is about a statically indeterminate (hyperstatic) beam with length l, with constant rectangular cross-section, clamped at the left end and simply supported at the right end, subjected to a point force at the mid-span. The objective is to evaluate the maximum force that the beam can support. Provided ideal plasticity is assumed (such as one of the springs of Figure 3.6b), Figure 3.8a shows the evolution of the stress within the cross-section, namely, at first yielding (and the consequent bending moment M_y), elastoplastic (intermediate situation) and fully plasticized (M_p). It is well known that the flexural stiffness (related to the product of Young's modulus E and moment of inertia I) of the three states is different, moving from a maximum value for the elastic stiffness (i.e. EI_e), to the intermediate elastoplastic one (i.e. EI_{ep}), until zero at the ultimate state. The related moment–curvature diagram is also depicted in Figure 3.8a.

Considering monotonic loading, the structure behaves elastically (Figure 3.8b) until the diagram of the bending moment is between $-M_y$ and M_y. As a consequence (compare Figure 3.8a), the stiffness is constant throughout the beam and equal to EI_e, as shown on the bottom of the figure. Once the structure reaches the yielding moment (Figure 3.8c), the stiffness starts reducing (according to the diagram of Figure 3.8a), and a

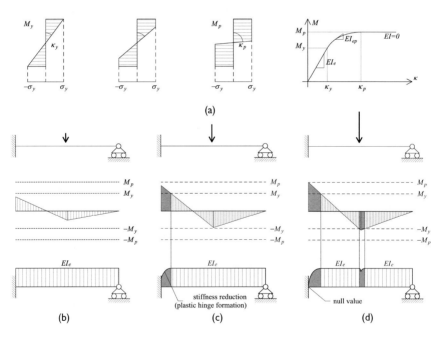

Figure 3.8 Statically indeterminate (hyperstatic) beam with a point force at mid-span. (a) Stress distributions at the section level and respective moment–curvature diagram. (b–d) Evolution of the structural response of the beam under monotonic loading, with bending moment diagram evolution (middle row) and flexural stiffness evolution (lower row).

plastic hinge starts to develop at the clamp. A subsequent increment of the external force (Figure 3.8c) leads the moment at the clamp to reach the plastic value, thus zero stiffness, unlimited rotation and a fully developed hinge. At the same time, for this particular case, a plastic hinge becomes evident in correspondence of the applied concentrated force, but still some additional force can be applied. In fact, the analysis only stops when a singular stiffness matrix is detected (i.e. a fully developed structural mechanism occurs, as shown ahead).

Even if the analysis may appear complex, it is worth underlining that the discussion related to Figure 3.8 is not totally correct and was kept schematic for the sake of clarity. The behaviour of any structure depends on its stiffness, or better, on the variation of the stiffness within the structure (generally, stiffer elements *attract* more load). In fact, according to basic continuum mechanics, the bending moment distribution is not affected by the stiffness magnitude in absolute terms (e.g. being steel or timber, large or small section) if the stiffness is constant throughout the beam. If stiffness changes, so does the moment distribution. Looking closely to Figure 3.8b–d, the distribution of the bending moment was kept constant, and the diagram incremented homothetically. In reality, there is an interdependence: bending distribution depends on the stiffness variation and vice versa. Thus, there is a need of an iterative approach, and some sort of discretization (such as FEM), to reach the equilibrium with the external loads.

When nonlinear constitutive laws are considered, it is evident that superposition does not hold because the change of stiffness modifies the mechanical characteristics at every step. In this case too, the detailed structural response can be studied only thanks to an incremental analysis, keeping track of the load history.

For the sake of completeness, the same problem can be addressed thanks to an incremental approach in the framework of limit analysis, which is based on a rigid-perfectly plastic relationship between moment and curvature. According to Figure 3.9a, it is assumed that free rotation occurs when the bending moment reaches M_p (i.e. a fully developed plastic hinge), while the rest of the structure is undeformed (compare to Figure 3.8a). First, according to linear elastic theory and assuming constant stiffness throughout the beam, the bending moment diagram is the one shown below the structural scheme of Figure 3.9b. Knowing the moment M_p that the beam section can resist, it is possible to calculate the maximum force F before the occurrence of the first plastic hinge. The consequent moment distribution is shown in Figure 3.9c where all the quantities are expressed in terms of M_p. In particular, for $F = 16M_p/3l$, the bending moment at mid-span is $M_p/6$ away from the plastic moment (a sort of *reserve* given by the structure hyperstatic conditions).

As the external force increases, the clamp behaves as a fully developed hinge (indefinite rotation without opposing additional resistance, i.e. null flexural stiffness). This is illustrated in Figure 3.9d showing that the original

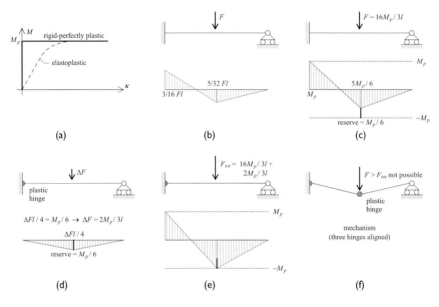

Figure 3.9 Statically indeterminate (hyperstatic) beam with a point force at mid-span. (a) Rigid-perfectly plastic moment–curvature diagram; (b) linear elastic analysis; (c) evaluation considering elastic behaviour and the plastic moment M_p; (d) increment of force ΔF after the first plastic hinge; (e) total force and (f) second plastic hinge (the structure is now statically indeterminate or a mechanism). Note that the bending moment diagrams are not drawn to scale.

structure became a simply supported beam. As the structure is now statically determined, it is still able to bear an increment of external force ΔF. According to Figure 3.9d, this increment cannot lead the mid-span section to have a bending moment larger than the plastic one. In other words, the maximum increment of force ΔF is the one able to develop a bending moment at the mid-span equal to its reserve (i.e. $M_p/6$). The necessary calculations are given in Figure 3.9d, whereas Figure 3.9e shows the overall results with $F_{tot} = 6\ M_p/l$. A subsequent increment of force is not possible as a second fully developed plastic hinge has been formed, leading to a statically indeterminate structure, thus a mechanism (Figure 3.9f).

As noticeable, the present example was solved with an event-driven approach considering three conditions for the beam, namely, (1) a hyperstatic beam with a clamped end and a simply supported end; (2) an isostatic simply supported beam and (3) a mechanism. In the second condition a stress redistribution was evident: the entire beam took some extra bending moment except for the clamped section that has already exhausted its capacity (and, obviously, the same for hinge at the right extremity). It is noted that, for the given example, in the case a rectangular section is considered, the ultimate force that may be applied to the beam is about 70%

higher than the force for which the first yielding occurs at the clamped section. This further demonstrates the relevance of adopting nonlinear structural analysis in engineering applications.

Finally, it must be stressed that, if the entire structural response is sought (usually given in terms of applied force vs. midspan deflection in this case), hand calculations are rather impractical even for such a simple problem. To account for stiffness reduction after yielding and the consequent updated moment distribution, FEM is the most used structural analysis tool and the solution, for the beam problem, would be rather straightforward.

3.2 ELASTOPLASTICITY MODELS

In Table 3.1 the *elastic limit* was indicated with the yield stress σ_y and was defined as the stress at which plastic (irreversible) deformation starts. Although the phenomenon of irreversible deformations can be related to specific processes inside the material (e.g. crystal slipping in metals), in a broader context, the mathematical formulation of plasticity can be applied to any material showing irreversible deformations, such as soil or concrete. The theory is developed here for perfect plasticity and hardening, historically associated with the behaviour of metals, which are the basis for the theory. Still, the approach can be used for softening materials also.

Looking at the elastoplastic case with hardening of Table 3.1, the unloading–reloading paths are always elastic. In the case of reloading, in particular, the behaviour is elastic until the previously attained maximum stress that becomes the current elastic limit. In other words, according to the stress history, the material exhibits a behaviour whose elastic limit evolves following the curve of monotonic loading. The new elastic limit is defined as the *plasticity threshold*: if the stress is below this value, the behaviour is elastic; plastic deformation occurs only for values of stress equal or larger than this limit (the latter only in the case of hardening plasticity). In turn, the growth of plastic deformation develops under the name of *plastic flow* (due to the irreversible "slip" movements at atomic level). The numerical description of hardening processes and the evolution of plastic deformations (the so-called *flow rule*) are addressed next.

Let us consider simple uniaxial loading tests and discuss their behaviour in the case of cyclic loading. According to Figure 3.10a, in the case of a perfectly elastoplastic material, the elastic limit σ_y is assumed equal for both tension and compression, and any cyclic load follows a path within these two boundaries (dashed lines). *Isotropic hardening* is sketched in Figure 3.10b where the hardening effects, either in tension or in compression, symmetrically reflect themselves on the other side of the graph. In other words, if the elastic limit moves from σ_y^0 to σ_y^1, due to a hardening process in tension, the new elastic limit in compression will be $-\sigma_y^1$ (the superscript indicates the subsequent hardenings, where 0 refers to the

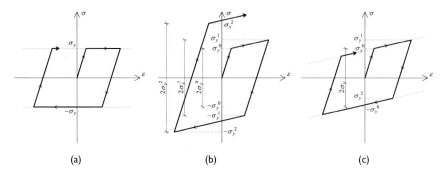

Figure 3.10 Simple uniaxial elastoplastic behaviour involving cyclic response. (a) Perfect elastoplasticity; (b) isotropic hardening and (c) kinematic hardening.

original uniaxial monotonic response). A subsequent hardening in compression moves the elastic limit for tension too (similarly to a spiral). The elastic domain, which, for the uniaxial case, can be pictured as the distance between the two opposite elastic limits in tension and compression, increases from $2\sigma_y^0$ to $2\sigma_y^1$ and $2\sigma_y^2$ and so on. It is worth underlining that, as hardening in metals is due to dislocation of atoms and crystals at the microscopic scale, anisotropy tends to occur in reality. However, despite this phenomenological aspect, isotropic hardening is very often used in numerical analysis because of its computational simplicity and overall accuracy in treating *proportional* and *monotonic* loads. More in detail, a proportional load refers to a single load or to a load set that depends only on a scalar factor that affects all the magnitudes simultaneously, while monotonic load means that the load factor evolves entirely increasing without unloading. Finally, Figure 3.10c shows the case of *kinematic hardening*. This schematization can be used to model anisotropic hardening. Conversely to the isotropic case, the elastic domain is kept constant (i.e. $2\sigma_y$ for the uniaxial case) and follows the hardening occurrence. Looking at the diagram, any cyclic load draws a path within the two inclined dashed lines. From the mechanical point of view, according to the sketched scheme, hardening in tension leads to softening in compression, i.e. plasticity in compression starts for an absolute value of stress smaller than σ_y, meaning that the absolute value of σ_y^1 in tension is different from the absolute value of σ_y^2 in compression; a subsequent hardening in compression leads to softening in tension, and so on. This phenomenon, also obtained in metals for higher levels of dislocations, goes under the name of *Bauschinger effect*, of which the kinematic hardening is the simplest approximation.

Finally, it is worth noticing the importance of *load history* and parameters to keep track of how load evolves. As already stressed, when dealing with nonlinear analysis, the unicity of the solution is lost. This aspect reflects itself in the constitutive laws too: in all the graphs of Figure 3.10

more than one strain leads to the same stress. In fact, knowing the strain alone provides no information about the stress level, which can be any, and vice versa.

3.2.1 Elastic domain and perfect elastoplasticity

Uniaxial loading is nothing but a particular case of loading of a three-dimensional continuum, which is characterized by six components of stress, i.e. $\boldsymbol{\sigma} = [\sigma_x, \sigma_y, \sigma_{zs}, \tau_{xy}, \tau_{xz}, \tau_{yz}]^T$. Intuitively, there are infinite combinations of the six stress components that lead to plastic deformations. An experimental campaign aimed at individuating all of them is not possible, due to the overall number of tests and the technical difficulties in reproducing complex stress combinations. However, by considering a few selected test configurations, it is possible to define the boundary of the elastic stress states, i.e. all the stress combinations that, once reached, lead to permanent deformations. Essentially, these combinations mark the onset of plastic strains and they compose the *yield surface* (the meaning of the term "surface" is explained ahead): every stress combination within this envelop is representative of an elastic state. With perfect elastoplasticity, no values outside the yield surface are admissible. How to practically define this envelop has been discussed in the past and usually a *yield function* $f(\boldsymbol{\sigma})$ is adopted, which furnishes a scalar according to the six components of stress. Accordingly, instead of handling the six stress components independently, the nature of the elastic or plastic state is indicated by a single scalar. This dramatically simplifies the problem because infinite stress combinations lead to the same scalar value.

The yield surfaces are defined such as $f(\boldsymbol{\sigma}) = 0$ and Eq. (3.2) provides more insight. According to differential calculus, this function can be increasing, decreasing or stationary based on the sign of the yield function rate $\dot{f}(\boldsymbol{\sigma})$, derived in Eq. (3.3). In the present section, rate measures the variability of a certain function with respect to time. In this case, $df = \dot{f}(\boldsymbol{\sigma})dt$, where dt is the differential of time. The reader should not be misled because time is here introduced only with the purpose of ordering the events and no dynamic effects are considered. From the computational point of view, the differential quantities indicate, in reality, small increments associated with the load step of an increment loading process, as discussed in Chapter 2. Looking at Eq. (3.4), from the mechanical point of view, $\dot{f}(\boldsymbol{\sigma}) > 0$ means that the value of $f(\boldsymbol{\sigma})$ is increasing and approaching the yielding surface $f(\boldsymbol{\sigma}) = 0$, i.e. the material is undergoing a *loading* process. Analogously, $\dot{f}(\boldsymbol{\sigma}) < 0$ is indicative of *unloading*, and a null value represents the permanency of the elastic or plastic state.

The most important consequence of Eqs. (3.2) and (3.4) is that, once $f(\boldsymbol{\sigma}) = 0$, the rate cannot be positive, namely, $\dot{f}(\boldsymbol{\sigma})$ cannot be positive. This aspect goes under the name of *consistency condition*, synthetically illustrated in Eq. (3.5).

$f(\sigma) < 0$ elastic state

$f(\sigma) = 0$ plastic state   (3.2)

$f(\sigma) > 0$ not admissible

$$\dot{f}(\sigma) = \left(\frac{\partial f}{\partial \sigma}\right)^T \dot{\sigma}$$ (3.3)

$\dot{f}(\sigma) > 0$ loading

$\dot{f}(\sigma) = 0$ permanency of the elastic or plastic state   (3.4)

$\dot{f}(\sigma) < 0$ unloading

when $f(\sigma) = 0$

$\dot{f}(\sigma) > 0$ violates the consistency condition

                                                                              (3.5)

$\dot{f}(\sigma) = 0$ plastic flow

$\dot{f}(\sigma) < 0$ elastic unloading

To explain the conditions of Eqs. (3.2), (3.4) and (3.5), let us consider a simple laboratory test depending only on the stress components σ and τ. Figure 3.11a shows a thin tubular metal specimen tested considering the simultaneous presence of axial load N (providing σ when divided by the specimen area) and torsional moment M_t (providing τ when divided by thickness and twice the specimen area enclosed by median line, according to the Bredt's formula). Different specimens were subjected to several tests, of which only the stresses corresponding to the onset of plastic deformation are marked with circles in the graph of Figure 3.11b. This outcome can be reached either by applying increasing loads with fixed ratio between σ and τ (i.e. between N and M_t), or by fixing one of the stresses and increasing the other one. The consequent *load directions* can be represented by means of vectors (which, in turn, define the *load histories*), as shown in Figure 3.11b.

From the geometrical point of view, a generic vector provides a fixed ratio between the components (e.g. $\tau = -0.25 \, \sigma$ in Figure 3.11b), and a scalar (positive or negative) can increase or reduce the magnitude of the components (keeping the same ratio). Accordingly, once a load direction is fixed, the intersection between the straight line (on which the vector lays) and the yield surface furnishes the yield stress combinations for the given direction. The consequent yielding points are two (the generic load direction line in

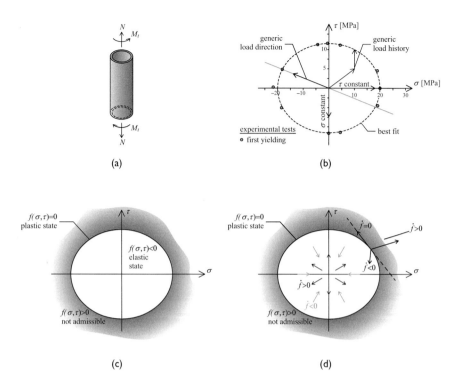

Figure 3.11 Tension–torsion test on a thin tubular metal specimen, after Lemaitre and Chaboche (1994). (a) Overall description of the test; (b) yield surface as the best fit curve and generic load directions; (c) description of the yield function sign and (d) description of yield function rate and consistency condition.

Figure 3.11b) if the surface is convex and closed, as explained in Section 3.2.3.

The best fit of the first yielding points is an ellipse whose equation is shown in Eq. (3.6) and drawn with dashed line in Figure 3.11b. According to Eq. (3.2), this curve is the yield surface and represents the threshold of the elastic domain. For this particular case, the domain is symmetric with respect to σ and τ, i.e. no differences arise if the specimen is compressed or tensioned, or if the torque moment changes the sign. As illustrated in Figure 3.11c, assuming ideal elastoplastic behaviour, the points within the ellipse are representative of elastic states. Points right on the ellipse are representative of plastic states, and states outside the ellipse are not admissible. The meaning of the yield function rate of Eq. (3.4) is shown in Figure 3.11d. Loading and unloading are illustrated by means of outward or inward directions with respect to the origin, respectively. In turn, if the stress state is on the yield surface, Eq. (3.5), Figure 3.11d shows that the only admissible possibilities are elastic unloading (going back to the elastic domain) or moving on the ellipse, i.e. on its tangent for small changes in the stress state (i.e. null yield function rate).

$$f(\sigma, \tau) = \sigma^2 + 3\tau^2 - 400 = 0 \qquad [\text{MPa}^2] \qquad (3.6)$$

Looking at the entire discussion from the reverse point of view, once the function $f(\sigma) = 0$ is known, any given stress state can be checked if elastic or plastic. This can be done either from the graphical point of view (not easy for complex three-dimensional stress states) or analytically. For the test shown in Figure 3.11, the yield function was given in Eq. (3.6). Accordingly, the generic stress state $\sigma = 10\,\text{MPa}$ and $\tau = 5$ MPa leads to $f(\sigma) = -225$ MPa^2 (note that the units of the function are being dropped hereafter). According to Eq. (3.2), the stress values are indicative of an elastic state. Analogously, maintaining the same normal stress σ, the values $\tau = 10$ MPa and $\tau = 15$ MPa lead to a plastic $(f(\sigma) = 0)$ and a not admissible state $(f(\sigma) = 375)$, respectively. These results are confirmed by the visual inspection of Figure 3.11b. According to perfect elastoplasticity (see also Table 3.1), the last result implies that the tubular metal specimen will never reach the values of stresses $\sigma = 10$ MPa and $\tau = 15$ MPa. As soon as the load history moves on the plastic domain, the specimen will deform indefinitely within the boundary of elastoplasticity.

In turn, from the practical point of view, by assuming a certain shape of the yield surface (see Section 3.2.2), for isotropic materials with symmetric response (meaning the sign of the stress is not determinant) such as metals, the yield function can be determined according to a single uniaxial test. For instance, considering Eq. (3.6), the coefficients of σ^2 and τ^2 are indicative of the shape of the ellipse (namely, stretched or flattened), calculated once for all on the basis of mathematical and mechanical considerations. The known term is the square of the uniaxial compressive/tensile yield stress $(\sigma_y = \pm 20$ MPa). This approach has a remarkable mechanical meaning that it is worth highlighting. Eq. (3.6) is rewritten in Eq. (3.7), showing that the yield function became a formula to derive an equivalent stress (underlined term) to be directly compared with the uniaxial test response. Intuitively, the material response is elastic if the equivalent stress is smaller than the yield limit of the uniaxial test. From this point of view, the yield functions assume the role of a *criterion*. The same reasoning can be applied to all the yield functions described next as long as it is possible to isolate σ_y.

$$\sigma^2 + 3\tau^2 = 400$$

$$\underline{\sqrt{\sigma^2 + 3\tau^2}} = 20 \text{ MPa} \qquad \text{plastic state}$$

$$(3.7)$$

$$\underline{\sqrt{\sigma^2 + 3\tau^2}} < 20 \text{ MPa} \qquad \text{elastic state}$$

3.2.2 Yield functions for isotropic materials

Although the example shown in Figure 3.11 has a didactic purpose, yield functions are usually chosen a priori according to the knowledge about material behaviour and a limited number of experimental tests or data from literature. In the following, the most used yield functions are introduced.

One of the simplest assumptions in evaluating the yield functions regards the material *isotropy*. Accordingly, it is possible to limit the discussion to the three principal stresses (σ_I, σ_{II}, σ_{III}), regardless of their orientation, as shown in Eq. (3.8). The convention $\sigma_I > \sigma_{II} > \sigma_{III}$ is widely accepted and used here. Isotropy guarantees also that the functions are invariant with respect to the principal stresses, i.e. there are no preferential principal directions, as described in Eq. (3.9). One of the great advantages of material isotropy is that it is possible to obtain a graphical description of the yield functions in the three-dimensional space of the principal stresses. Consequently, all the points representative of the elastic envelop create a surface (the so-called *yield surface*). The consideration of the principal stresses is of value also for analytical studies and hand calculations because the expressions shown next are rather simple, even if the principal stress space is less used in computational implementations due to the need to keep track of the orientation of the principal stresses and the need to sort them according to the magnitude.

More in detail, experiments showed that materials such as metals keep behaving elastically even under high level of hydrostatic stresses (i.e. when all the principal stresses assume the same value). This aspect leads to a twofold consideration: (1) in the tension space, the hydrostatic line $\sigma_I = \sigma_{II} = \sigma_{III}$ is part of the elastic domain and (2) the yield criteria for metals do not depend on the hydrostatic stress (i.e. they are pressure independent), but on the deviatoric part only (this is the part that tends to distort a body, which can be understood, in a very simplistic way, as the shear stress).

$$f(\boldsymbol{\sigma}) = f(\sigma_I, \sigma_{II}, \sigma_{III}) \tag{3.8}$$

$$f(\sigma_y, 0, 0) = f(0, \sigma_y, 0) = f(0, 0, \sigma_y) = 0 \tag{3.9}$$

Tresca yield function assumes that the elastic domain depends only on the shear stress. Considering Eq. (3.10), the material exhibits plastic deformation when the maximum shear stress τ_{max} (depending on σ_I, σ_{II} and σ_{III}) achieves the threshold value τ_y. This assumption, supported by experimental evidence, can be derived according to continuum mechanics, as shown in Eqs. (3.11) and (3.12). The latter is used to evaluate the value of τ_y from a uniaxial test, where σ_y is the yielding normal stress. Finally, Tresca yield function is derived in Eq. (3.13) as a function of $\boldsymbol{\sigma} = [\sigma_I, \sigma_{II}, \sigma_{III}]^T$.

$$\tau_{max}\left(\sigma_{I},\sigma_{II},\sigma_{III}\right)=\tau_{y} \tag{3.10}$$

$$\tau_{max}=\frac{1}{2}max\left(\left|\sigma_{I}-\sigma_{II}\right|,\left|\sigma_{I}-\sigma_{III}\right|,\left|\sigma_{II}-\sigma_{III}\right|\right) \tag{3.11}$$

$$\tau_{y}=\frac{1}{2}\sigma_{y} \tag{3.12}$$

$$f\left(\sigma\right)=max\left(\left|\sigma_{I}-\sigma_{II}\right|,\left|\sigma_{I}-\sigma_{III}\right|,\left|\sigma_{II}-\sigma_{III}\right|\right)-\sigma_{y}=0 \tag{3.13}$$

The graphical representation of Tresca yield function, shown in Figure 3.12a, is an orthogonal prism of hexagonal base with its axis along the hydro-static line. It should be noted that, for the sake of clarity and comparison with the yield functions provided next, the three-dimensional axes are oriented towards the negative side to highlight the compressive part of the yield surface. A plane perpendicular to the hydrostatic line is called *deviatoric plane*. In this case, the projection of the yield surface on this plane is constant, as illustrated in Figure 3.12b, whereas Figure 3.12c shows the two-dimensional case with $\sigma_{III}=0$.

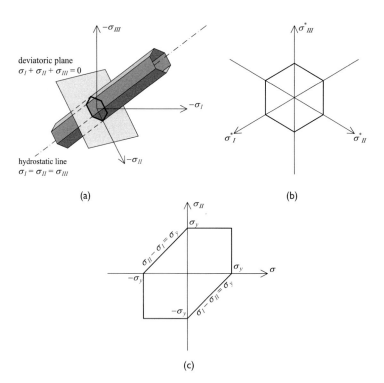

Figure 3.12 Tresca yield function. (a) Three-dimensional representation with deviatoric plane at the origin; (b) projection on the deviatoric plane (* projected axes) and (c) in-plane domain ($\sigma_{III}=0$).

The *Mohr–Coulomb yield surface* represents reasonably well concrete-like (also designated as cohesive-frictional) materials such as rock, brick or concrete. For the well-known two-dimensional case, the elastic domain is illustrated in Figure 3.13a and is defined by the lines in Eq. (3.14), where φ is the angle of internal friction (or, simply, friction angle) and c is the cohesion. On closer inspection, the shear stress is not limited by a threshold value, but it can grow indefinitely as long as an increasing compressive stress "confines" its effect. In Eq. (3.14), the absolute value brackets indicate that the sign of shear stress is indifferent.

Mohr–Coulomb yield function is often adopted for materials that exhibit different yield stresses in compression and tension, σ_y^- and σ_y^+, respectively. According to the Mohr's circles in Figure 3.13a relative to uniaxial compression stress states, simple trigonometry, or the use of Eq. (3.14), allows expressing tensile or compressive yield stresses by means of cohesion and friction angle, as shown in Eq. (3.15). In turn, based on Figure 3.13b, the generic yield stress combination can be calculated from principal stresses, as derived in Eq. (3.16). Combining Eqs. (3.14) and (3.16), the generic Mohr–Coulomb yield function for the two-dimensional case is derived in Eq. (3.17). By developing the absolute value and considering Eq. (3.15), the two equations of Eq. (3.18) are derived.

$$|\tau| + \sigma \tan \varphi - c = 0 \qquad (3.14)$$

$$\sigma_y^- = -\frac{2\cos\varphi}{1-\sin\varphi} c$$

$$\qquad (3.15)$$

$$\sigma_y^+ = \frac{2\cos\varphi}{1+\sin\varphi} c$$

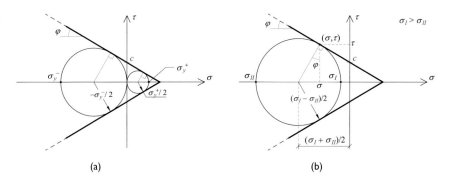

(a)   (b)

Figure 3.13 Mohr–Coulomb yield function for two-dimensional stress state. (a) Uniaxial load tests (tension and compression) and (b) generic yield stress combination.

$$\sigma = \frac{1}{2}(\sigma_\mathrm{I} + \sigma_\mathrm{II}) + \frac{1}{2}(\sigma_\mathrm{I} - \sigma_\mathrm{II})\sin\varphi$$

$$\tau = \frac{1}{2}(\sigma_\mathrm{I} - \sigma_\mathrm{II})\cos\varphi$$

$$\text{(3.16)}$$

$$|\sigma_\mathrm{I} - \sigma_\mathrm{II}|\cos^2\varphi + (\sigma_\mathrm{I} - \sigma_\mathrm{II})\sin^2\varphi + (\sigma_\mathrm{I} + \sigma_\mathrm{II})\sin\varphi - 2c\cos\varphi = 0 \qquad \text{(3.17)}$$

$$\sigma_y^+\sigma_\mathrm{I} + \sigma_y^-\sigma_\mathrm{II} - \sigma_y^+\sigma_y^- = 0$$

$$\sigma_y^-\sigma_\mathrm{I} + \sigma_y^+\sigma_\mathrm{II} - \sigma_y^+\sigma_y^- = 0$$

$$\text{(3.18)}$$

The extension to the three-dimensional case can be obtained by simply permuting the principal stresses of Eq. (3.17), including σ_III. The synthetic form of the yield function according to cohesion and friction angle is derived in Eq. (3.19). The same permutation can be applied to Eq. (3.18) to obtain the yield function in terms of yield stresses σ_y^- and σ_y^+, synthetically illustrated in Eq. (3.20). In the space of the principal stresses, the yield function is represented by the six planes drawn in Figure 3.14a. Its projection on the deviatoric plane and the two-dimensional case are shown in Figure 3.14b and c. As already highlighted, conversely to Tresca yield function, Mohr–Coulomb allows different limit values for compression and tension. The Tresca prism became an irregular pyramid whose axis still lays along the hydrostatic line: the apex limits the hydrostatic stress in tension (which does not coincide with the tensile yield stress, see Figure 3.13a), whereas the base of the pyramid increases according to the hydrostatic stress in compression. Practically, the projection of the yield function on deviatoric planes traces homothetic irregular hexagons (getting larger with higher confining stresses). In other words, the elastic domain is unlimited only for compression, allowing increasing values of shear stress along the compression part of the hydrostatic axis.

$$f(\sigma) = (\sigma_i - \sigma_j) + (\sigma_i + \sigma_j)\sin\varphi - 2c\cos\varphi = 0 \qquad \text{with } i,j = \mathrm{I, II, III} \quad \text{(3.19)}$$

$$f(\sigma) = \sigma_y^+\sigma_i + \sigma_y^-\sigma_j - \sigma_y^+\sigma_y^- = 0 \qquad \text{with } i,j = \mathrm{I, II, III} \quad \text{(3.20)}$$

Both Tresca and Mohr–Coulomb yield functions describe the elastic domain by means of multiple-equation surfaces, which is not computationally appealing. According to Figure 3.11b, every load direction is associated to a vector and the yield stress combinations along that direction are the intersection between the straight line (on which the vector lays) and the yield surface. If the yield surface is made by multiple planes (such as Tresca and Mohr–Coulomb yield functions), the intersection will lead to several incorrect stress states. Figure 3.15a explains this concept considering the

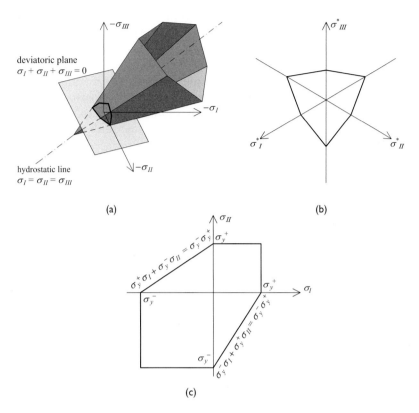

Figure 3.14 Mohr–Coulomb yield function. (a) Three-dimensional representation with deviatoric plane at the origin; (b) projection on the deviatoric plane (* projected axes) and (c) in-plane domain ($\sigma_{III} = 0$).

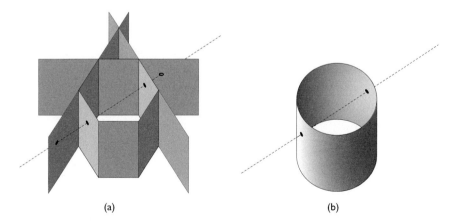

Figure 3.15 Intersection between load directions and yield function. (a) Multiple-equation and (b) single-equation surfaces.

Tresca yield function, which is represented by an orthogonal prism of hexagonal base. For the sake of clarity, only three of the six planes are displayed extended. As it is possible to see, the two external intersections fall outside of the surface and cannot be accepted. The extension to the other planes not shown in the figure is straightforward. The mentioned disadvantage can be avoided if a single-equation surface is chosen. Looking at Figure 3.15b, if the orthogonal prism of hexagonal base is substituted with a circular cylinder, the intersections are only two and both of them belong to the yield surface. In addition, multiple-equation surfaces define edges in which properties cannot be easily defined, as shown in Section 3.2.3.

In this regard, yield functions such as von Mises and Drucker–Prager are preferred. Without entering the merit of the dissertation and the mechanical reasoning, Eq. (3.21) shows the squared and square root formulations of the von Mises yield function. This is based on the elastic shear energy and its three-dimensional representation is a circular cylinder of infinite length, whose axis coincides with the hydrostatic line (Figure 3.16a). Figure 3.16b shows the projection of the yield surface on the deviatoric plane, namely, a circle that circumscribes the Tresca surface projection. A similar relationship is found for the two-dimensional case ($\sigma_{III} = 0$) of Figure 3.16c, displaying a rotated ellipse.

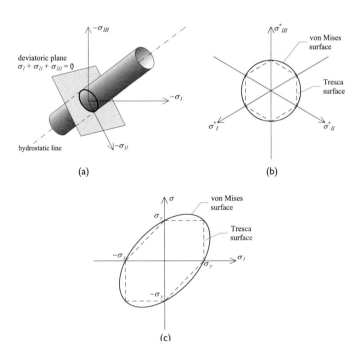

Figure 3.16 Von Mises yield function, compared with Tresca. (a) Three-dimensional representation with deviatoric plane at the origin; (b) projection on the deviatoric plane (* projected axes) and (c) in-plane domain ($\sigma_{III} = 0$).

$$f(\sigma) = (\sigma_I - \sigma_{II})^2 + (\sigma_I - \sigma_{III})^2 + (\sigma_{II} - \sigma_{III})^2 - 2\sigma_y^2 = 0$$

$$f(\sigma) = \sqrt{\frac{1}{2}\left[(\sigma_I - \sigma_{II})^2 + (\sigma_I - \sigma_{III})^2 + (\sigma_{II} - \sigma_{III})^2\right]} - \sigma_y = 0 \qquad (3.21)$$

With no surprise, the pyramid of the Mohr–Coulomb function becomes a cone with the Drucker–Prager approach (Figure 3.17a). This is a family of criteria in the shape of Eq. (3.22), where the constants A and B can be tuned according to experimental outcomes (or qualitative data). For instance, considering the yield stresses of uniaxial loading tests, both in tension and compression, the relationships in Eq. (3.23) must hold. It is worth highlighting that the left-hand-side term of the former must be positive (due to second power and square root), thus the minus before the negative σ_y^-. Solving the system, the constants A and B are derived in Eq. (3.24). Similar equations can be found from the values of the angle of internal friction φ and the cohesion c.

In general, this yield surface being the "smooth" version of the Mohr–Coulomb, it is possible to set the constants differently. Looking at Figure 3.17b, the two limits of the cone of Drucker–Prager yield surface can be inscribed or circumscribed to the Mohr–Coulomb one. It is stressed that Eqs. (3.22)–(3.24) provide the circumscribed solution. In turn, the intersection of the cone with the plane $\sigma_{III} = 0$ produces an ellipse (Figure 3.17c) whose centre does not coincide with the origin. In this case too, the ellipse

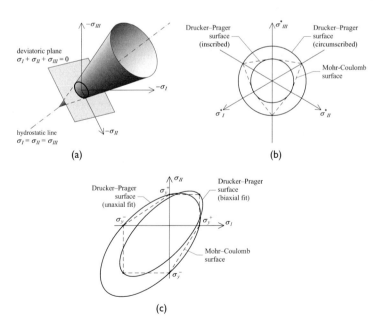

Figure 3.17 Drucker–Prager yield function, compared with Mohr–Coulomb. (a) Three-dimensional representation with deviatoric plane at the origin; (b) projection on the deviatoric plane (* projected axes) and (c) in-plane domain ($\sigma_{III} = 0$).

can be set according to the Mohr–Coulomb function, interpolating the uniaxial yield stresses (as done with Eq. 3.24) or the biaxial ones. Note that continuous yield surfaces that better approximate Mohr–Coulomb are available in literature, with increasing complexity.

$$f(\sigma) = \sqrt{\frac{1}{6}\left[(\sigma_{\mathrm{I}} - \sigma_{\mathrm{II}})^2 + (\sigma_{\mathrm{I}} - \sigma_{\mathrm{III}})^2 + (\sigma_{\mathrm{II}} - \sigma_{\mathrm{III}})^2\right]} - A - B(\sigma_{\mathrm{I}} + \sigma_{\mathrm{II}} + \sigma_{\mathrm{III}}) = 0$$

(3.22)

$$\frac{-\sigma_y^-}{\sqrt{3}} = A + B\sigma_y^- \qquad \frac{\sigma_y^+}{\sqrt{3}} = A + B\sigma_y^+ \qquad (3.23)$$

$$A = -\frac{2}{\sqrt{3}}\left(\frac{\sigma_y^- \sigma_y^+}{\sigma_y^+ - \sigma_y^-}\right) \qquad B = \frac{1}{\sqrt{3}}\left(\frac{\sigma_y^+ + \sigma_y^-}{\sigma_y^+ - \sigma_y^-}\right) \qquad (3.24)$$

At this point, the reader should have developed the idea that yield functions/surfaces represent a mathematical abstraction that, based on theoretical insights (e.g. maximum shear stress or elastic shear energy), approximates experimental outcomes. In this regard, it is easy to picture that two different formulations may be adopted simultaneously for the same material to address, for instance, the different behaviour in tension and compression. This is the case of the Rankine yield function, often used in combination with von Mises or Drucker–Prager ones.

Rankine yield function (or the maximum stress criterion) is such that the material is assumed to behave elastically if the principal stresses lay between the positive and negative yield stresses, as shown in Eq. (3.25). The graphical representation is illustrated in Figure 3.18a and b, namely, an eccentric cube (whose side is equal to $\sigma_y^+ - \sigma_y^-$) in the space of the principal stresses and an eccentric square in the case $\sigma_{\mathrm{III}} = 0$. The Rankine yield surface in tension can be combined with e.g. von Mises or Drucker–Prager's in compression to create a model whose graphical illustrations are shown in Figure 3.18c and Figure 3.18d for the two-dimensional space.

$$\sigma_y^- \le \sigma_{\mathrm{I}} \le \sigma_y^+ \qquad \sigma_y^- \le \sigma_{\mathrm{II}} \le \sigma_y^+ \qquad \sigma_y^- \le \sigma_{\mathrm{III}} \le \sigma_y^+ \qquad (3.25)$$

To gain familiarity with the yield functions, let us consider an application case of a concrete cube deformed homogeneously in compression. Assuming $\varphi = 30°$ and $c = 10\,\mathrm{MPa}$, let us calculate the value of the principal stress σ_{III}, according to the Mohr–Coulomb yield surface so that yielding is obtained, when $\sigma_{\mathrm{I}} = -10\,\mathrm{MPa}$ and $\sigma_{\mathrm{II}} = -20\,\mathrm{MPa}$. The solution is here attained independently thanks to the two strategies of Eqs. (3.19) and (3.20), in terms of $\varphi{-}c$ and $\sigma_y^+{-}\sigma_y^-$, respectively. According to the former, the given principal stresses should be substituted in each of the six formulas of Eq. (3.20). The possible results are $\sigma_{\mathrm{III}} = -94.64, -64.64, 4.88$ and $8.21\,\mathrm{MPa}$, plus two impossible solutions (due to the fact that $\sigma_{\mathrm{I}} = -10\,\mathrm{MPa}$ and $\sigma_{\mathrm{II}} = -20\,\mathrm{MPa}$ do not belong to any of the two planes perpendicular to $\sigma_{\mathrm{II}}{-}\sigma_{\mathrm{III}}$ plane).

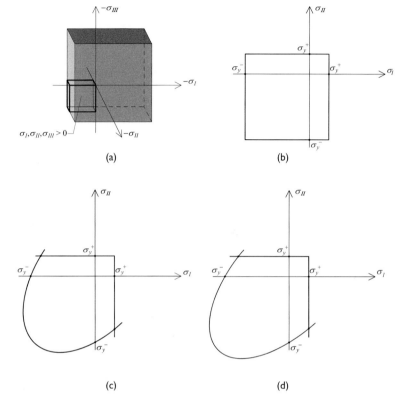

Figure 3.18 Rankine yield surface. (a) Three-dimensional representation; (b) in-plane domain ($\sigma_{III} = 0$); in combination with (c) von Mises and (d) Drucker–Prager's formulations (for the compression part).

It should be noted that the order $\sigma_I > \sigma_{II} > \sigma_{III}$ would request the renumbering of the principal stresses for each trial, here ignored. The possibility of finding multiple solutions was already discussed in Figure 3.15. For the given example, only two solutions are possible, namely, $\sigma_{III} = -64.64\,\text{MPa}$ and $\sigma_{III} = 4.88\,\text{MPa}$, as the other values violate (fall outside) some of the prism planes with a value of the yield function larger than zero. Figure 3.19 shows the relative Mohr's circles, centred on the σ axis and passing through two of the principal stresses. To find the missing one, it is possible to consider the ultimate Mohr's circle, i.e. the outer one tangent to the yield function. Without entering the merit of the discussion, only σ–τ combinations that lie on these circles or within the shaded area enclosed by them are admissible. The graphical construction of Figure 3.19a confirms that the minimum (negative, thus compressive) principal stress cannot be smaller than $-64.64\,\text{MPa}$; Figure 3.19b is inherent to the maximum (positive, thus tensile) value equal to $4.88\,\text{MPa}$, as expected. Both solutions lead the concrete specimen to yield, whereas intermediate values keep the response elastic,

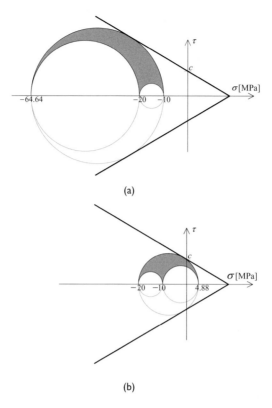

(a)

(b)

Figure 3.19 Mohr's construction for the case of $\sigma_{\mathrm{I}} = -10\,\mathrm{MPa}$ e $\sigma_{\mathrm{II}} = -20\,\mathrm{MPa}$. (a) $\sigma_{\mathrm{III}} = -64.64\,\mathrm{MPa}$, representative of a fully compressed state, and (b) $\sigma_{\mathrm{III}} = 4.88\,\mathrm{MPa}$.

namely, none of the Mohr's circles achieves the yield line. As the cube is deformed in compression, the correct solution is $\sigma_{\mathrm{III}} = -64.64\,\mathrm{MPa}$.

In turn, based on Eq. (3.15), the given values of friction angle and cohesion allow calculating $\sigma_y^+ = 11.55\,\mathrm{MPa}$ and $\sigma_y^- = -34.64\,\mathrm{MPa}$. The use of Eq. (3.19) leads to the same results seen earlier. However, it is interesting to discuss further the result $\sigma_{\mathrm{III}} = -64.64\,\mathrm{MPa}$. This value is much smaller (much larger in magnitude) than the yielding compressive stress and may be considered (erroneously) inaccurate. On the contrary, this value is the result of the confining effect of the other principal stresses that, by laterally compressing the specimen, enlarges the elastic domain (Figure 3.14a).

To provide a better insight on the confining stress, let us consider a column made of the same material. The confining effect of the stirrups is assumed $\sigma_{\mathrm{I}} = \sigma_{\mathrm{II}} = -2\,\mathrm{MPa}$ (uniform in the cross-section). It is interesting to evaluate the increase of compressive yield stress following this confinement. According to Eq. (3.19), $\sigma_{\mathrm{III}} = 10.88\,\mathrm{MPa}$ and $\sigma_{\mathrm{III}} = -40.64\,\mathrm{MPa}$, of which, the latter regards compression. If compared with the yielding compressive stress, the gain is of about 17%, with respect to the uniaxial compressive strength of $-34.64\,\mathrm{MPa}$.

Finally, for the sake of clarity, the same problems are discussed according to *Drucker–Prager yield surface*, according to the formulation of Eqs. (3.22)–(3.24). Regarding the concrete specimen under compression, this leads to $\sigma_{III} = -93.69$ and $10.60\,\text{MPa}$, being much larger in magnitude for compression with respect to Mohr–Coulomb formulation (see also Figure 3.17c). Considering the second problem of the concrete column, the results are $\sigma_{III} = 11.55\,\text{MPa}$ and $\sigma_{III} = -42.64\,\text{MPa}$. According to Drucker–Prager yield surface, the increment of compressive yield stress is of about 23%.

3.2.3 Constitutive law with associated plastic flow and perfect elastoplasticity

Let us now collect the main assumptions of elastic perfectly plastic materials:

1. A generic strain ε is the summation of elastic (reversible) ε_e and plastic (irreversible) ε_p part, namely, $\varepsilon = \varepsilon_e + \varepsilon_p$. The same relation holds for the three-dimensional case ($\boldsymbol{\varepsilon} = \boldsymbol{\varepsilon}_e + \boldsymbol{\varepsilon}_p$).
2. The yield surface is the geometric representation of all the yield stress states.
3. The yield surface is *continuous* (even if nonsmooth).
4. The yield surface is *closed*. This means it marks the threshold between elastic (within) and plastic states (on the boundary). Partly unlimited surfaces can be considered closed at infinite distance.

This information alone is not able to describe the relationship between stress and strain when the material attains its elastic limit. For this purpose, the *Drucker's stability first* and *second postulates* are usually implemented, and the consequent implication in the constitutive model takes the name of *associated plastic flow*. The postulates are derived from energetic considerations not discussed here. The main outcomes of stability postulates are: (1) the yield surface is convex and (2) the direction of plastic strain rate $\dot{\varepsilon}_p$ is parallel to the outward-pointing normal to the yield surface (**n**). In the case one of the previous two points does not hold, the plastic flow is *nonassociated*, and a simple example is discussed ahead.

First, it is worth reminding that the outward-pointing normal **n** is, mathematically, the gradient of the yield function with respect to the stress components, as shown in Eq. (3.26); **n** is a vector that indicates a given direction and $\dot{\varepsilon}_p$ is calculated according to Eq. (3.27), where the magnitude of the increment is defined by a scalar parameter $\dot{\lambda} \geq 0$.

$$\mathbf{n} = \frac{\partial f}{\partial \sigma} \tag{3.26}$$

$$\dot{\varepsilon}_p = \dot{\lambda}\,\frac{\partial f}{\partial \sigma} \tag{3.27}$$

Given the strain decomposition $\boldsymbol{\varepsilon} = \boldsymbol{\varepsilon}_e + \boldsymbol{\varepsilon}_p$, it is possible to detail the constitutive law as shown in Eq. (3.28), where most of the quantities have been already introduced. On closer inspection, the third part of Eq. (3.28) is inherently the consistency condition illustrated in Eq. (3.5). For elastic perfectly plastic materials, in the case numerical analyses or experimental tests are performed under force control, the constitutive law does not provide any information about the magnitude of $\dot{\lambda}$ that results arbitrary (see also Table 3.1), confined only by possible boundary conditions or adjacent parts of the structural element that still behave elastically. Therefore, to keep track of the current state of the material for a given equilibrium condition, a state variable is needed and this can be derived from $\dot{\boldsymbol{\varepsilon}}_p$ or $\dot{\boldsymbol{\varepsilon}}$.

$$\dot{\boldsymbol{\varepsilon}} = \dot{\boldsymbol{\varepsilon}}_e + \dot{\boldsymbol{\varepsilon}}_p \qquad \rightarrow \qquad \dot{\boldsymbol{\varepsilon}}_e = \dot{\boldsymbol{\varepsilon}} - \dot{\boldsymbol{\varepsilon}}_p$$

$$\dot{\boldsymbol{\sigma}} = \mathbf{D}\dot{\boldsymbol{\varepsilon}}_e \qquad \rightarrow \qquad \dot{\boldsymbol{\sigma}} = \mathbf{D}\left(\dot{\boldsymbol{\varepsilon}} - \dot{\boldsymbol{\varepsilon}}_p\right)$$

$$\dot{\boldsymbol{\varepsilon}}_p = \begin{cases} 0 & \text{if } f(\boldsymbol{\sigma}) < 0 & \text{elastic state} \\[2mm] 0 & \text{if } f(\boldsymbol{\sigma}) = 0, \ \dot{f}(\boldsymbol{\sigma}) < 0 & \text{elastic unloading} \\[2mm] \dot{\lambda}\dfrac{\partial f}{\partial \boldsymbol{\sigma}} & \text{if } f(\boldsymbol{\sigma}) = 0, \ \dot{f}(\boldsymbol{\sigma}) = 0 & \text{plastic flow} \end{cases} \tag{3.28}$$

An example makes these concepts clearer. Let us consider a brick couplet with a masonry joint under compression normal to the joint and shear load, where the yield function is calculated according to the Mohr–Coulomb formulation. The overall scheme is depicted in Figure 3.20a, where $\sigma_n = F_n/A$, $\tau = F_s/A$, where A is the area of the brick. The example is limited to the plane case where the directions n and s are normal and parallel to the masonry joint, respectively. As a first step, the derivative of Eq. (3.27) is expanded in its scalar shape in Eq. (3.29) and, for the sake of clarity, Eq. (3.14) is rewritten in Eq. (3.30), assuming a positive shear stress τ. According to the two equations, the two nonnull components of the plastic strain rate are reported in Eq. (3.31). As expected, they both depend on $\dot{\lambda}$, which is defined by the applied load or displacement increment to the material (or structure). Combining the two formulas, it is possible to obtain the ratio between the two components of plastic strain, as shown in Eq. (3.32). If $\varphi = 30°$, for instance, the ratio between normal and shear strain is equal to 0.58. This value provides information about the *direction of plastic flow*: no matter the magnitude of $\dot{\lambda}$, the normal component is of the same sign and 0.58 as large as the shear one.

$$\dot{\varepsilon}_{p,n} = \dot{\lambda}\frac{\partial f}{\partial \sigma_n} = \dot{\lambda}\frac{\partial f}{\partial \sigma} \qquad\qquad \dot{\varepsilon}_{p,s} = \dot{\lambda}\frac{\partial f}{\partial \sigma_s} = 0$$

$$\dot{\varepsilon}_{p,ns} = \dot{\lambda}\frac{\partial f}{\partial \tau_{ns}} = \dot{\lambda}\frac{\partial f}{\partial \tau} \tag{3.29}$$

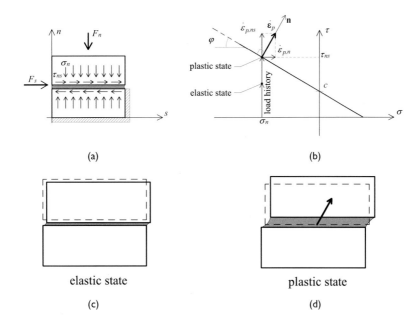

(a)   (b)

elastic state   plastic state

(c)   (d)

Figure 3.20 Brick couplet under compression F_n and shear F_s load. (a) Overall scheme; (b) Mohr–Coulomb yield formulation; (c) elastic and subsequent (d) plastic deformation.

$$f(\sigma) = f(\sigma, \tau) = \tau + \sigma \tan\varphi - c = 0 \tag{3.30}$$

$$\dot\varepsilon_{p,n} = \dot\lambda \tan\varphi$$
$$\tag{3.31}$$
$$\dot\varepsilon_{p,ns} = \dot\lambda$$

$$\dot\varepsilon_{p,n} = \tan\varphi\, \dot\varepsilon_{p,ns} \tag{3.32}$$

Let us consider this outcome from a geometrical point of view and a load history with a fixed normal force F_n (or stress σ_n), as shown in Figure 3.20b. As long as the stress state lays below the yield line, the specimen behaves elastically, and the expected deformation is shown in Figure 3.20c. According to linear elasticity, there is no coupling between normal and shear deformation. However, if the stresses achieve the yield line, the specimen exhibits plastic flow whose direction is perpendicular to the yield line. According to Figure 3.20b, and in line with Eqs. (3.31) and (3.32), both $\dot\varepsilon_{p,n}$ and $\dot\varepsilon_{p,ns}$ are present. Their relative ratio is equal to $\tan\varphi$, evident also from the graph. They both are positive, i.e. clockwise distortion and longitudinal expansion, as shown in Figure 3.20d.

This, seemingly, surprising and anti-intuitive outcome is a type of deformation that produces an increment of volume and a negative work for the compressive action. In reality, this is to be expected because the surfaces are not smooth and plastic work in the joint involves, usually, an increase of volume, which is commonly referred to as *dilatancy* (see Chapter 4). The same occurs in concrete-like materials in which plastic deformation leads to volume increase, in opposition to metals, in which there is no volume increase upon plastic deformation. According to experimental tests, either in concrete-like homogenous materials or masonry joints, the volume generation accounted for by associated plasticity is too large: the correct result should be smaller than the one given by the friction angle. This violates Drucker's stability postulates because the plastic flow is not perpendicular to the yield surface, and *nonassociated plastic flow* should be considered instead.

Going back to the consistency condition, the third part of Eq. (3.28) can be synthetically rewritten in Eq. (3.33). From the third part of Eq. (3.33), it is clear that plastic flow (i.e. $\dot{\lambda} \neq 0$) can occur only if $f(\sigma) = 0$ and $\dot{f}(\sigma) = 0$. Figure 3.21a–c shows the three possibilities of $\dot{f}(\sigma)$ when $f(\sigma) = 0$. In this regard, for the consistency condition, $\dot{f}(\sigma) > 0$ is inadmissible. Additionally, in correspondence of a singular point, such as the edges of the Tresca yield surface, a unique normal to the surface does not exist and the plastic strain increment is calculated according to Eq. (3.34), where n_p is the number of planes concurrent in a singular point. A visual explanation of Eq. (3.34) is shown in Figure 3.21d.

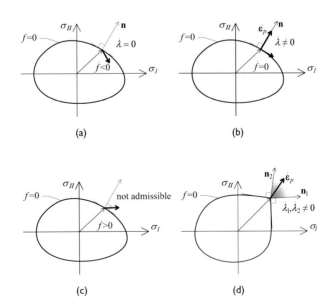

Figure 3.21 Plastic flow for perfectly plastic materials. (a) Unloading case; (b) load increment on the yield surface; (c) inadmissible load increment and (d) plastic strain for a singular point of the yield function.

$$\dot{\varepsilon}_p = \lambda \frac{\partial f}{\partial \sigma} = \lambda \mathbf{n}$$

$$f(\sigma) \le 0 \qquad \dot{\lambda} \ge 0$$

$$\lambda f(\sigma) = 0 \qquad \lambda \dot{f}(\sigma) = 0 \qquad\qquad (3.33)$$

$$\dot{\varepsilon}_p = \sum_{i=1}^{n_p} \lambda_i \frac{\partial f_i}{\partial \sigma} = \sum_{i=1}^{n_p} \lambda_i \mathbf{n}_i \qquad\qquad (3.34)$$

The reader should pay attention to the peculiarity of Figure 3.21b. A null yield function rate means the load path is moving on the tangent line. As already highlighted in the present text, the tangent line is "coincident" with the curve only with small increments (rates, in this case) and the final value might easily depart from the yield surface. This has a twofold consequence. On the one hand, as the yield surface is assumed concave, the final stress state always lays outside of it, violating the condition $f(\sigma) \le 0$. On the other hand, the normal to the surface is calculated at the beginning of the load increment, and it may not be coincident with the normal at the end of the increment. For these reasons, an iterative procedure is necessary, and the same will be briefly discussed in Section 3.2.5.

3.2.4 Constitutive law with associated plastic flow and hardening

In the previous subsection, the elastic domain was modelled by means of theories that produced different yield surfaces in the space of the principal stresses. The region within the surface represents the elastic domain and any point in this region is indicative of an elastic state (i.e. reversible deformation). In the framework of perfect plasticity, these surfaces are fixed and, once the load history achieves the envelop, plastic flow occurs with unlimited magnitude. If this assumption might be reasonable for many engineering simplified applications, it is often a limitation for existing structures.

Let us consider the same test on the tubular specimen discussed in Figure 3.11 and shown again in Figure 3.22a. As no information was provided for possible hardening, perfect plasticity was assumed. It seems now interesting to discuss what happens if stresses exceed the elastic limit, as it occurs in many metals. With this purpose, looking at Figure 3.22b, after reaching the initial yield curve, the tensile load was increased according to the horizontal vector. For the present case, the fact that the specimen can sustain a larger stress than the initial yield value means that, essentially, the material exhibits *hardening* (under uniaxial tensile load, in this case).

The description of hardening by the dislocation of atoms and crystals at the microscopic scale was already introduced at the beginning of Section 3.1.3. This means that the overall behaviour of the specimen has

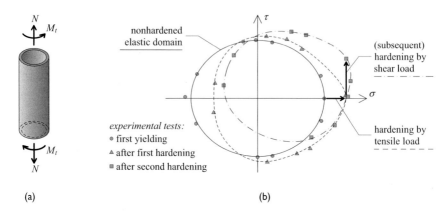

(a)   (b)

Figure 3.22 Tension–torsion test on a thin tubular metal specimen, after Lemaitre and
Chaboche (1994). (a) Overall description of the test and (b) elastic domain
evolution after subsequent hardening processes.

changed irreversibly, and it is interesting to perform other tests to investigate the new yield stress states. Looking at Figure 3.22b, these are marked
with triangles and the best fit curve is drawn with a dashed line. As it is
possible to notice, the new curve is not symmetric with respect to the τ axis,
and the hardening in tension produced a softening in compression. From
the geometrical point of view, it is possible to state that, according to the
stress history, the elastic surface can *move* and *deform*, paving the way to
a new elastic domain. As expected, the subsequent hardening by pure shear
load produces similar results: the tests are marked with squares and the best
fit curve is drawn with dot-dashed line.

As hardening produces an alteration of the yield surface, the one inherent
to the first yield (i.e. the one discussed in Figure 3.11) is usually referred to
as *nonhardened elastic domain* (or initial elastic domain). Looking at the
figure, conversely to perfect plasticity, when hardening is assumed, stress
states can fall outside of the nonhardened domain. However, consistently
with the previous discussion, and in line with the experimental outcomes,
every stress state is either within the new elastic domain (elastic) or on the
domain itself (plastic).

Although no information has been given about plastic strain, from
Figure 3.22b it is clear that, in the σ–τ plane, and, by analogy, in the space
of the principal stresses or any other stress space, the elastic domain moves
and deforms. Knowing the exact hardening evolution is not a simple task,
also because stress history plays a significant role. However, simplified models, also supported by experimental evidence, can be adopted with a certain
confidence. This is the case of *isotropic* and *kinematic hardening.* They have
already been introduced for the case of uniaxial load at the beginning of
Section 3.2 and discussed in Figure 3.10. Let us now expand the definitions
for the two-dimensional case in the plane of the principal stresses σ_I–σ_{II}.
Figure 3.23a shows the schematic understanding of isotropic hardening,

according to which, the evolution of the elastic domain is homothetic. More concisely, the increment of yield stresses is proportionally the same for any direction passing through the origin. In turn, kinematic hardening in Figure 3.23b assumes a translation of the elastic domain according to the applicable stress increment. The combination of both models leads to isotropic–kinematic hardening. For the sake of comparison, perfect elastoplasticity is illustrated in Figure 3.23c. In all the cases of Figure 3.23, the final stress state is elastic, as a consequence of unloading.

From the computational point of view, it is necessary to introduce an internal variable that describes the hardening process, understood as how the yield stress evolves through the loading process. This goal is addressed with the *hardening parameter* κ. To have an intuitive insight, let us consider the Tresca yield function as shown in Eq. (3.13). In the case of hardening, the function is updated in Eq. (3.35), where the yield stress is now dependent on κ. Accordingly, the new function $\sigma_y (\kappa)$ should provide the nonhardened yield stress when hardening is not activated ($\kappa = 0$), and provide the current value of yield stress as κ evolves. Let us consider the structural response of a certain material under uniaxial monotonic test, such as illustrated in the diagram on the left side of Figure 3.24. As noticeable, the abscissa is the total strain and it is intuitive to set $\kappa = \varepsilon_p$ so that the diagram on the right can be derived.

Two points need to be noticed. First, recalling Figure 3.1, the area under the σ–ε diagram is the energy per unit volume dissipated during the test. Being null the elastic energy when the stress approach the ultimate value, the area under the two graphs of Figure 3.24 is the same. The shape of

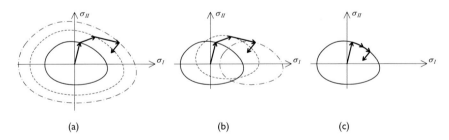

(a)   (b)   (c)

Figure 3.23 Simple models for the elastic domain evolution. (a) Isotropic and (b) kinematic hardening; (c) perfect elastoplasticity.

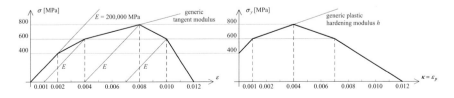

Figure 3.24 Structural response under uniaxial monotonic test. Stress–total strain diagram (left) and hardening diagram with $\kappa = \varepsilon_p$ (right).

the diagrams being different, another feature can be derived. Whereas the tangent to the curve on the left is defined as *tangent modulus* (of which the initial value is Young's modulus E), the equivalent on the right-side graph is named *plastic hardening modulus*, generally indicated with the letter h.

$$f(\sigma,\kappa) = \max\left(|\sigma_1 - \sigma_{II}|, |\sigma_1 - \sigma_{III}|, |\sigma_{II} - \sigma_{III}|\right) - \sigma_y(\kappa) = 0 \qquad (3.35)$$

One of the advantages of having introduced the internal variable κ is that the same mathematical approach can be used for both hardening and softening. By updating σ_y, the yield function keeps the form $f(\sigma, \kappa) = 0$ but it becomes larger or smaller according to hardening or softening, respectively. Considering the Tresca yield elastic domain in two-dimensional space ($\sigma_{III} = 0$), its evolution according to the diagrams of Figure 3.24 is shown in Figure 3.25. As it is possible to notice, both hardening and softening are isotropic. Despite the discrete values of κ, the reader should imagine the yield surface evolving throughout the numerical analysis, i.e. expanding and reducing, until (in this case) it degenerates into the origin with no stress and no stiffness.

The same approach discussed for the Tresca yield function can be extended to any yield function. Nonetheless, the richer is the model, the more parameters may be considered. As an example, for Mohr–Coulomb yield function, both friction angle and cohesion (or, analogously, the yield stresses) must depend on κ. This is evident from Eqs. (3.19) and (3.20). In other words, the main mechanical parameters evolve during the loading process and this evolution must be described by the internal variable κ.

Besides $\kappa = \varepsilon_p$ related to uniaxial case, similar to the approach used for perfect plasticity, also for the internal different κ hypotheses must be made for its evolution. The most common formulas are reported in Eq. (3.36), whereas the rate of the yield function is calculated in Eq. (3.37). Finally, in analogy with Eq. (3.28), the constitutive law for associated plastic flow and hardening is given in Eq. (3.38): the only change regards the introduction of the hardening parameter κ. The reader can extend Eqs. (3.33) and (3.34) to the cases with hardening.

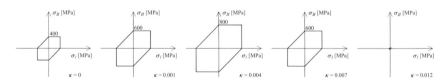

Figure 3.25 Evolution of Tresca yield surface according to the structural response of Figure 3.24.

$$\dot{\kappa} = \mathbf{\sigma}^T \dot{\mathbf{\varepsilon}}_p \qquad\qquad \text{work hardening}$$

$$\dot{\kappa} = \sqrt{\frac{2}{3} \dot{\mathbf{\varepsilon}}_p^{\ T} \dot{\mathbf{\varepsilon}}_p} \qquad\qquad \text{strain hardening}$$

(3.36)

$$\dot{f}(\mathbf{\sigma},\kappa) = \left(\frac{\partial f}{\partial \mathbf{\sigma}}\right)^T \dot{\mathbf{\sigma}} + \frac{\partial f}{\partial \kappa}\dot{\kappa}$$

(3.37)

$$\dot{\mathbf{\varepsilon}} = \dot{\mathbf{\varepsilon}}_e + \dot{\mathbf{\varepsilon}}_p \qquad \rightarrow \qquad \dot{\mathbf{\varepsilon}}_e = \dot{\mathbf{\varepsilon}} - \dot{\mathbf{\varepsilon}}_p$$

$$\dot{\mathbf{\sigma}} = \mathbf{D}\dot{\mathbf{\varepsilon}}_e \qquad \rightarrow \qquad \dot{\mathbf{\sigma}} = \mathbf{D}\left(\dot{\mathbf{\varepsilon}} - \dot{\mathbf{\varepsilon}}_p\right)$$

$$\dot{\mathbf{\varepsilon}}_p = \begin{cases} 0 & \text{if } f(\mathbf{\sigma},\kappa) < 0 & \text{elastic state} \\[2mm] 0 & \text{if } f(\mathbf{\sigma},\kappa) = 0, \ \dot{f}(\mathbf{\sigma},\kappa) < 0 & \text{elastic unloading} \\[2mm] \dot{\lambda}\dfrac{\partial f}{\partial \mathbf{\sigma}} & \text{if } f(\mathbf{\sigma},\kappa) = 0, \ \dot{f}(\mathbf{\sigma},\kappa) = 0 & \text{plastic flow} \end{cases}$$

(3.38)

From the computational point of view, conversely to perfect plasticity, hardening allows calculating the magnitude of the plastic strain rate. Let us consider the consistency condition, according to which plastic flow occurs when the yield function rate is null, as shown in Eq. (3.39). For the sake of brevity, the gradient of the yield function is substituted by **n**. Eqs. (3.38) and (3.39) are combined in Eq. (3.40), where subsequent manipulation is also described. Finally, the relationship between $\dot{\lambda}$ and $\dot{\mathbf{\varepsilon}}$ is shown in Eq. (3.41). The parameter h depends on $\dot{\lambda}$ but simplifications apply for simple yield functions and hardening laws. In a simple way, h can be defined as the *plastic hardening modulus* of Figure 3.24 (right side).

$$\left(\frac{\partial f}{\partial \mathbf{\sigma}}\right)^T \dot{\mathbf{\sigma}} + \frac{\partial f}{\partial \kappa}\dot{\kappa} = 0 \qquad \rightarrow \qquad \mathbf{n}^T \dot{\mathbf{\sigma}} + \frac{\partial f}{\partial \kappa}\dot{\kappa} = 0$$

(3.39)

$$\mathbf{n}^T \mathbf{D}\left(\dot{\mathbf{\varepsilon}} - \dot{\lambda}\mathbf{n}\right) + \frac{\partial f}{\partial \kappa}\dot{\kappa} = 0$$

$$\mathbf{n}^T \mathbf{D}\dot{\mathbf{\varepsilon}} - \dot{\lambda}\mathbf{n}^T \mathbf{D}\mathbf{n} + \frac{\partial f}{\partial \kappa}\dot{\kappa} = 0$$

(3.40)

$$\mathbf{n}^T \mathbf{D}\dot{\mathbf{\varepsilon}} + \dot{\lambda}\left(\underbrace{-\mathbf{n}^T \mathbf{D}\mathbf{n} + \frac{1}{\dot{\lambda}}\frac{\partial f}{\partial \kappa}\dot{\kappa}}_{-h} \right) = 0$$

$$\dot{\lambda} = \frac{\mathbf{n}^T \mathbf{D} \dot{\boldsymbol{\varepsilon}}}{\mathbf{n}^T \mathbf{D} \mathbf{n} + h} \tag{3.41}$$

Finally, the relationship between stress and strain rates is developed in Eq. (3.42), where $\mathbf{D}_{ep}$ is the *elastoplastic stiffness*, the equivalent counterpart of the elastic stiffness now relating stress rate and strain rate. From the numerical point of view, this equation must be integrated through the load history by considering an incremental procedure with small increments of strain $\Delta\boldsymbol{\varepsilon} = \dot{\boldsymbol{\varepsilon}}\Delta t$, as shown in the following section.

$$\dot{\boldsymbol{\sigma}} = \mathbf{D}\left(\dot{\boldsymbol{\varepsilon}} - \mathbf{n}\dot{\lambda}\right)$$

$$\dot{\boldsymbol{\sigma}} = \mathbf{D}\left(\dot{\boldsymbol{\varepsilon}} - \mathbf{n}\frac{\mathbf{n}^T \mathbf{D} \dot{\boldsymbol{\varepsilon}}}{\mathbf{n}^T \mathbf{D} \mathbf{n} + h}\right)$$

$$\dot{\boldsymbol{\sigma}} = \underbrace{\left(\mathbf{D} - \frac{\mathbf{D}\mathbf{n}\mathbf{n}^T \mathbf{D}}{\mathbf{n}^T \mathbf{D} \mathbf{n} + h}\right)}_{\text{elastoplastic stiffness } \mathbf{D}_{ep}}\dot{\boldsymbol{\varepsilon}} \quad \rightarrow \quad \dot{\boldsymbol{\sigma}} = \mathbf{D}_{ep}\dot{\boldsymbol{\varepsilon}} \tag{3.42}$$

Finally, for the sake of completeness, it is interesting to provide a few comments on nonassociated plastic flow. As mentioned, this applies when the normality between yield function and strain rate direction is lost. Accordingly, the new direction of plastic flow can be labelled with $\mathbf{m}$ and can be regarded as the gradient of a new function g defined as *plastic potential*. Consequently, the plastic strain rate is calculated with Eq. (3.43). Providing the same passages, Eq. (3.42) becomes Eq. (3.44). Note that for generic vectors $\mathbf{m}$ and $\mathbf{n}$, the product $\mathbf{m}\mathbf{n}^T$ provides a nonsymmetric matrix, meaning that the elastoplastic stiffness loses symmetry.

$$\dot{\boldsymbol{\varepsilon}}_p = \dot{\lambda}\frac{\partial g}{\partial \boldsymbol{\sigma}} = \dot{\lambda}\mathbf{m} \tag{3.43}$$

$$\dot{\boldsymbol{\sigma}} = \left(\mathbf{D} - \frac{\mathbf{D}\mathbf{m}\mathbf{n}^T \mathbf{D}}{\mathbf{n}^T \mathbf{D} \mathbf{m} + h}\right)\dot{\boldsymbol{\varepsilon}} \tag{3.44}$$

3.2.5 Euler Forward and Euler Backward incremental methods with perfect elastoplasticity

Similar to what was shown in Chapter 2, Eq. (3.45) shows the iterative procedure for the *Euler Forward method*. Basically, moving from Eq. (3.42) and by considering a single integration step for increments of time Δt (i.e. assuming that the quantity is constant within the load step), it is possible to directly replace $\dot{\varepsilon}$ with $\Delta\varepsilon$. A closer inspection of Eq. (3.45) shows that the elastoplastic stiffness is calculated by means of quantities relative to the previous step $n - 1$, thus known. Additionally, it is worth reminding that Eqs. (3.42) and (3.45) provide the relationship between stress and strain once the load history has reached the elastic limit, i.e. with σ^{n-1} on the yield function and $\Delta\varepsilon^n$ the consequent increment of strain.

$$\sigma^n = \sigma^{n-1} + \Delta\sigma^n$$

$$\Delta\sigma^n = \underbrace{\left(D - \frac{Dn^{n-1}\left(n^{n-1}\right)^T D}{\left(n^{n-1}\right)^T Dn^{n-1} + h^{n-1}} \right)}_{\text{elastoplastic stiffness at step } n-1} \Delta\varepsilon^n \qquad (3.45)$$

With the aim of illustrating the method for the integration of the elastoplastic relations, a simple problem with a metal plate is considered. The analysis is performed, in a simplified way, considering the principal stresses assuming plane stress conditions. Young's modulus is $E = 210,000\,\text{MPa}$ (for the sake of simplicity the Poisson's coefficient ν is assumed as zero) and the von Mises yield function is assumed. Two plates, initially unloaded and with null stresses $\sigma_{\text{initial}} = [0,0]^T$ MPa, are subjected to different strain increments, namely, $\Delta\varepsilon = [0.001, 0.001]^T$ for plate 1 and $\Delta\varepsilon = [0.004, 0.002]^T$ for plate 2.

The constitutive matrix is evaluated in Eq. (3.46). To understand if a generic strain increment leads to an elastic or plastic state, let us assume that the elastic relationship in Eq. (3.47) holds. Once the trial value of stress is calculated in Eq. (3.48), it is possible to check it according to the von Mises yield function, as shown in Eq. (3.49), see also Eq. (3.21). The calculation is given in Eq. (3.50). The yield function being negative, the stress state is elastic, and the stresses calculated in Eq. (3.48) are the correct ones. The graphical comparison between yield function and actual stress state is shown in Figure 3.26, where the square marker within the ellipse is indicative of an elastic stress state.

The same reasoning can be applied to the second plate, achieving a final stress of $\sigma_{\text{trial}} = [840, 420]^T$ MPa. By means of the von Mises yield function, this stress state leads to $f(\sigma_{\text{trial}}) = 369,200$, not acceptable because it is positive. In the following, this value will be addressed as the reference value for assessing the accuracy of the incremental procedure. Graphically, it is possible to see in Figure 3.26 that the cross marker falls outside of the ellipse, in contradiction with the feature of perfect elastoplasticity. However, there is no reason to think the steel plate cannot undergo the given increment of strain, as long as the final stress state is confined by the yield surface (Figure 3.23c). The calculation of this value is discussed next.

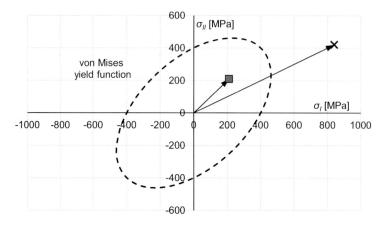

Figure 3.26 Graphical check of the elastic (square) and not admissible state (cross marker) according to von Mises yield function.

$$D = \frac{E}{1-v^2}\begin{bmatrix} 1 & v \\ v & 1 \end{bmatrix} = \begin{bmatrix} 210,000 & 0 \\ 0 & 210,000 \end{bmatrix} MPa \qquad (3.46)$$

$$\Delta\sigma_{trial} = D\Delta\varepsilon = \begin{bmatrix} 210,000 & 0 \\ 0 & 210,000 \end{bmatrix}\begin{bmatrix} 0.001 \\ 0.001 \end{bmatrix} = \begin{bmatrix} 210 \\ 210 \end{bmatrix} MPa \qquad (3.47)$$

$$\sigma_{trial} = \sigma_{initial} + \Delta\sigma_{trial} = \begin{bmatrix} 0 \\ 0 \end{bmatrix} + \begin{bmatrix} 210 \\ 210 \end{bmatrix} = \begin{bmatrix} 210 \\ 210 \end{bmatrix} MPa \qquad (3.48)$$

$$f(\sigma) = \sigma_I^2 + \sigma_{II}^2 - \sigma_I\sigma_{II} - \sigma_y^2 = 0 \qquad (3.49)$$

$$f(\sigma) = 210^2 + 210^2 - 210 \cdot 210 - 400^2 \approx -10^5 < 0 \qquad (3.50)$$

According to Figure 3.26, it seems reasonable to analyse the problem "chronologically". Moving from null stress, the two components of stress increase along the load direction with a certain fixed ratio (i.e. the tangent of the vector slope), following the given load path, see also Figure 3.11b. The steel plate behaves elastically until the load vector reaches the yield function (the circle in Figure 3.27a). This limit stress σ_y can be calculated according to Eq. (3.51), where α is a scalar value that scales the trial stresses σ_{trial} calculated earlier. Imposing that the components of $\sigma_y = \alpha\sigma_{trial}$ satisfy the yield function, Eq. (3.52) leads to $\alpha = 0.55$. This is an important outcome because it allows decomposing the total strain increment in a purely elastic part $\Delta\varepsilon_e$ (55%) and subsequent remaining elastoplastic part $\Delta\varepsilon_p$ (45%), as shown in Eq. (3.54).

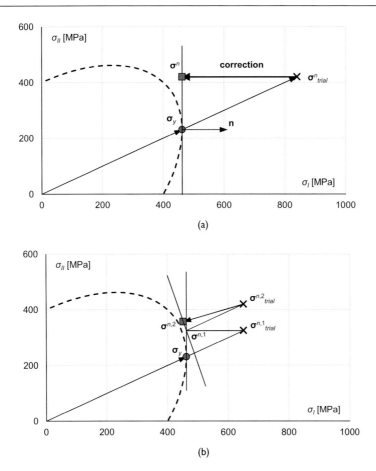

(a)

(b)

Figure 3.27 Explicit Euler Forward method with null Poisson's ratio. (a) Single and (b) two subdivisions of plastic strain increment.

$$f(\sigma) = \alpha^2 \left[\sigma_\mathrm{I}^2 + \sigma_\mathrm{II}^2 - \sigma_\mathrm{I}\sigma_\mathrm{II} \right] - \sigma_y^2 = 0 \tag{3.51}$$

$$\alpha = \sqrt{\frac{400^2}{840^2 + 420^2 - 840 \cdot 420}} = 0.55 \tag{3.52}$$

$$\sigma_y = \alpha \begin{bmatrix} 840 \\ 420 \end{bmatrix} = \begin{bmatrix} 462 \\ 231 \end{bmatrix} \mathrm{MPa} \tag{3.53}$$

$$\Delta\varepsilon = \alpha \underbrace{\begin{bmatrix} 0.004 \\ 0.002 \end{bmatrix}}_{\Delta\varepsilon_e} + (1-\alpha) \underbrace{\begin{bmatrix} 0.004 \\ 0.002 \end{bmatrix}}_{\Delta\varepsilon_p} = \underbrace{\begin{bmatrix} 0.0022 \\ 0.0011 \end{bmatrix}}_{\Delta\varepsilon_e} + \underbrace{\begin{bmatrix} 0.0018 \\ 0.0009 \end{bmatrix}}_{\Delta\varepsilon_p} \tag{3.54}$$

Following the chronological description of the phenomenon, of the imposed strain increment, 55% is used to move the material to the elastic limit, achieving $\boldsymbol{\sigma}_y = [462,231]^T$ MPa, and 45% is left for the subsequent phase. Considering $\boldsymbol{\sigma}_y$ as the starting point of the iterative procedure (namely, $\boldsymbol{\sigma}^0$), it is possible to use Eq. (3.45), expanded and decomposed in Eq. (3.55), where the term h is neglected because no hardening is considered in the current example. It is worth noticing that the equation is maintained in its full form to allow the reader extending the discussion to the case of nonassociated plastic flow. Additionally, the superscript n refers to a generic step n of a structural analysis.

It is worth noticing that, in correspondence of $\boldsymbol{\sigma}^0$, $\varepsilon_p = 0$. With reference to Figure 3.27a, the trial stress σ_{trial}^n is calculated assuming the increment of strain $\Delta\varepsilon_p^n$ as elastic, applying the highlighted correction subsequently. Looking again at Figure 3.27a, the correction projects the trial stress σ_{trial}^n on the tangent to the yield curve (calculated at the previous step). This is in line with the consistency condition and Figures 3.21b and 3.23c. However, being an approximated (linearized) approach, $\boldsymbol{\sigma}^n$ does not lie on the yield curve and the lack of accuracy is evident. The relative calculations are detailed in Table 3.3, where the last column reports the ratio between the current value of $f(\boldsymbol{\sigma}^n)$ and $f(\boldsymbol{\sigma}_{\text{trial}})$, the latter referring to the first trial and equal to 369,200.

$$\boldsymbol{\sigma}^n = \boldsymbol{\sigma}^{n-1} + \Delta\boldsymbol{\sigma}^n$$

$$\Delta\boldsymbol{\sigma}^n = \mathbf{D}\Delta\boldsymbol{\varepsilon}_p^n - \frac{\mathbf{D}\mathbf{n}^{n-1}\left(\mathbf{n}^{n-1}\right)^T \mathbf{D}}{\left(\mathbf{n}^{n-1}\right)^T \mathbf{D}\mathbf{n}^{n-1}}\Delta\boldsymbol{\varepsilon}_p^n \tag{3.55}$$

$$\boldsymbol{\sigma}^n = \underbrace{\boldsymbol{\sigma}^{n-1} + \mathbf{D}\Delta\boldsymbol{\varepsilon}_p^n}_{\boldsymbol{\sigma}_{\text{trial}}^n} \underbrace{- \frac{\mathbf{D}\mathbf{n}^{n-1}\left(\mathbf{n}^{n-1}\right)^T \mathbf{D}}{\left(\mathbf{n}^{n-1}\right)^T \mathbf{D}\mathbf{n}^{n-1}}\Delta\boldsymbol{\varepsilon}_p^n}_{\text{correction } \boldsymbol{\sigma}_c^n}$$

An intuitive method to reduce the error is to subdivide the strain increment $\Delta\varepsilon_p^n$ in smaller subincrements i. As a matter of fact, the smaller are the load increments, the smaller are the displacement increments (here translated

Table 3.3 Step-by-step calculations for the Forward Euler method (stress values in MPa)

σ^{n-1}	ε_p^{n-1}	$\Delta\varepsilon_p^n$	$\varepsilon_p^n = \varepsilon_p^{n-1} + \Delta\varepsilon_p^n$	n^{n-1}	σ_{trial}^n	σ_c^n	$\sigma^n = \sigma_{\text{trial}}^n + \sigma_c^n$	$f(\sigma^n)$	$f(\sigma^n)/f(\sigma_{\text{trial}})$
	$[10^{-3}]$	$[10^{-3}]$	$[10^{-3}]$	$\partial f(\sigma)/\partial\sigma$	Eq. (3.55)	Eq. (3.55)	Eq. (3.55)	Eq. (3.49)	
462	0	1.80	1.80	693	840	−378	462	35,744	0.09681
231	0	0.90	0.90	0	420	0	420		

Table 3.4 Step-by-step calculations for the Euler Forward method with two subincrements ($i = 1, 2$) (stress values in MPa)

i	$\sigma^{n,i-1}$	$\varepsilon_p^{n,i-1}$	$\Delta\varepsilon_p^{n,i}$	$\varepsilon_p^{n,i} = \varepsilon_p^{n,i-1} + \Delta\varepsilon_p^{n,i}$	$n^{n,i-1}$	σ_{trial}^n	$\sigma_c^{n,i}$	$\sigma^{n,i} = \sigma_{trial}^{n,i} + \sigma_c^{n,i}$	$f(\sigma^{n,i})$	$f(\sigma^{n,i})/f(\sigma_{trial})$
	$[10^{-3}]$	$[10^{-3}]$	Eq. (3.56)	$\partial f(\sigma)/\partial\sigma$	Eq. (3.56)	Eq. (3.56)	Eq. (3.56)	Eq. (3.49)		
1	462	0	0.90	0.90	693	651	−189	462	8,936	0.02420
	231	0	0.45	0.45	0	325	0	325		
2	462	0.90	0.90	1.80	598	651	−199	452	10,352	0.02804
	325	0.45	0.45	0.90	189	420	−63	357		

to smaller strain increments), meaning also, the better is the approximation between tangent lines and yield curve (thus, the lower is the error in the integration of the elastoplastic relations). Considering subincrements, Eq. (3.55) becomes Eq. (3.56). It is noted that the first of Eq. (3.56) highlights the subdivision of the plastic strain increment $\Delta\varepsilon_p^n$ in small subincrements $\Delta\varepsilon_p^{n,i}$, whereas $\varepsilon_p^{n,i}$ is the current plastic strain (calculated as a summation of the previous increments). The example with two equal subincrements ($i = 1, 2$) of the elastoplastic part $\Delta\varepsilon_p^n$ (45%) is discussed in Table 3.4 and shown Figure 3.27b. As it is possible to notice, the error is sensibly reduced. However, the solution being based on the result of the previous iteration, there is no possibility to limit the error that keeps increasing. Accurate results can be achieved only with a dramatic increase of the number of subincrements. In this regard, the number of subincrements should be chosen such as the final value of $f(\sigma^n)$ is in the range of 10^{-4} to 10^{-6} of the first violation (i.e. last column of the mentioned tables). For the present case, around 400 subincrements are needed to reach an error lower than 10^{-4}, achieving a final value $\sigma^{n,400} = [445,331]^T$ MPa. The final value $f(\sigma^{n,400}) = 35$ (which is still not zero!) vs. the initial $f(\sigma_{trial}) = 369,200$. The incremental procedure basically "follows" the yield surface (see ahead Figure 3.28a).

$$\Delta\varepsilon_p^n = \sum_{i=1}^{} \Delta\varepsilon_p^{n,i} \quad \text{with} \quad \varepsilon_p^{n,i} = \varepsilon_p^{n,i-1} + \Delta\varepsilon_p^{n,i}$$

$$\sigma^{n,i} = \sigma^{n,i-1} + \Delta\sigma^{n,i}$$

$$\Delta\sigma^{n,i} = D\Delta\varepsilon_p^{n,i} - \frac{Dn^{n,i-1}\left(n^{n,i-1}\right)^T D}{\left(n^{n,i-1}\right)^T Dn^{n,i-1}}\Delta\varepsilon_p^{n,i} \qquad (3.56)$$

$$\sigma^{n,i} = \underbrace{\sigma^{n,i-1} + D\Delta\varepsilon_p^{n,i}}_{\sigma_{trial}^{n,i}} - \underbrace{\frac{Dn^{n,i-1}\left(n^{n,i-1}\right)^T D}{\left(n^{n,i-1}\right)^T Dn^{n,i-1}}\Delta\varepsilon_p^{n,i}}_{\text{correction } \sigma_c^{n,i}}$$

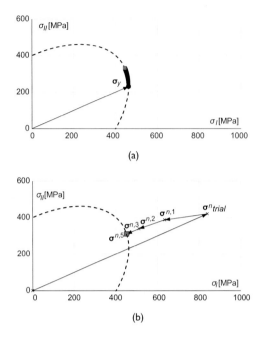

Figure 3.28 Comparison between explicit and implicit Euler methods. (a) Euler Forward method with 400 subincrements and (b) Euler Backward method with five iterations.

The reader may have recognized the similarities between the Euler Forward method used for the integration of the elastoplastic relations with what was shown in Chapter 2. Additionally, as all quantities are evaluated at the previous load step where all the information is known, the approach is defined as *explicit*. In general, the method can be understood as a two-step process, with an initial calculation of a trial stress (made by assuming elastic behaviour) and the subsequent projection towards the yield curve (or surface for three-dimensional problems). For this reason, the term *return mapping* is also quite popular for this procedure.

The error made with the Euler Forward method increases with the curvature of the yield surface and the numerical instability of the method for large load steps is not appealing for a robust implementation in a finite element code. For these reasons, an *implicit method*, e.g. *Euler Backward*, is preferred. The method evaluates all quantities at the end of the load increment, resulting rather efficient and accurate. With this goal, a mathematical manipulation of the formulas seen earlier must be made.

The last of Eq. (3.55) is rewritten in Eq. (3.57) where the terms are slightly rearranged and the normal to the yield surface is evaluated at the current step n (namely, $\mathbf{n}^n$). Recalling that the normal vector is the gradient to the yield surface, Eq. (3.57) can be rewritten in Eq. (3.58), where the dependency of

σ^n from $\Delta\lambda^n$ is also highlighted. Without entering the merit of the mathematical calculations, it is possible to state that the goal of the analysis is to seek for a certain $\Delta\lambda^n$ for which Eq. (3.59) holds. At this point, to refine the solution, the Newton–Raphson method can be adopted where the iterations are indicated with i (see Chapter 2). The method is synthetically illustrated in Eq. (3.60), whose symbol meaning was already discussed. The first derivative of $f(\Delta\lambda^n)$ is calculated according to the chain rule, as shown in Eq. (3.61). Operatively, the starting point for the method is $\Delta\lambda^{n,0} = 0$, thus $\sigma^{n,0} = \sigma^n_{trial}$. Without detailing all the mathematical manipulations, the vector containing the gradient of σ with respect to λ at step $n, i-1$ (denominator of the Newton–Raphson method) is calculated thanks to Eq. (3.62). Furthermore, the vector of the stresses at the end of the current iteration i is calculated with Eq. (3.63), provided the updated $\Delta\lambda^n$ of Eq. (3.60).

The method can be easily extended to the case of nonassociated flow and hardening, as long as all the quantities (namely $\mathbf{m}^n$ and κ^n) are expressed in terms of $\Delta\lambda^n$.

$$\sigma^n = \underbrace{\sigma^{n-1} + \mathbf{D}\Delta\varepsilon^n_p}_{\sigma^n_{trial}} - \mathbf{D}\mathbf{n}^n \underbrace{\frac{\left(\mathbf{n}^n\right)^T \mathbf{D}}{\left(\mathbf{n}^n\right)^T \mathbf{D}\mathbf{n}^n} \Delta\varepsilon^n_p}_{\Delta\lambda^n} \tag{3.57}$$

$$\sigma^n = \sigma^n_{trial} - \Delta\lambda^n \mathbf{D}\mathbf{n}^n$$

$$\sigma^n = \sigma^n_{trial} - \Delta\lambda^n \mathbf{D} \left.\frac{\partial f}{\partial \sigma}\right|_{\sigma=\sigma^n} \qquad \rightarrow \qquad \sigma^n = \sigma^n\left(\Delta\lambda^n\right) \tag{3.58}$$

$$f\left(\sigma^n\right) = 0 \qquad \rightarrow \qquad f\left(\Delta\lambda^n\right) = 0 \tag{3.59}$$

$$\Delta\lambda^{n,i} = \Delta\lambda^{n,i-1} - \frac{f\left(\Delta\lambda^{n,i-1}\right)}{f'\left(\Delta\lambda^{n,i-1}\right)} \tag{3.60}$$

$$f'(\Delta\lambda) = \frac{\partial f}{\partial \lambda} = \frac{\partial f}{\partial \sigma}\frac{\partial \sigma}{\partial \lambda} \tag{3.61}$$

$$\left(\frac{\partial \sigma}{\partial \lambda}\right)^{n,i-1} = \begin{bmatrix} \dfrac{E\left(\sigma^{n,0}_{II} - 3E^2\sigma^{n,0}_{II}\left(\Delta\lambda^{n,i-1}\right)^2 - 2\sigma^{n,0}_{I}\left(1+3E\Delta\lambda^{n,i-1}\left(1+E\Delta\lambda^{n,i-1}\right)\right)\right)}{\left(1+E\Delta\lambda^{n,i-1}\right)^2\left(1+3E\Delta\lambda^{n,i-1}\right)^2} \\[3ex] \dfrac{E\left(\sigma^{n,0}_{I} - 3E^2\sigma^{n,0}_{I}\left(\Delta\lambda^{n,i-1}\right)^2 - 2\sigma^{n,0}_{II}\left(1+3E\Delta\lambda^{n,i-1}\left(1+E\Delta\lambda^{n,i-1}\right)\right)\right)}{\left(1+E\Delta\lambda^{n,i-1}\right)^2\left(1+3E\Delta\lambda^{n,i-1}\right)^2} \end{bmatrix} \tag{3.62}$$

Table 3.5 Step-by-step calculations for the Euler Backward method (stress values in MPa)

i	$\sigma^{n,i-1}$	$f(\sigma^{n,i-1})$ Eq. (3.49)	$\Delta\lambda^{n,i-1}$ $[10^{-6}]$	$(\partial f(\sigma)/\partial\sigma)^{n,i-1}$	$(\partial\sigma/\partial\Delta\lambda)^{n,i-1}$ $[10^8]$ Eq. (3.62)	$(\partial f(\sigma)/\partial\Delta\lambda)^{n,i-1}$ $[10^{11}]$ Eq. (3.61)	$\Delta\lambda^{n,i}$ $[10^{-6}]$ Eq. (3.60)	$\sigma^{n,i}$ Eq. (3.63)	$f(\sigma^{n,i})$ Eq. (3.49)	$f(\sigma^{n,i})/f(\sigma_{trial})$
1	840 420	369,200	—	1,260 0	−2.65 0.00	−3.33	1.11	635 387	147,163	0.39860
2	635 387	147,163	1.11	882 140	−1.33 −0.41	−1.23	2.30	510 339	42,345	0.11469
3	510 339	42,345	2.30	682 168	−0.82 −0.38	−0.62	2.98	460 314	6,053	0.01640
4	460 314	6,053	2.98	606 169	−0.66 −0.34	−0.46	3.11	452 310	167	0.00045
5	452 310	167	3.11	594 168	−0.63 −0.33	−0.43	3.12	452 310	0	0.00000

$$\boldsymbol{\sigma}^{n,i} = \begin{bmatrix} \dfrac{\sigma_{\mathrm{I}}^{n,0} + E\Delta\lambda^{n,i}\left(\sigma_{\mathrm{II}}^{n,0} + 2\sigma_{\mathrm{I}}^{n,0}\right)}{\left(1 + E\Delta\lambda^{n,i}\right)\left(1 + 3E\Delta\lambda^{n,i}\right)} \\[4mm] \dfrac{\sigma_{\mathrm{II}}^{n,0} + 2E\Delta\lambda^{n,i}\sigma_{\mathrm{II}}^{n,0} + E\Delta\lambda^{n,i}\sigma_{\mathrm{I}}^{n,0}}{1 + 4E\Delta\lambda^{n,i} + 3E^2\left(\Delta\lambda^{n,i}\right)^2} \end{bmatrix} \tag{3.63}$$

The application of the Euler Backward method to the mentioned example is described in Figure 3.28b, whereas the calculations are shown in Table 3.5. As it is possible to notice, only five iterations are needed for the final iteration to reach $f(\boldsymbol{\sigma}^{n,5}) < 10^{-4} f(\boldsymbol{\sigma}_{\mathrm{trial}})$. The two graphs in Figure 3.28 show the profound difference of the two methods in terms of strategy. Euler Forward method reaches the yield surface and then very small subincrements of plastic strain are needed to stay in the neighbourhood of the surface. Conversely, Euler Backward method does not need to define the intersection between the load path and the yield surface, and the solution is refined by subsequent iterations (not subincrements!).

In terms of results, the Euler Forward and Euler Backward methods lead to different values of $\boldsymbol{\sigma}^n = [445,331]^T$ MPa and $\boldsymbol{\sigma}^n = [452,310]^T$ MPa, respectively. This is because the quantities are evaluated at different points and the integration of the varying quantities in "time" is made in a single step assuming a constant value through the load step. Note that both solutions found lie in the yield surface.

3.3 DAMAGE MODELS

Many materials subjected to unfavourable mechanical and environmental conditions undergo microchanges in their inner structure, which, consequently, affect their macroscopic behaviour. Since these changes impair the mechanical properties, the term *damage* is generally used. Damage is particularly relevant for brittle and quasi-brittle materials such as concrete, rocks, mortar or other concrete-like materials, exhibiting smaller or larger cracks upon loading.

More in detail, damage represents a progressive and irreversible deterioration of the material from the so-called *virgin state* (i.e. the material in its undamaged conditions) until the occurrence of macroscopic cracks. With the aim of providing more qualitative insight, let us consider the simplified schematization of the continuum shown in Figure 3.29a, where a generic material element is subjected to uniaxial tension. Once the elastic limit stress is reached, atomic dislocations may occur in the shape of microscopic cracks and voids (Figure 3.29b). Consequently, the amount of material effectively reacting to the external load changes throughout the test. Due to propagating and interacting defects, in fact, the initial cross-section area A has become an effective cross-section area A^*, with a consequent macroscopic reduction of strength and stiffness (Figure 3.29b and c).

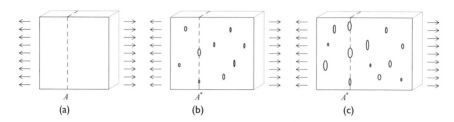

Figure 3.29 Schematic damage evolution for the continuum under uniaxial loading. (a) Undamaged configuration with cross-section area A; (b) and (c) microcracks evolution with reduction of the cross-section area to an effective area A^*.

As the mechanical properties are related to A^*, knowing its evolution might be a logical approach. However, the measurement of A^* implies the physical measurement of microcracks and cavities, which is impracticable. Instead of getting a direct measurement of the cross-section reduction, it is possible to interpret the global mechanical behaviour (such as strength and stiffness reduction) according to damage. In other words, being impossible to measure A^* directly, all the physical quantities can be still related to the undamaged configuration, but their variation provides indirect information on the evolution of A^*.

For the sake of completeness, it is intuitive to think that microcracks are going to close upon full load removal or load inversion. If the original configuration is restored, upon load removal, the cross-section area A is fully restored, and the material behaves as undamaged, as discussed in Section 3.3.3. In turn, damage may occur for high compressive actions too, with dramatic differences at the microscopic level. Due to the more intuitive insight, in the following, damage is mainly referred to tension, but the computational approach is equivalent also for compression.

3.3.1 Uniaxial loading

Let us now extend the example of Figure 3.29 to a fictitious specimen made of three parallel bars as illustrated in Figure 3.30a, where the cardinal number indicates each bar. They represent the microstructure composing the continuum and it is worth investigating how the behaviour of the bars influences the macroscopic behaviour of the specimen under uniaxial loading. The bars are similar and follow a linear elastic constitutive law, with Young's modulus E^* (Figure 3.30b). In the following, * indicates the quantities related to microscopic structure, assumed as the *effective* ones. With the aim of simulating microscopic damage, they break at a certain increasing strain (where $\varepsilon_2 = 1.25$ and $\varepsilon_3 = 1.5$, normalized with respect to ε_1). A is the cross-section area of the specimen, which is equally shared by the three bars (i.e. $A/3$ each). The comparison is formulated by equating the effective force ($\sigma^* A^*$) with the macroscopic (or nominal, as related to the initial geometry) measurement (σA). Additionally, since the problem deals with

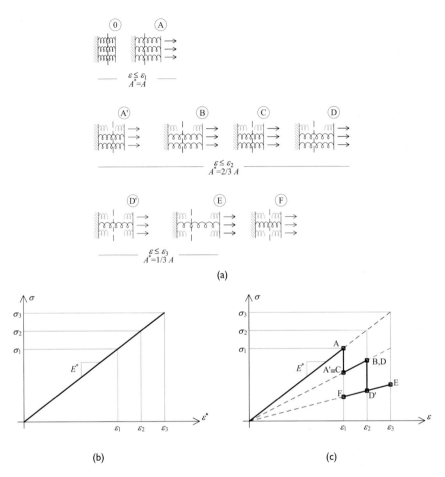

Figure 3.30 Schematic approach of damage by means of parallel bars. (a) Representation at the microscopic level; (b) effective and (c) macroscopic stress–strain relationships.

parallel bars, it is possible to assume that the effective and the macroscopic strains are the same.

When $\varepsilon \leq \varepsilon_1$, i.e. until point A of Figure 3.30, the entire cross-section is active, thus $A = A^*$, and effective and macroscopic stresses coincide, as well as Young's moduli. If strain keeps increasing, the bar 1 breaks and the structure jumps to point A'. The effective area is now 2/3 of A (as 1/3 is damaged) and, consequently, the stress is 2/3 of the effective one. In other words, due to the breakage of one of the bars, although the other bars are basically working with the same stress as earlier, a reduction of capacity is clear at the macroscopic level. As damage represents a permanent deterioration of the continuum, a partial loading–unloading–reloading process leads to points B-C-D, without travelling through the previous (elastic) path.

Essentially, the response is still elastic but with a smaller cross-section, lead-ing to an apparent lower macroscopic Young's modulus. This is the nov-elty of damage model with respect to elastoplasticity, according to which Young's modulus is never updated. A subsequent increment of strain from point D leads to the breakage of the second bar and to a jump to D', with analogous consequences. It is not difficult to follow the structural response through a possible loading–unloading to points E and F. As clearly notice-able from Figure 3.30c, despite the same strain ε_1, the macroscopic stresses of points A, A', C and F are different. They depend on the load history, so does the damage level (no damage for point A, 1/3 damage for point A' and C and 2/3 damage for point F).

In order not to explicitly consider the effective area A^*, let us intro-duce the damage variable $\omega(\varepsilon)$, whose general meaning is illustrated in Eq. (3.64). More in detail, ε_d is the value of strain at which damage starts (ε_1 in Figure 3.30), whereas ε_u is the value in correspondence of which the cross-section is totally damaged (ε_3 in Figure 3.30). Defining a priori the evolution of $\omega(\varepsilon)$ provides the stress–strain law in the postelastic phase, either loading or unloading. In the following, for the sake of clarity, the dependence from ε is dropped.

Considering the previous example, the values of ω in the graph of Figure 3.31a lead to the same macroscopic behaviour of Figure 3.30c. As it is possible to see, the jumps of the damage level are marked with vertical segments. In turn, looking at points C and F of Figure 3.31a, the reduction of strain does not produce a reduction of the damage variable, in line with the assumption of damage as an irreversible process. Computationally, this reflects itself on the need of keeping track of the damage variable from the past load history, just as done in plasticity with the internal variable κ. A possible smoother func-tion for the damage variable (to be preferred as more realistic) is shown in Figure 3.31b, whose equation is discussed ahead in Eq. (3.70).

$$A^* = (1-\omega)A \quad \begin{array}{lll} \omega = 0 & \text{undamaged state} & \varepsilon \le \varepsilon_d \\ \omega = 1 & \text{null resisting area} & \varepsilon \ge \varepsilon_u \\ 0 < \omega < 1 & \text{damaged state} & \varepsilon_d < \varepsilon < \varepsilon_u \end{array} \quad (3.64)$$

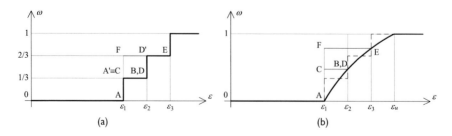

Figure 3.31 Examples of damage variables ω. (a) Stepwise function for the case of Figure 3.30 and (b) smooth function.

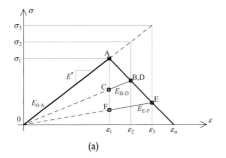 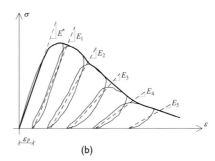

(a)   (b)

Figure 3.32 Constitutive laws with damage. (a) Following the damage variable of Figure 3.31b and (b) typical uniaxial load test of a concrete specimen subjected to cyclic compressive loading showing residual (plastic) strain and reducing (damaged) stiffness.

The example of Figure 3.30 can be discussed according to Eq. (3.64), leading to Eq. (3.65). As the macroscopic (elastic) strain is assumed equal to the effective one of the bars, Eq. (3.66) shows the relationship between the effective and macroscopic Young's moduli. It is worth noticing that, for undamaged material, effective and macroscopic Young's moduli are equivalent (when $\omega = 0$, $E \equiv E^*$). Eq. (3.66) has another important interpretation: after manipulation, it leads to Eq. (3.67), according to which it is possible to calculate the variable ω knowing the evolution of E^*, e.g. from experimental tests (see Figure 3.32b).

$$\sigma^* A^* = \sigma A \qquad A^* = (1-\omega)A \qquad \sigma = (1-\omega)\sigma^* \tag{3.65}$$

$$\varepsilon = \frac{\sigma^*}{E^*} = \frac{\sigma}{E} \qquad \frac{\sigma^*}{E^*} = \frac{(1-\omega)\sigma^*}{E} \qquad E = (1-\omega)E^* \tag{3.66}$$

$$\omega = 1 - \frac{E}{E^*} \tag{3.67}$$

For the sake of completeness, the effect of the damage variable illustrated in Figure 3.31b is shown in Figure 3.32a. The σ–ε curve is obtained following Eq. (3.65), whereas Young's modulus is calculated according to Eq. (3.66). It is worth mentioning that the latter represents the slope of the unloading phase (i.e. the line between the origin and the points on the curve), and it is often referred to as *secant elastic modulus*, indicated with E_{sec} in the following.

As it is possible to notice, the damage variable was calculated to obtain linear softening. This goal was achieved isolating ω in Eq. (3.68) from Eqs. (3.65) and (3.66). The expected softening branch is a straight line passing through the elastic limit point $(\varepsilon_d, \sigma_y)$ and the ultimate one $(\varepsilon_u, 0)$, whose equation is derived in Eq. (3.69). Noting that $E^* = \sigma_y/\varepsilon_d$, the damage function is calculated in Eq. (3.70). For the current example, $(\varepsilon_d, \sigma_y) \equiv (\varepsilon_1, \sigma_1)$ and $\varepsilon_u = 1.5 \, \varepsilon_d$.

$$\sigma = (1-\omega)\sigma^* \qquad \rightarrow \qquad \sigma = (1-\omega)E^*\varepsilon$$

$$\omega = 1 - \frac{\sigma}{E^*\varepsilon} \tag{3.68}$$

$$\sigma = \frac{\varepsilon_u - \varepsilon}{\varepsilon_u - \varepsilon_d}\sigma_y \qquad \varepsilon_d \leq \varepsilon \leq \varepsilon_u \tag{3.69}$$

$$\omega = 1 - \frac{\varepsilon_d(\varepsilon_u - \varepsilon)}{\varepsilon(\varepsilon_u - \varepsilon_d)} \qquad \varepsilon_d \leq \varepsilon \leq \varepsilon_u \tag{3.70}$$

As already mentioned, the variation of Young's modulus (following e.g. cyclic tests) may be used to evaluate the damage variable. Figure 3.32b shows the case of uniaxial cyclic test on concrete specimen whose softening behaviour is highlighted by the envelope curve (bold line). Upon unloading–reloading, the stress–strain curve follows new paths highlighted with dashed lines. The reduction of their slope is the macroscopic evidence of internal damage. In Figure 3.32b, the subscripts indicate the unloading–loading cycle and, as noticeable, the unloading branch modulus reduces at every cycle, meaning that the damage level is increasing.

Looking carefully at Figure 3.32, a great difference between the two diagrams is evident. Whereas pure damage leads to null residual strain (Figure 3.32a), structural materials usually exhibit also plastic deformation (Figure 3.32b). For the sake of brevity, the combination of damage and elastoplasticity is not discussed in the present text, but marginally dealt with in Section 3.6.4.

3.3.2 Isotropic and anisotropic damage models

Let us now consider a more rigorous approach assuming that the damage variable ω depends on a new internal variable $r(t)$, as shown in Eq. (3.71). The need of a new variable is related to the load history. As shown in Figure 3.31, the damage variable ω must be evaluated according to the largest previously reached strain (a simplification valid for simple one-dimensional problems) and not from the current one. The goal of r is to take note of this aspect. In this regard, as stated before, the variable t does not necessarily indicate time, but it describes the evolution of the loading process.

Accordingly, the variable $r(t)$ indicates either the value of strain ε_d at which damage starts or the maximum value in the previous history of the material up to a given load step t (the variable τ is a dummy one). As noticeable, $r(t)$ can either increase or maintain the same value. Synthetically, its rate is $\dot{r}(t) \geq 0$. The consequent computation of the damage variable is shown in Eq. (3.72). In the following, the dependences are dropped when evident from the context.

$$r(t) = \max\left(\varepsilon_d, \max_{\tau \le t} \varepsilon(\tau)\right)$$

$$\dot{r}(t) \ge 0 \tag{3.71}$$

$$\omega(r) = \begin{cases} 0 & \text{if } r \le \varepsilon_d \\ 0 < \omega \le 1 & \text{if } r > \varepsilon_d \end{cases} \tag{3.72}$$

With the aim of defining the constitutive law for pure damage, let us introduce the *damage function* $f(\varepsilon, r)$ and the uniaxial loading–unloading conditions, Eqs. (3.73) and (3.74), respectively. The reader should be able to recognize the similarity with elastoplasticity (discussed in Section 3.2) and Eq. (3.33). In this case, $f = 0$ represents the boundary of the elastic domain in the strain space. Eq. (3.74) represents the equivalent of the consistency condition shown for elastoplasticity in Eq. (3.33). This means r can increase ($\dot{r} > 0$) only when $f = 0$ (thus, $\varepsilon = r$) and $\dot{f} > 0$. For the sake of completeness, the σ–ε relationship is recalled in Eq. (3.75) by means of the secant modulus E_{sec}.

$$f(\varepsilon, r) = \varepsilon - r \tag{3.73}$$

$$f \le 0 \qquad \dot{r} \ge 0$$

$$\dot{r}f = 0 \qquad \dot{r}\dot{f} = 0 \tag{3.74}$$

$$\sigma = (1-\omega)\sigma^* = \underbrace{(1-\omega)E^*}_{E_{\text{sec}}}\varepsilon = E_{\text{sec}}\varepsilon \tag{3.75}$$

The previous equations provide the necessary information to describe the damage evolution and the consequent macroscopic constitutive law for uniaxial loading–unloading. Before extending these outcomes to three-dimensional cases, it is worth discussing the new formulas according to the example of the previous subsection. In Figure 3.33a, $f = 0$ is achieved when the strain is equal to ε_d, and it represents the threshold of the current elastic domain. No matter loading or unloading process, in this elastic phase, stresses and strains are univocally determined according to the bold line. Subsequently, ε moves to ε_2 and to ε_3 and, by updating r, they represent the new elastic limit (where $f = 0$).

The reader should be able to notice a certain analogy with Figure 3.10b. More in detail, dealing with stress, hardening describes the evolution of the elastic limit stress. Damage, instead, describes the evolution of the

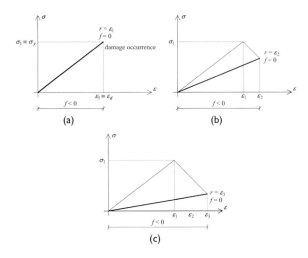

Figure 3.33 Evolution of the elastic domain for uniaxial loading and damage. (a) Undamaged case and (b and c) subsequent increments of damage.

elastic limit strain, and the slope of the bold lines in Figure 3.33b and c. Accordingly, in analogy with yield surfaces, *damage surfaces* are the three-dimensional representation of the elastic domain in the space of the principal strains.

The simplest assumption to address three-dimensional problems is by means of an *isotropic damage model*. In this case, the term isotropic has a twofold meaning. As expected, the damage variable is independent from the direction of the external loads. Additionally, damage is assumed to affect all the orientations simultaneously, i.e. the originally isotropic material keeps being isotropic. Looking at Figure 3.29, the inner defects are not ellipses but circles: the reduction of the cross-section area does not depend on the orientation of the cutting plane considered. Intuitively, this approach does not match the reality exactly. As already underlined when introducing hardening at the beginning of Section 3.2, the occurrence of anisotropy is rather evident.

However, isotropic damage is often adopted in engineering applications thanks to its simplicity and computational efficiency. First and foremost, despite the six components of strain in three-dimensional problems, the damage state is completely defined by a single scalar ω, as shown in Eq. (3.76). The similarity with one-dimensional problems is clear if compared with Eq. (3.75).

$$\sigma = (1-\omega)\sigma^* = \underbrace{(1-\omega)\mathbf{D}\varepsilon}_{\mathbf{D}_{sec}} = \mathbf{D}_{sec}\varepsilon \qquad (3.76)$$

If the computation of ω was easy and evident for the previous example, this statement does not hold for three-dimensional problems. With the aim of

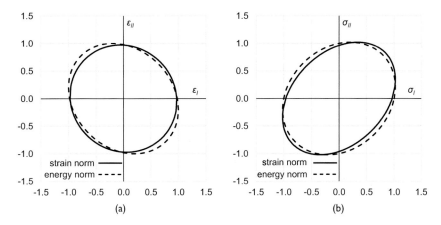

Figure 3.34 Damage surface considering unitary ε_{eq} as strain and energy norm. Graphical representation for the two-dimensional case in the (a) principal strain and (b) principal stress planes.

maintaining the same formalism, let us introduce the equivalent strain $\varepsilon_{eq}(\boldsymbol{\varepsilon})$ as a function of the six components of strain. Basically, the strain state of a complex three-dimensional problem is synthetically indicated by a scalar. Consequently, Eqs. (3.71) and (3.73) are updated with Eqs. (3.77) and (3.78), respectively. Several formulations have been proposed in literature to calculate the equivalent strain. In the following, only a few are introduced limiting the discussion to the two-dimensional case.

$$r(t) = \max\left(\varepsilon_d, \max_{\tau \leq t} \varepsilon_{eq}(\tau)\right) \tag{3.77}$$

$$f(\varepsilon_{eq}, r) = \varepsilon_{eq} - r \tag{3.78}$$

The simplest choice is to define the equivalent strain as the Euclidean norm of the strain vector or as the energy norm, Eqs. (3.79) and (3.80), respectively. The former is also expanded in terms of principal strains. In the plane of the principal strains $\varepsilon_I - \varepsilon_{II}$, the given equations lead to the graphs of Figure 3.34a. It is worth stressing that the curves are based on plane stress with $E = 210,000\,\mathrm{MPa}$ and $v = 0.2$ with $\varepsilon_{eq} = 1$. Additionally, although plane stress, $\varepsilon_{III} \neq 0$ and must be computed, as explicitly stated in Eq. (3.79). This component of strain can be derived according to the linear elastic constitutive law, as a function of ε_I and ε_{II}. The consequent domain in the plane of principle stress is shown in Figure 3.34b, derived as $\boldsymbol{\sigma} = \mathbf{D}\boldsymbol{\varepsilon}$.

$$\varepsilon_{eq} = \sqrt{\boldsymbol{\varepsilon}^T \boldsymbol{\varepsilon}} \qquad \rightarrow \qquad \varepsilon_{eq} = \sqrt{\varepsilon_I^2 + \varepsilon_{II}^2 + \varepsilon_{III}^2} \tag{3.79}$$

$$\varepsilon_{eq} = \sqrt{\frac{\boldsymbol{\varepsilon}^T \mathbf{D}\boldsymbol{\varepsilon}}{E}} \tag{3.80}$$

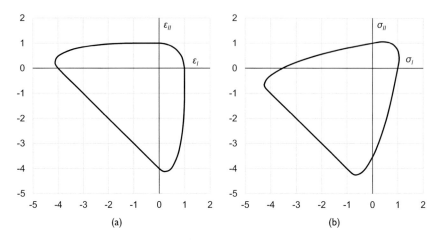

Figure 3.35 Damage and yield surfaces considering unitary ε_{eq} as positive strain norm. Graphical representation for the two-dimensional case in the (a) principal strain and (b) principal stress planes.

The norms described above lead to symmetric damage and response, i.e. the elastic limits are the same (in absolute terms) for tension and compression. However, several materials (such as masonry and concrete) often show a non-symmetric damage surface with a different behaviour in tension and compression. With the aim of getting a more accurate model, the definition of the equivalent strain should be updated. In general, as microcracks mainly appear when materials are stretched, it seems rather intuitive to consider only the positive (tensile) strains, neglecting the negative ones (compressive), when considering tensile failure. Reminding the meaning of the Macaulay brackets "$\langle\rangle$" (also known as positive part operator) in Eq. (3.81), Eqs. (3.79) and (3.80) are updated in Eqs. (3.82) and (3.83), defined as positive strain norm and positive energy norm, respectively. The former is illustrated in Figure 3.35.

$$\langle x \rangle = \begin{cases} x & \text{if } x > 0 \\ 0 & \text{if } x \leq 0 \end{cases} \tag{3.81}$$

$$\varepsilon_{eq} = \sqrt{\langle \varepsilon \rangle^T \langle \varepsilon \rangle} \tag{3.82}$$

$$\varepsilon_{eq} = \sqrt{\frac{\langle \varepsilon \rangle^T \mathbf{D} \langle \varepsilon \rangle}{E}} \tag{3.83}$$

The graphical representations of damage in the principal strain space allow also a clear understanding of the effects of damage evolution. Figure 3.36 shows how the damage surfaces become larger and larger according to

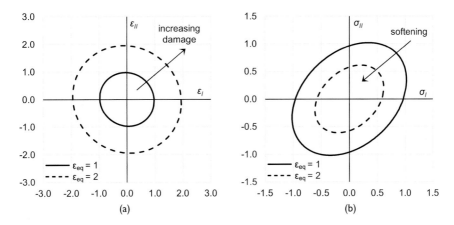

Figure 3.36 Graphical representation of damage evolution for the two-dimensional case according to equivalent strain norm. (a) Principal strain and (b) principal stress space.

the level of damage. The reader may recognize the strong analogy with Figure 3.23a. In turn, as damage is generally associated with softening, the surface defining the stress elastic domain gets smaller, while damage and elastic limit strain increase.

An important advantage of the isotropic damage model is that the stress evaluation algorithm is usually explicit and there is no need to use an iterative solution for nonlinear equations. Thus, given ε^{n-1}, r^{n-1}, and a certain increment of strain $\Delta\varepsilon^n$, the expressions in Eq. (3.84) hold. From the strain ε^n it is possible to calculate the equivalent strain ε_{eq}^n, thus the state variable r^n and the damage variable ω^n (for which a certain law must be set, in the fashion of Eq. (3.70)). Accordingly, the effective stress tensor $\boldsymbol{\sigma}^{*,n}$ is adjusted to provide the current macroscopic stress $\boldsymbol{\sigma}^n$.

$$\varepsilon^n = \varepsilon^{n-1} + \Delta\varepsilon^n$$

$$\varepsilon_{eq}^n = \varepsilon_{eq}^n\left(\varepsilon^n\right)$$

$$r^n = \begin{cases} r^{n-1} & \text{if } \varepsilon_{eq}^n \leq r^{n-1} \\ \varepsilon_{eq}^n & \text{if } \varepsilon_{eq}^n > r^{n-1} \end{cases}$$

(3.84)

$$\omega^n = \omega^n\left(r^n\right)$$

$$\boldsymbol{\sigma}^{*,n} = \mathbf{D}\varepsilon^n$$

$$\boldsymbol{\sigma}^n = \left(1-\omega^n\right)\boldsymbol{\sigma}^{*,n}$$

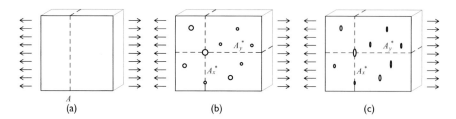

Figure 3.37 Schematic damage evolution under uniaxial loading. (a) Undamaged configuration; (b) isotropic and (c) anisotropic damages.

Thanks to the orientation independence, a unique scalar (damage) variable was sufficient to describe the deterioration of the material properties in three-dimensional problems. With the aim of introducing the *anisotropic damage model*, let us consider the simple scheme of Figure 3.37 (based on Figure 3.29). Figure 3.37b shows the case of isotropic damage, where the material properties are degraded along every direction even under a uniaxial loading. In this case, the inner microcracks and flaws are pictured with circles. In turn, for anisotropic damage (Figure 3.37c), the material behaves differently according to the load direction (or possible preferential orientations). In this case, damage is pictured by ellipses, showing the different effects on the effective areas A_x^* and A_y^*. As evident, a single damage variable is not enough.

In the most general scenario, it appears natural to assign a damage variable to each stress component, thus, replacing the scalar ω with the tensor $\boldsymbol{\Omega}$, and introducing $\mathbf{I}$ as the identity matrix. The new relationship is shown in Eq. (3.85) and the similarity with Eq. (3.76) is evident. It must be stressed that the relationship in Eq. (3.85) may lead to nonsymmetric secant stiffness matrix $\mathbf{D}_{sec}$ (inadmissible from the thermodynamic point of view) and mathematical manipulations are often implemented to avoid this drawback. This aspect is not covered in the present text.

$$\boldsymbol{\sigma} = (\mathbf{I} - \boldsymbol{\Omega})\boldsymbol{\sigma}^* = \underbrace{(\mathbf{I} - \boldsymbol{\Omega})\mathbf{D}}_{\mathbf{D}_{sec}}\boldsymbol{\varepsilon} = \mathbf{D}_{sec}\boldsymbol{\varepsilon} \tag{3.85}$$

3.3.3 Damage deactivation

As mentioned in the introduction to the section, damage has been discussed considering only the material behaviour under tensile actions. This approach was motivated with the more intuitive understanding for structural materials such as concrete or masonry. However, the different behaviour in tension and compression was already highlighted by considering the positive energy norm. It is of interest now to understand what happens if the material experiences different stress states that include tension and compression.

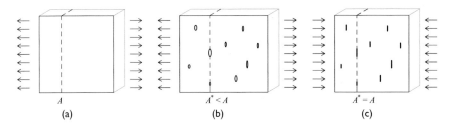

Figure 3.38 Schematic damage evolution for the continuum under uniaxial loading. (a) Undamaged configuration with cross-section area A; (b) microcracks opening under tensile actions (reduced cross-section area A^*) and (c) microcracks closing under compression (cross-section area $A^* = A$).

Considering the example of Figure 3.29, let us imagine that the specimen undergoes damage in tension and is subsequently loaded in compression (Figure 3.38). According to the previous subsection, the damage of Figure 3.38b (in terms of microcracks and voids) is an irreversible process of deterioration with loss of stiffness and strength. However, if the external load is reversed, the microcracks close and the two crack faces can interact again (Figure 3.38c). This is an important outcome because, although the "pristine" continuity is lost, this new contact is sufficient to transmit compressive stresses (any tensile action would make the crack reopen). On closer inspection, the effective stress coincides with the macroscopic one, i.e. the entire cross-section is reactive ($A^* = A$).

Trying to summarize the phenomenon described in Figure 3.38, when microcracks close (i.e. null strain) the damage produced by tensile actions should be ignored. In other words, microcracks behave as unilateral contact constraint allowing separation but preventing interpenetration. Mathematically, this aspect can be studied according to Eq. (3.86), where the Heaviside step function H is detailed. In particular, H is a discontinuous function whose value is zero for negative arguments and one for positive arguments, thus, a sort of switch for the damage variable. It is worth mentioning that, in the case of new tensile stress state, the damage variable recovers the last values reached before the inversion of loading.

$$\sigma = \left[1 - \omega H(\varepsilon)\right]\sigma^*$$

$$H(\varepsilon) = \begin{cases} 0 & \text{if } \varepsilon < 0 \\ 1 & \text{if } \varepsilon \geq 0 \end{cases} \quad \begin{array}{l} \rightarrow \quad \sigma = \left[1 - 0\right]\sigma^* \\ \rightarrow \quad \sigma = \left[1 - \omega\right]\sigma^* \end{array} \qquad (3.86)$$

Whereas the simple Heaviside function works accurately for uniaxial loading, this is not the case for multiaxial models. A more complex approach must be based on the overall stress state, whose dissertation falls out of the scope of the present text.

3.4 SMEARED CRACK MODELS

3.4.1 Main concept and strain decomposition

Intuitively, a crack is a geometrical discontinuity that effectively separates a structural element into two distinct solids (Figure 3.39a), or surfaces in the case of plane elements. This displacement discontinuity does not fit the nature of the finite element displacement method (if standard continuum finite elements are used). A discrete crack would request that a node, originally shared between two (or more elements), "splits" and becomes two independent nodes (Figure 3.39b). As shown in Chapter 1, this aspect may be tackled by interface elements, which are defined by originally overlapped nodes that can move (independently) according to a defined interface constitutive law. These elements must be localized a priori, thus, constraining the crack propagation to follow predetermined paths. Other strategies to address this are the automatic update of the mesh at each iteration (*remeshing* as crack propagates, with some computational effort), the use of special finite elements with embedded discontinuities or other approaches, which will not be addressed here.

Alternatively, despite some loss of realism, cracks can be easily described with FEM if idealized as *smeared* (Figure 3.39c). Basically, rather than computing the real irregular crack, FEM evaluates the average effects of cracks on the structural behaviour. The reader should not be misled by the term smeared. In a wider sense, if cracks are associated to the weakening of the material properties, any softening continuum model (such as plasticity or damage discussed earlier) could be understood as "smeared". Like plasticity and damage, also smeared crack models lead the "cracked" element (although no real fracture is visible) to large values of strains (thus displacements) consistent with a real physical crack (Figure 3.39c). However,

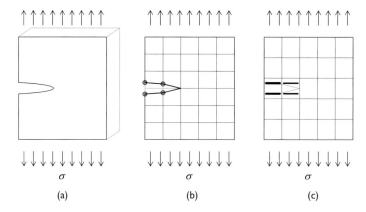

Figure 3.39 Computational approach to crack analysis. (a) Real specimen; (b) interface element and discrete crack and (c) schematic approach of the smeared crack model.

if compared with plasticity, smeared crack model does not postulate a yield condition and a flow rule, but the crack opening is directly related to the traction transmitted across the crack plane. In turn, damage models should be anisotropic to describe the oriented nature of cracking.

In this regard, smeared crack models appear particularly suited for quasi-brittle materials that exhibit densely distributed cracks, such as large-scale reinforced concrete structures (at least if the scale of the representative continuum is large compared to the crack spacing). For the sake of simplicity, the term "crack" is here used to mean a direction in which cracking is detected at the single integration point: the closest physical meaning is that there exists a continuum of microcracks around that point, oriented as determined by the model. From the conceptual point of view, the model is based on a two-step procedure. First, a criterion must be enforced to detect the crack initiation. This can be based on the maximum principal stress or any other failure criterion, such as the ones seen in Section 3.2.2. Second, the mechanical properties of the material are lowered along the directions defined by the crack plane. In other words, if the model deals with isotropic materials, cracks change it to orthotropic.

In particular, once the crack is individuated (Figure 3.40a), it is necessary to describe how the crack direction evolves, i.e. how the crack propagation affects the structural behaviour of finite elements. In this regard, two main concepts are usually introduced, namely, *fixed* and *rotating smeared crack models*. The former assumes a fixed orientation of the crack during the entire computational process (Figure 3.40b), and possible change of load direction activates shear Mode II (in-plane, as shown in the figure) and Mode III (in the perpendicular out-of-plane direction). The permanent memory of crack orientation is the prominent feature of the fixed smeared crack concept, which coincides with its main drawback. The idea of preserving the original crack direction may represent an intrinsic constraint for the material, possibly resulting in overstiff and overstrength responses around and

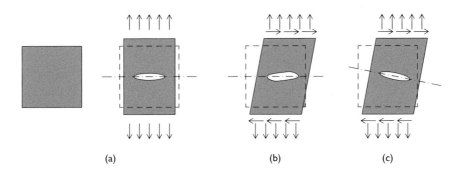

Figure 3.40 Plane element subjected to tensile and subsequent shear loading. (a) Crack initiation; (b) fixed and (c) rotating smeared crack schematization.

after peak, in the case of unreinforced materials. This is essentially due to the misalignment of principal directions and crack directions. In turn, rotating smeared crack models (Figure 3.40c) allow the orientation of the crack to co-rotate with the axes of principal strain. The rotating crack concept does not preserve permanent memory of the damage orientation. This implies that inactive defects cannot be re-activated during a subsequent stage of the loading process. In general, the choice of the model to use has consequences when the loading process is nonproportional or in the postpeak response.

Smeared crack models usually share with plasticity the concept of strain decomposition, according to which the total strain is here decomposed in elastic and crack parts, $\varepsilon = \varepsilon_e + \varepsilon_{cr}$, respectively: the similarity with the first of Eq. (3.28) is evident. A generic stress–strain diagram with softening under tension can be decoupled according to Figure 3.41. When the material behaves elastically, $\varepsilon_{cr} = 0$ and only the first diagram holds, where $\sigma = E\varepsilon_e$. When the total strain exceeds the elastic threshold, $\varepsilon_{cr} > 0$ and cracking occurs. Conceptually, the decomposition of Figure 3.41 is the same of Figure 3.24. Also, the area under the crack stress–strain diagram is equal to the one of the total stress–strain diagram (because in correspondence of the ultimate strain, the elastic energy per unit volume is null).

In analogy with damage, cracking is an irreversible process. In the case of unloading, the term "inactive" should be preferred to "closed" because cracking re-activates as soon as the stress conditions are suitable. Figure 3.42 shows the two extreme cases of unloading–reloading, namely, elastic and secant. The former assumes that the crack becomes inactive immediately upon a strain reversal and the latter assumes that the crack becomes inactive when the normal strain goes back to a null value. Elastic unloading is computationally appealing because the strain is totally elastic and strain

Figure 3.41 Strain decomposition between elastic and crack state for Mode I elastic-softening model.

Figure 3.42 Strain decomposition between elastic and crack state. Secant and elastic unloading–reloading in Mode I.

decomposition is not necessary (the crack strain is null). In turn, secant unloading accounts for a stiffness reduction of the cracked material, even though the crack is not active. Additionally, it is worth reminding that, according to a fixed crack model, crack orientation is fixed also during a possible unloading–reloading process. This means that possible re- and de-activations occur along the same direction.

Certainly, the presented approaches are approximations of the reality. Judgement should guide the user towards the correct choice, also with reference to shear effects. Upon unloading–reloading, a crack can be active in Mode I and inactive in Mode II, inactive in Mode I and active in Mode II, active or inactive in both modes. This topic also falls out of the scope of the book.

3.4.2 Implementation of the fixed and rotating smeared crack model

In fixed smeared crack models, Mode II and III inelastic responses are generally assumed to start after Mode I crack is detected (e.g. when tensile stress reaches the tensile strength f_t). Mode I response, already introduced in Figure 3.4, is presented again in Figure 3.43a for clarity. The figure shows the relationship between the stress and the measure of the longitudinal displacement u. As already mentioned in the description of Figure 3.4, strain becomes meaningless in the case of a discrete, localized, crack. The reason lays on the fact that the strain measure dramatically changes if measured in the immediate neighbourhood of the crack or including some elastic material. However, dealing with FEM, a σ–ε relationship is more appealing. To avoid mesh sensitivity (i.e. to avoid measures that depend on the element size) it is possible to address the problem from the energetic point of view. Considering Figure 3.43b, the area under the σ–ε curve represents the energy per unit volume g_f. To guarantee that a single element does not dissipate more energy than its size can allow, g_f is set equal to the ratio between the fracture energy G_f^I and the average size of the element. As a geometrical

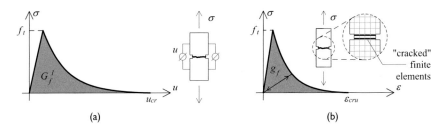

(a)   (b)

Figure 3.43 Mode I description according to total strain crack model. (a) Idealized σ–u diagram (with u total displacement) and (b) σ–ε diagram and secant stiffness (providing g_f is equal to G_f^I divided by the average element size h).

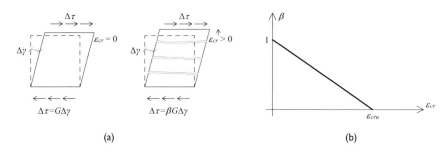

Figure 3.44 Mode II and shear retention factor β. (a) Schematization of shear stiffness reduction and (b) linear variation of β according to tensile strain after crack.

quantity is involved, the σ–ε relationship is evidently *mesh dependent*, and so is the ultimate cracked strain ε_{cru}. This aspect will be discussed further in Section 3.5.

Based on observation, when a certain crack is activated, the material is still able to resist to actions parallel to the crack faces. Indicatively, regardless of the horizontal crack, the specimen sketched in Figure 3.43 would still perform soundly to horizontal tension. This phenomenon can be pictured with another uniaxial stress–strain relationship along the horizontal direction: a second crack (perpendicular to the first) may occur only when the tensile stress reaches the tensile strength in this direction. This concept can be further refined by allowing more than two cracks per integration points (e.g. at every 60°, thus, nonorthogonal cracks), leading to the so-called *fixed multi-directional crack concept*. Accordingly, besides the threshold represented by the tensile strength, a new crack initiates also when the angle of inclination between the existing crack(s) and the current direction of principal stress exceeds a certain threshold angle.

Moving to shear (Mode II), a phenomenological approach is usually adopted, as schematized in Figure 3.44a. Basically, cracks deteriorate the continuity of the material, affecting the structural response to shear. In other words, the larger is the crack, the lesser is the contact between crack faces (and possible interlocking) to transmit shear. This concept can be modelled by reducing the shear modulus and setting it to a proportion of the elastic value, as $G_{cr} = \beta G$, where G is the elastic shear stiffness and β the *shear retention factor*. Basically, reducing the shear stiffness leads to small resistance and large deformations, in line with the smeared crack model approach. Different strategies are available in literature for evaluation of β. A variable shear retention model depending linearly on the tensile strain after crack ε_{cr} is detailed in Figure 3.44b, where $0 \leq \beta \leq 1$, simulating the progressive loss of shear stiffness when crack opens. In particular, when $\varepsilon_{cr} = 0$, crack width is null and $G_{cr} = G$; in turn, when $\varepsilon_{cr} = \varepsilon_{cru}$, Mode I crack is fully open and the contact between the two crack faces is lost, together with the possibility of transmitting shear actions (i.e. $G_{cr} = 0$). It is noted that this approach may introduce some mesh dependency in the response

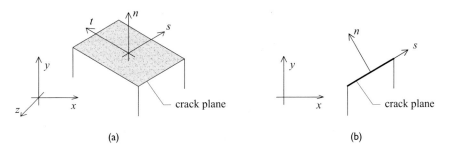

Figure 3.45 Global and local coordinate systems (according to the crack plane) for (a) three- and (b) two-dimensional configurations.

as the notion of strain is lost in presence of a discontinuity. However, it is still acceptable for reinforced concrete structures in which crack width is limited by the reinforcement. Other relations, such as exponential, between β and ε_{cr} are also feasible. Alternatively, a variable shear retention model can be used in which the shear stiffness gradually reduces to zero for a crack width of a certain length (e.g. half the average aggregate size for concrete). Constant values of β are also discussed in literature. However, constant shear retention models are not advisable for unreinforced structures, or should at least be accompanied with thorough postanalysis checks of spurious principal tensile stresses.

Figure 3.45a shows the general case of a crack plane for three-dimensional configurations, further detailed in Figure 3.45b for the two-dimensional one. Limiting the discussion to the latter, according to above and mathematical manipulation, Eq. (3.87) illustrates the incremental relation between stress and total strain for plane stress problems. As Eq. (3.87) does not make distinction between elastic strain and strain due to cracking, the procedure is usually referred to as *total strain crack model*. Besides the known symbols, μ is the *reduction factor for the Mode I stiffness* and is related to the stress–strain relationship, such as in Figures 3.41 and 3.43b. In the case of softening, μ turns negative, i.e. a positive increment of strain leads to a negative increment of stress. Additionally, as Eq. (3.87) represents a linearized form (by means of tangent stiffness), it assumes the constitutive relations calculated at previous step as constant during the current strain increment. An iterative approach is thus needed to refine the solution.

$$
\begin{bmatrix} \Delta\sigma_n \\ \Delta\sigma_s \\ \Delta\tau_{ns} \end{bmatrix} = \begin{bmatrix} \dfrac{\mu E}{1-v^2\mu} & \dfrac{v\mu E}{1-v^2\mu} & 0 \\[2mm] \dfrac{v\mu E}{1-v^2\mu} & \dfrac{E}{1-v^2\mu} & 0 \\[2mm] 0 & 0 & \beta G \end{bmatrix} \begin{bmatrix} \Delta\varepsilon_n \\ \Delta\varepsilon_s \\ \Delta\gamma_{ns} \end{bmatrix} \qquad (3.87)
$$

Moving to the rotating crack model, instead of assuming stress–strain relationships along certain directions defined by the initiation of cracking, the response is always evaluated along the principal directions of the strain vector (i.e. *coaxial stress–strain concept*), continuously updated throughout the analysis. As the reference system is updated continuously, there is no need to explicitly model shear Mode II and III: for every stress state, the final effect is a Mode I crack whose direction is not fixed and *follows* the structural behaviour. This model shares with the fixed crack concept the same mathematical framework and is recommended in the case of unreinforced masonry and concrete structures.

More in detail, to guarantee the co-axiality between principal stress and strain, the shear modulus G must satisfy the relation in Eq. (3.88). In the formula, 1 and 2 represent the principal directions of stress, aligned with principal directions of strain and the crack reference system. The reader should be not misled by the use of shear modulus when dealing with principal directions. The conditions in Eq. (3.88) guarantee the co-axiality after a small increment of shear and strain. In other words, if the tangent stiffness calculated at the previous step (in the previous reference system) includes Eq. (3.88), then the current step will maintain the co-axiality between principal stress and strain. That explains also why the principal directions are numbered with Arabic and not with Roman numbers.

$$G_{12} = \frac{\sigma_1 - \sigma_2}{2(\varepsilon_1 - \varepsilon_2)} \tag{3.88}$$

3.5 STRAIN LOCALIZATION AND MESH SENSITIVITY

Within a loaded structure, *strain localization* refers to the development of narrow bands where material degradation and deformation concentrates, while the areas outside these bands tend to unload elastically. This phenomenon is expected when softening is considered, unless some embedded reinforcement is active, whose ductile behaviour mitigates the problem (at least before its own yielding) and dominates the global behaviour. Examples of concrete structures where localization effects can have a significant influence on behaviour include punching shear, shear in members without transverse reinforcement, over-reinforced columns and beams, and large unreinforced concrete structures such as dams. In each of these cases the behaviour at failure is more controlled by that of plain concrete than that of the reinforcement. The same holds for (unreinforced) masonry structures.

In structures made of quasi-brittle materials, strain localization bands are expected to coincide with cracks (or crushing phenomena), which are generally responsible of local or global collapse. In these cases, a careful definition of the constitutive stress–strain relationships is needed, as it should account for the average size of the element (as already mentioned in Section 3.4.2 and Figure 3.43). The mesh size itself, in fact, introduces a fixed length scale that can bias the results of localization sensitive problems. Consequently, *lack of objectivity* will arise, according to which, upon mesh refinement, the results may not converge to a meaningful solution (conversely to the general FEM concept). Mesh refinement will lead to narrower crack bands but lower energy dissipation. For this reason, the use of terms such as *mesh sensitivity* or *mesh dependency* is popular.

To have a physical understanding, let us consider the simple bar in Figure 3.46a discretized by means of m equal rectangular elements and subjected to axial tensile load. The constitutive law is elastic with linear softening. According to equilibrium conditions, the stress must be uniform throughout the element and equal to F/A. Due to numerical approximation, it may happen that, in proximity of the peak load, a certain element jumps on the softening branch, whereas the rest of the structure remains on the elastic one, particularly if explicit solution methods are used to find the global solution. In the case of using the Regular Newton–Raphson solution method, a bifurcation is found with m possible solutions, i.e. as much as the number of elements, and divergence is to be expected, as no decision can be made on which element will undergo failure.

For the problem at hand, a possibility to enforce localization of deformation in a single finite element is to introduce a defect, such as a slightly lower strength (or slightly smaller width). When the specimen is subjected to tensile force, all the elements behave the same until the weaker element (the shaded element of Figure 3.46a) reaches its maximum strength; for increasing strain, this element moves on the softening branch (Figure 3.46b) while the rest of the structure unloads to guarantee equilibrium. Accordingly, the generic stresses marked with a circle in both

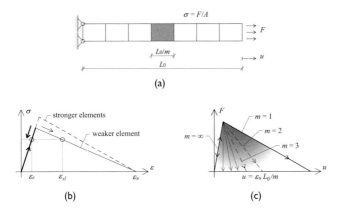

Figure 3.46 Schematic representation of strain localization. (a) Bar subjected to tensile loading with weaker (shaded) element; (b) stress–strain diagrams for weaker and stronger elements and (c) overall response of the bar in terms of force–displacement.

diagrams correspond to ε_{sl} and ε_e for the weaker and stronger elements, respectively. In other words, the stress is the same throughout the beam, but only the element that previously reached the maximum strength faces softening. Consequently, the inelastic deformation occurs solely in one element that results inelastic, while the rest of the structure unloads elastically (thus, the name strain localization).

Proceeding with the analysis, the external force ultimately goes to zero as the weaker element follows the softening branch up to the ultimate strain ε_u, while the rest of the structure displays null strain (Figure 3.46b). Integration of strains throughout the bar furnishes its elongation, calculated according to Eq. (3.89). At ultimate stage, the calculation can be limited to the damaged element only, whose length is L_0/m. A closer look to the Eq. (3.89) shows that the bar elongation depends on the number of elements m in which the bar is subdivided. This means the solution is mesh-sensitive, i.e. the same load process leads to (dramatically) different solutions.

$$u = \varepsilon L_0 = \varepsilon_u \frac{L_0}{m} \tag{3.89}$$

Figure 3.46c shows the overall response of the bar in terms of force–longitudinal displacement ($F-u$) for different discretizations. If $m = 1$, the entire bar is damaged and the ultimate displacement is equal to $\varepsilon_u L_0$; in turn, $m = 2$ leads to a value half as large as the previous one, and so on. At this point, it is interesting to notice that a large number of subdivisions ($m \rightarrow \infty$) means the damage is localized in a small element, with no energy being dissipated, which is physically unrealistic.

Despite the easiness of the uniaxial case, strain localization occurs in multiaxial cases too. More advanced models may be used to solve this issue, but this is outside the scope of this text. In practice, a technique to avoid mesh size sensitivity is to make the stress–strain relation a function of the element size. The simplest approximation is to consider the strain uniformly distributed within the element so that the displacement u can be calculated as $u \approx \varepsilon h$, where h represents the average size of the finite element. Accordingly, the fracture energy can be computed with Eq. (3.90), see also Eq. (3.1). Essentially, each finite element can store only the energy per unit volume g_f, which is the ratio between the fracture energy G_f and the average size of the element h. Along similar lines, g_c is the volumetric energy in compression.

$$G_f = \int_0^{u_{cr}} \sigma_{(u)} du = \int_0^{\varepsilon_u} \sigma_{(\varepsilon)} h \, d\varepsilon = h \underbrace{\int_0^{\varepsilon_u} \sigma_{(\varepsilon)} d\varepsilon}_{g_f} = g_f h \quad \rightarrow \quad g_f = \frac{G_f}{h} \tag{3.90}$$

3.6 EXAMPLES OF CONSTITUTIVE MODELS IN COMMERCIAL SOFTWARE CODES

The present section provides a brief description of typical material libraries offered by commercial software codes. Far from being exhaustive, it is recommended for the analyst to always check the user manual to be aware of the actual potentialities of the software code or of the most up-to-date features. Additionally, given the greater easiness in dealing with ductile materials, the discussion mainly focuses on quasi-brittle (i.e. concrete-like) materials. In this regard, constitutive models intended for concrete or masonry may be interchangeable, at least with some degree of simplification. As far as steel reinforcement is considered, the use of embedded elements is suggested, adopting standard metal plasticity models.

3.6.1 Smeared crack model (in tension) and plasticity in compression

This approach is intended for applications in which the material is essentially subjected to monotonic loading under fairly low confining pressures. In this case, cracking is assumed to be the most crucial aspect of the material behaviour, which controls the structural response. Being a smeared crack approach, the presence of cracks enters the calculation by modifying the constitutive relationships in correspondence of the integration points, introducing, consequently, anisotropy. In particular, cracking is assumed when the stresses reach a limiting condition: in this case the definition *crack detection surface* results rather appropriate. With this aim, a Rankine-type criterion is often employed, sometimes referred to as constant and linear tension cut-off. These are shown in Figure 3.47, where f_t and f_c are the tensile and compressive strengths, respectively.

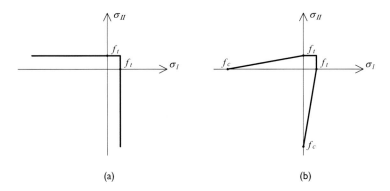

Figure 3.47 Typical crack detection surfaces in smeared crack models. (a) Rankine (constant) criterion and (b) Rankine combined with a linear criterion.

Once the crack is detected (e.g. when the principal stress reaches the tensile strength or crack detection surface), the softening behaviour of the material can be described by means of uniaxial stress–strain relationships. The most used ones are shown in Figure 3.48 where nonlinear diagrams are also depicted. In the case concrete is considered, it is worth noticing that the tensile behaviour is independent from the likely presence of rebar, whose effect on concrete may be assessed by means of tension stiffening (see Chapter 4). In the case of load removal, it is assumed that there is no residual strain, i.e. crack closure occurs following the secant stiffness and null residual strain is found upon load removal (see Figure 3.42).

Looking closely at Figure 3.48, the postpeak relationship can be defined by means of fracture energy G_f^1 and the finite element dimension h, or by explicitly stating the ultimate cracked strain ε_{cru}. As explained in Section 3.5, the second option disregards the finite element dimensions, leading to a loss of objectivity with regard to mesh refinement. For this reason, it should be used judiciously in unreinforced concrete or masonry structures.

After describing the tension cut-off criterion and tension softening, shear behaviour must also be defined. In the fixed crack model, this is accounted for by means of the shear retention factor β, which defines the evolution of the shear modulus according to the crack strain or opening. In the case of rotating crack model, there is no need to define the shear retention factor.

In turn, when the principal stress components are dominantly compressive, the material response is defined according to an elastoplastic theory (e.g. Mohr–Coulomb, Drucker–Prager, Willam–Warnke or Menetrey–Willam yield functions), with associated or nonassociated flow, and isotropic hardening. In this case, the evolution of the mechanical parameters in the postelastic phase must be defined in terms of a hardening parameter (see Figure 3.24). Examples are depicted in Figure 3.49. Regarding concrete-like materials, one of the main drawbacks of this approach is that the elastic stiffness in the unloading response is too stiff when compared with experimental observation (characterized by a combination of irreversible damage and irreversible deformation). However,

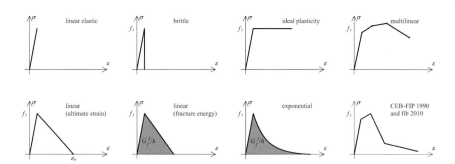

Figure 3.48 Examples of tension-softening relationships.

the consequent algorithm is rather robust, and the model provides useful predictions for a variety of problems involving inelastic loading.

Synthetically, Figure 3.50a shows the one-dimensional material response, where compressive and tensile behaviours are highlighted. As noticeable, crushing produces irreversible (plastic) deformation while cracks are represented by loss of stiffness (and strength) with totally reversible deformation. In the space of the principal stresses, the elastic domain is delimited by two individual surfaces, as depicted in Figure 3.50b, where the most significant strengths are also indicated.

3.6.2 Plasticity models in tension and compression

As an alternative to the previous approach (where two distinct types of constitutive models are used), it is possible to use plasticity models for both tension and compression. Rankine, Rankine/von Mises and Rankine/Drucker–Prager yield functions are usually adopted (Figure 3.51, see also Figure 3.18). In all cases, the Rankine function bounds the tensile

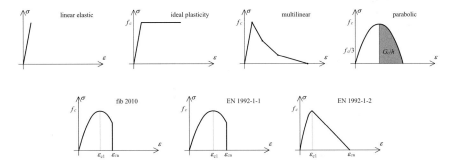

Figure 3.49 Examples of compression stress–strain relationships.

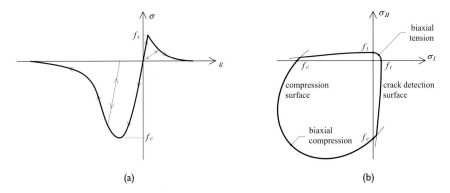

Figure 3.50 Typical constitutive relationships for smeared crack model and plasticity in compression. (a) One-dimensional material response and (b) elastic domain in the principal stress space.

stresses while either von Mises or Drucker–Prager function can be adopted in the compression region (here it is noted that von Mises is only adequate in the two-dimensional stress space, because of the lack of the hydrostatic pressure dependency). Each of these models can be combined with hardening/softening models to better predict the observed response. Similar hardening/softening models to Figures 3.48 and 3.49 are possible. Being plastic models, the unloading leads to residual deformations (i.e. plastic strains).

It is worth comparing Figures 3.50b and 3.51b and c. Despite the same description of the compressive behaviour by means of a plasticity model, the tensile behaviour follows two different approaches. In the previous section, the surface in the tension region was used as crack detection, i.e. a crack occurs when the principal stress combination reaches the surface. From this point, the material behaviour at the integration point becomes anisotropic and follows the uniaxial stress–strain relationship such as one of Figure 3.48. In turn, the Rankine surface in Figure 3.51 is used in the framework of plasticity, i.e. inelastic deformations are governed directly by the plastic flow with possible softening.

As a final remark, as both von Mises and Drucker–Prager criteria describe isotropic materials, the natural evolution of the model is to consider orthotropic plasticity. The most used approach is based on multisurface plasticity, comprising of an anisotropic Rankine-type yield function combined with an anisotropic Hill function for compression. A typical application

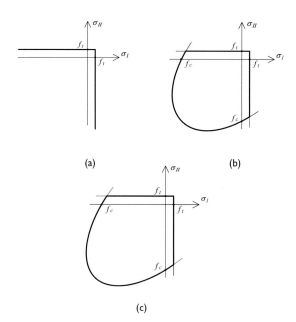

Figure 3.51 Typical yield functions for plasticity models in tension and compression. (a) Rankine, (b) Rankine/von Mises and (c) Rankine/Drucker–Prager.

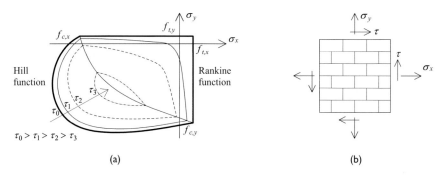

Figure 3.52 Rankine–Hill anisotropic model for masonry. (a) Multisurface elastic domain with iso-shear stress lines and (b) description of the three stress components involved in plane stress, aligned with the material axes (horizontal and vertical joints).

for this model is masonry having different strengths parallel and perpendicular to the bed joints (Figure 3.52). Other orthotropic materials, like fibre-reinforced composites, fractured rock and other layered or stacked materials can be also modelled with the Rankine–Hill model.

3.6.3 Total strain crack model

Still following the smeared approach, a total strain crack model describes both tensile and compressive behaviour with one-dimensional stress–strain relationships. Four types of cracking models are usually available in commercial software codes, including: (1) fixed crack model (constitutive relations are evaluated in a coordinate system that is fixed upon cracking), (2) rotating crack model (constitutive relations are evaluated in the principal directions of the strain vector, ensuring that the crack remains normal to the direction of the maximum principal strain), (3) rotating to fixed crack model (combination of the previous ones, rotating model switches to fixed crack direction once a critical strain is reached) and (4) nonorthogonal model (crack directions are assumed to be not orthogonal). Neglecting the two last models, in the fixed formulation, as stated earlier, a shear retention parameter is required for the definition of the shear behaviour, while, in the rotating model, shear softening occurs implicitly as a result of the principal stress and strain co-axiality. Examples of constitutive models in tension and compression have been shown in Figures 3.48 and 3.49, respectively. Regarding the compressive behaviour, these models can be enhanced by adding an increase in compressive strength due to lateral confinement or a reduction in compressive strength due to lateral cracking. The reader is referred to Chapter 4 for further details.

In a total strain crack model, the deterioration of the material due to cracking and crushing is monitored by means of damage variables, with null residual strain upon loading (i.e. secant stiffness). It is assumed that damage recovery is not possible, which implies that the absolute values of the internal damage variables are increasing. Additionally, as total strain crack models are based on secant unloading and reloading, they underestimate the energy absorption under cyclic loading. For static nonlinear cyclic or transient dynamic nonlinear, other models are better suited, e.g. damaged plasticity (whose methodology for cyclic loading is marginally addressed in Section 3.6.4).

A similar approach is performed considering a biaxial strength failure criterion. According to smeared crack formulation, after crack initiation, which is controlled by a biaxial failure envelope (Figure 3.53a), the originally isotropic material formulation changes to an orthotropic one. The unloading moduli for each direction are determined from an equivalent uniaxial diagram shown in Figure 3.53b. In particular, the nonlinear behaviour of concrete-like materials in the biaxial stress state is described by means of the so-called effective stress and the equivalent uniaxial strain. The former is in most cases a principal stress; the latter is introduced in order to eliminate the Poisson's effect in the plane stress state. Looking at Figure 3.53b, unloading is a linear function that returns to the origin (i.e. secant stiffness) and, upon reloading, the stress–strain relation follows the unloading path until the last loading point is reached.

3.6.4 Damaged plasticity

The damaged plasticity model is based on isotropic damage, and it is intended for applications in which quasi-brittle materials are subjected to arbitrary loading conditions, including cyclic and/or dynamic loading. The model takes into consideration the degradation of the elastic stiffness induced by plastic straining in tension and compression independently. However, in the case of cyclic loading and load sign reversal, stiffness recovery effects

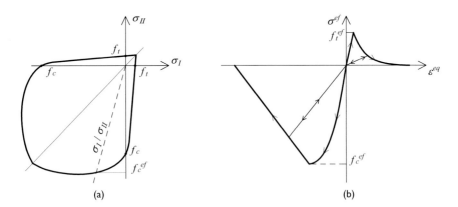

Figure 3.53 Equivalent constitutive relationship for total strain crack model. (a) Biaxial failure surface. (b) Tensile and compressive behaviour.

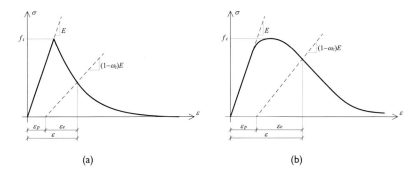

Figure 3.54 Typical constitutive relationships for damaged plasticity model. (a) Tensile and (b) compressive behaviour.

may be assumed. Being based on plasticity and damage, the model does not use the smeared crack approach of the previous models, but criteria can be adopted to visualize graphically the effective crack direction and width.

Let us consider the damaged plasticity model for the one-dimensional material response, as shown in Figure 3.54. Either in tension or compression, when the material is unloaded from any point on the softening branch, the unloading response is observed to be weakened. Basically, the elastic stiffness of the material appears to be damaged (or degraded), and this phenomenon is assumed to be different for tension and compression, i.e. two independent uniaxial damage variables govern the problem (ω_t and ω_c, respectively). However, conversely to what shown for elastic damage, plastic strain is also found (i.e. irreversible deformation upon load removal). Accordingly, the damage variables are assumed increasing functions of plastic strains, ranging from zero, for the undamaged material, to one, for the fully damaged material. Schematically, the stresses in tension and compression (σ_t and σ_c, respectively) can be evaluated according to Eq. (3.91), where the effect of the damage variables influences the elastic strains only (computed according to the strain decomposition). Like the previous two models, postcracking behaviour can be accounted for by directly specifying the stress–strain relationships or by applying the fracture energy approach.

$$\sigma_t = (1 - \omega_t) E (\varepsilon - \varepsilon_p) \quad \text{in tension}$$

$$\sigma_c = (1 - \omega_c) E (\varepsilon - \varepsilon_p) \quad \text{in compression}$$

(3.91)

The extension to the two- and three-dimensional behaviour can be made introducing functions about the elastic limit stress and hardening variables. For the two-dimensional case, the typical function is depicted in Figure 3.55a (compare with Figure 3.50b), while the projection on the

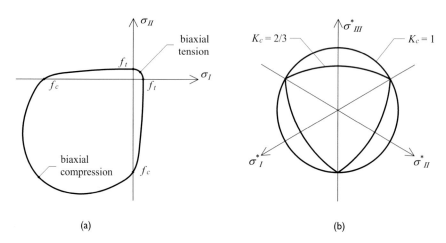

(a)   (b)

Figure 3.55 Typical elastic limit function for damaged plasticity. (a) Representation in the two-dimensional space of the principal stresses (plane stress) and (b) projection on the deviatoric plane.

deviatoric plane is shown in Figure 3.55b. The shape of the latter is governed by the parameter K_c. This is larger than 0.5 and $K_c = 2/3$ is generally recommended; when its value tends to 1, the projection of the failure surface on the deviatoric plane becomes a circle. As nonassociated plastic flow is assumed, so the plastic potential must be defined too.

Finally, although not discussed in the present text, a brief description of the damaged plasticity model with cyclic loading seems worthy. Limiting the analysis to uniaxial cyclic behaviour, the opening and closing of previously formed microcracks, as well as their interaction, represent a complex issue. However, from experimental observation, most quasi-brittle materials recover the compressive stiffness upon crack closure (as the load changes from tension to compression). On the other hand, the tensile stiffness is not recovered as the load changes from compression to tension once crushing microcracks have developed. Without entering the merit of the mathematical formulation, Figure 3.56 schematically shows the phenomenon by introducing two parameters w_c and w_t, for compressive and tensile behaviours, respectively. They range from 0 to 1, representative of null and total stiffness recovery, respectively. Figure 3.56 describes the typical case of total recover for crack closure ($w_c = 1$) and null recovery for crack reopening after crushing ($w_t = 0$). Intermediate values result in partial recovery of the stiffness.

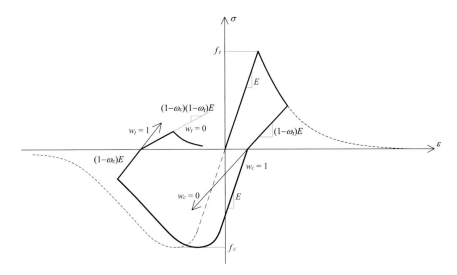

Figure 3.56 Typical uniaxial load cycle (tension-compression-tension) for damaged plasticity with $w_c = 1$ and $w_t = 0$.

BIBLIOGRAPHY AND FURTHER READING

Chen, W.-F. and Han, D.-J. (2007) *Plasticity for Structural Engineers*. Plantation, FL: J. Ross Publishing.

Jirasek, M. and Bazant, Z. P. (2001) *Inelastic Analysis of Structures*. West Sussex, UK: John Wiley & Sons Inc.

Lemaitre, J. and Chaboche, J.-L. (1994) *Mechanics of Solid Materials*. Cambridge, UK: Cambridge University Press.

Rots, J. G. and Blaauwendraad, J. (1989) 'Crack models for concrete, discrete or smeared? Fixed, multi-directional or rotating?', *HERON*, 34(1), p. 59.

Chapter 4

Recommended properties for advanced numerical analysis

INTRODUCTION

Based on the insights of Chapter 3, a numerical model should be able to incrementally describe the structural behaviour of a construction from the linear elastic stage, through cracking and crushing, until complete loss of strength or stability. Only then is it possible to control the serviceability limit state, fully understand the failure mechanism and assess safety. With this aim, material characterization should include a complete stress–strain behaviour description considering the observed experimental failure modes, stiffness degradation and energy dissipation. The goal of the present chapter is the description of the main materials encountered in existing civil engineering structures. Essential material properties are discussed according to their experimental behaviour. In this regard, it must be stressed that data are also available in codes of practice, although mainly focused on concrete and steel. Far less information is usually available for timber and masonry due to their heterogeneity and diversity of construction techniques, which makes generalization complex. The incomplete material characterization is particularly felt in the case of nonlinear analysis, which should be primarily considered for understanding the behaviour of existing structures. The recommendations presented in this chapter are not intended as a substitute to detailed material investigation and testing (in situ or on material sample, when possible), being the careful judgement of the analyst required at every level of the structural assessment. Sensitivity analyses should also be used to determine the influence of specific material parameters in the assessment outcome. This can be of support in judging the effectiveness of additional tests to refine material and structural characterization. Finally, information about safety assessment is introduced in the last section of the present chapter. Aspects such as cyclic behaviour, creep, shrinkage, moisture movement or thermal expansion are not discussed in the present text, despite their possible relevance in the evaluation of structural performance.

DOI: 10.1201/9780429341564-4   209

4.1 MASONRY

Masonry is a heterogeneous material that consists of units and joints. Typical units are clay bricks, adobes (sun dried earth blocks), ashlars (finely dressed stones), irregular stones and others. More recently, calcium silicate, aerated, lightweight and regular concrete represent common materials in the form of hollow of solid blocks (the latter defined as of larger size than bricks). Mortar for joints, used to ensure adequate bond and to accommodate dimensional tolerance of units, can be clay, lime/cement-based, polymer-based or other materials. In some cases, units are stacked without any mortar (so-called dry-joint masonry). The possible combinations generated by geometry, origin and arrangement of units, together with the intrinsic characteristics of mortars, are countless, but they all share a very low tensile strength for masonry. Despite this feature, masonry has revealed itself as a successful building material during thousands of years, justified by its simplicity and durability.

In the present section, masonry is considered as unreinforced, i.e. masonry not containing reinforcement, such as bars or a mesh embedded in the mortar joints or in a concrete infill, which acts together in resisting action effects. After a general introduction of masonry typologies and their overall behaviour, the so-called micro- and macro-modelling approaches are introduced, where the mechanical properties are discussed according to experimental outcomes. It is noted that several aspects arising during testing are not considered here, even if they may be relevant in specific applications, such as the aspect ratio of specimens (affecting the uniaxial compressive strength due to constraining effects of the testing platens), moisture content (affecting the uniaxial compressive strength differently for each material) or strain rate of testing (affecting all mechanical properties, if not quasi-static).

The discussion is aimed at describing the two-dimensional mechanical behaviour of masonry, which can satisfactorily represent plates (walls) and shells (domes, vaults, etc.), thus, sufficiently accurate for the analysis of most existing constructions. The concepts given can easily be extended for typical three-dimensional behaviour of masonry. Regarding the nomenclature, the subscripts b, m, i and M inform the reader about the interested component, namely, unit (e.g. brick or block), mortar, interface and masonry, respectively. For instance, Young's modulus and the compressive strength of unit and mortar are E_b and E_m, $f_{c,b}$ and $f_{c,m}$, respectively. To avoid an excess of notation, subscripts are limited to the cases where possible misinterpretations may arise. Finally, dealing with existing structures, if not explicitly stated, the values reported hereinafter are average values based on testing or expected average values according to building codes. In the case characteristic values are given, in the absence of more information from experimental tests, the average values may be computed according to the applicable European standard (EN 1052 series). This corresponds to conversion factors from characteristic to average values equal to 1.2 for compressive, 1.25 for shear and 1.5 for flexural strengths. Assuming a normal distribution, the related coefficients of variation are 10.2%, 12.2% and 20.3%, respectively. Section 4.1.7 collects a synthesis of the main material properties to be used in nonlinear analysis.

4.1.1 Masonry morphologies

According to the Italian code (Circolare, 2019), regardless of the level of complexity of the analysis to be performed, a careful and critical analysis of the masonry components and their interaction is indispensable for assessing the structural capacity of the material, thus, of the building at hand. Together with the geometric survey at the building scale (when e.g. connections between intersecting walls, floor beams and possible wooden or metallic ties are investigated), the characterization of masonry unit typology and arrangement, masonry joint degradation, and possible discontinuities along the length or the thickness of structural elements are essential aspects the analyst should check before performing a structural analysis. The structural response of masonry members is also affected by the so-called *rules of the art* (or *best practice*, resulting from proven experience), as described in Section 4.1.6.2.

Dealing with existing masonry, according to Giuffrè (1996), a first distinction can be made between public and monumental buildings (e.g. religious or military) and common buildings related to popular traditions (so-called *vernacular*). Regarding the first group, masonry may involve dressed stones with regular horizontal courses and good interlocking of units across the wall (Figure 4.1a). This suitable arrangement of the stones guarantees a monolithic behaviour without or with little mortar between the units. Figure 4.1a describes also the so-called *diatone*, that is a long stone or header unit (also called *through stone* or *bond stone*) that, by crossing the entire thickness of the wall, connects two *leaves* (also called *wythes*) together. In modern masonry, a through stone can be a concrete

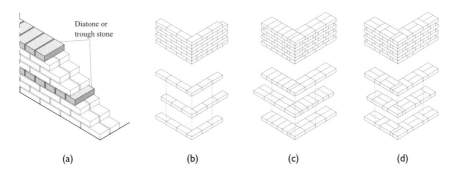

Figure 4.1 Different masonry arrangements. (a) Roman *opus quadratum*; (b) stretcher bond, (c) English bond and (d) Flemish bond.

block, a wood element or bars with hooked ends embedded in concrete that perform the same function. On closer inspection, the wall of Figure 4.1a is the archetype of the Roman *opus quadratum* and innumerable variants have followed ever since, some of them shown in Figure 4.1b–d. In this case, the term bond is used to describe a specific arrangement of units in regular patterns for a better response towards external actions, for aesthetic purposes or related to the construction technology and geometrical modularity.

As clearly noticeable, the horizontality of courses represents one of the basic rules of the art, together with the staggering of the units in consecutive courses so that the vertical joints are discontinuous with interlocking and no weakness plane is predefined. Additionally, a good interlocking across the thickness of the wall is essential to guarantee the integral performance of the wall, making masonry disintegration difficult. A schematic view of these two rules is shown in Figures 4.2 and 4.3, respectively. Figure 4.2 illustrates the structural response of a masonry wall subjected to in-plane (lateral) actions: the better is the level of interlocking, the more monolithic is the wall behaviour. Figure 4.3, in perfect analogy, depicts the behaviour of a wall subjected to out-of-plane loads. Three paradigmatic cases are shown, namely, perfect connection between the two leaves, null connection (the leaves are independent) and the case with one leaf composed by rubble masonry. The last case, even though good looking from one side, may be disastrous in the case of seismic events. The

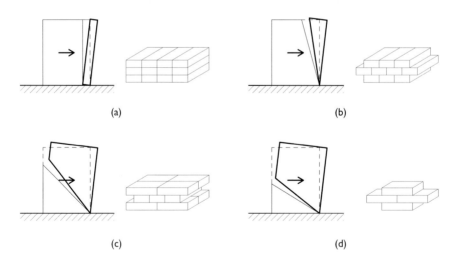

Figure 4.2 In-plane action for a masonry wall (shown as a front view) subjected to combined vertical and lateral in-plane loading with different levels of interlocking (after Giuffrè, 1996.) From (a) to (d), no, low, medium and high level of interlocking, respectively. The arrows in the middle indicate the lateral body forces in the case of a seismic action.

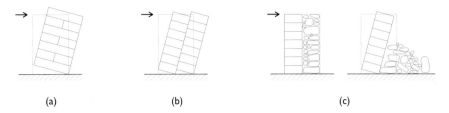

Figure 4.3 Effectiveness of transversal connections on a two-leaf regular masonry wall (shown as a cross-section in elevation) under combined vertical and out-of-plane actions. (a) With and (b) without through stone; (c) one leaf made of rubble stone. The arrows indicate the direction of the lateral body forces in the case of a seismic action.

rubble masonry may crumble away and trigger the collapse of the entire wall/building.

Conversely to Figures 4.2 and 4.3, in the case units are not squared (as for vernacular walls), friction is not as effective as in regular arrangements, especially when subjected to cyclic loading and progressive environmental deterioration processes. A good interlocking is still possible, but the wall capacity relies on the skill of the mason that built it and on the quality of the mortar used. Aiming at reproducing the quality of *opus quadratum*, other rules of the art are usually put in place. Examples include horizontal courses (in brickwork) placed regularly along the height, or large stones oblong-shaped disposed across the wall thickness as frequently as possible (similar to the through stones in Figure 4.1a). The inevitable voids among stones may be partly filled with small wedged stones used to level larger ones into position or with rubble to fill gaps. The result is a suitable base for the upper part of the wall with a uniform distribution of the stresses. In turn, the disrespect of the rules of the art, namely, the lack of horizontal courses and through stones across leaves, may lead to a poor-quality masonry, prone to disintegration.

4.1.2 Introduction to numerical modelling

Masonry can be described as a material with a visible structure, usually designated as *microstructure*. The last definition is borrowed from materials such as metals, polymers, concrete or ceramics, where microscopic arrangements of atoms and molecules, their bonds, possible inclusions, voids or defects have a significant influence on the macroscopic physical and mechanical properties, such as strength, ductility, hardness or corrosion resistance. Dealing with the structural analysis of masonry, the designation *mesostructure* would be more appropriate because the term *micro* is not referred to such a low level of scale, but to units and mortar, and their interaction. Accordingly, regular masonry can be approximated with an arrangement of units (either bricks or stones) separated by mortar joints.

The latter can be classified in *bed* and *head joints*, horizontal (typically continuous) and vertical (typically discontinuous), respectively. The presence of the joints (that act as planes of weakness) and the size of the units (usually comparable with the wall thickness in modern masonry but not in existing masonry) make masonry anisotropic. Additionally, the complex interaction between the elementary components makes difficult to foresee the mechanical behaviour of masonry (as a composite) from that of the components (see Section 4.1.5 for further details).

In this regard, Figure 4.4 shows examples of stress–strain curves of compressive tests performed independently on bricks, mortar and masonry prisms, after Binda et al. (1994a) who carried out an intense experimental campaign on units and mortars. Bricks, and most units in historical buildings, have a larger compressive strength than mortar. In turn, units may show an increasingly brittle behaviour (with a lower ultimate strain and ductility) as the compressive strength gets higher, which is in contrast with the ductile behaviour of traditional mortars. The consequent masonry exhibits a smaller capacity with respect to the strongest component (i.e. brick) but larger than the weakest one (i.e. mortar). This is the so-called *go-between* compressive behaviour of masonry when units are stronger than mortar. However, different behaviours are possible. Kaushik et al. (2007) performed uniaxial compression tests on prisms using a mortar with strength and stiffness comparable with bricks and found that the stress–strain curve of masonry falls on the lower side of those corresponding to bricks and mortar.

Furthermore, although not shown in Figure 4.4, both bricks and mortar have tensile strengths prominently lower than the compressive ones. Tensile strength of masonry, instead, is more related to the bond between units and

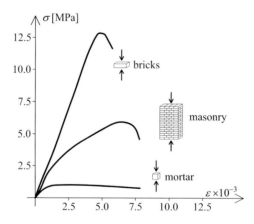

Figure 4.4 Synthetic representation of the compressive behaviour of brick, mortar and masonry, after Binda et al., 1994a.

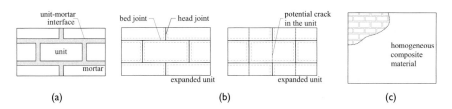

Figure 4.5 Modelling strategies for masonry structures. (a) Detailed micro-modelling, (b) simplified micro-modelling without and with potential crack in the units and (c) macro-modelling.

mortar, thus to the interface between the two materials and not to bricks and mortar themselves.

From the numerical point of view, according to the level of accuracy and simplicity desired, the description of masonry can follow three main computational strategies, namely (see Figure 4.5):

- **Detailed micro-modelling:** Units and mortar in the joints are represented by continuum elements, whereas the unit–mortar interface is represented by discontinuous elements. This strategy represents the most detailed one and the mechanical parameters must be obtained by a thorough experimental activity. This approach is suited for small structural elements with a particular interest in strongly heterogeneous states of stress and strain. The numerical model must reflect the actual geometry and arrangement of units and joints (also through the structural element thickness, if applicable), and their properties, differentiating bed and head joints, if necessary. It is noted that mortar behaves differently in the composite (due to the interaction with units and the related water migration) if compared to laboratory evidences of a mortar specimen tested independently. Due to the large amount of data and the relatively high computational effort, detailed micro-modelling is not recommended for professional applications.
- **Simplified micro-modelling:** Expanded units are represented by continuum elements whereas the behaviour of the mortar joints and unit–mortar interface is lumped in discontinuous elements. In other words, mortar joints (and unit–mortar interfaces) are substituted with a zero-thickness interface, while the units are expanded to keep the geometry unchanged. In its simplest approach, units are considered linear elastic, and all the material nonlinearities are lumped in the interface elements, that act as potential crack, slip or crushing planes. The main consequence of this approach is a small loss of accuracy since the Poisson's effect of the mortar is not included (see Section 4.1.5.1). A more accurate approach considers the occurrence of a possible crack within the units to reproduce the crack progress from one head joint to the other (immediately below or above), which is a typical masonry

characteristic. In this case too, a zero-thickness interface element is used. In the professional field, these approaches are recommended if an accurate analysis of a single structural elements is requested, such as slender piers that, due to door or window openings, may result of few units in length and whose behaviour influences the overall response of the entire wall. Applications to full buildings have been made, usually by creating some assemblage of masonry units into a larger equivalent unit.

- **Macro-modelling**: Units, mortar and unit–mortar interface are smeared out in the continuum. Different from the previous ones, this approach replaces the actual heterogeneity of masonry with an equivalent homogenous anisotropic material. Isotropic assumption can be also considered as a further simplification. A proper definition of the constitutive law is a key point of this strategy and different choices may be made. In fact, it is possible to distinguish between constitutive laws based on phenomenological or micro-mechanical approaches. Regarding the former, the identification of the characterizing law parameters is based on experimental results of masonry specimens of sufficient size, under homogeneous states of stress and strain. About the latter, homogenization techniques are often implemented, according to which masonry constitutive relations are established in terms of average strains and average stresses, through a mathematical process involving the geometry of the masonry arrangement and the properties of the individual components. Consequently, thanks to the reduced time and computational requirements, macro-modelling is more practice-oriented and user-friendly in the mesh generation, most valuable when a compromise between accuracy and efficiency is needed. Additionally, macro-modelling should be preferred when dealing with irregular masonry for which micro-modelling would be hardly adequate.

As expected, the extent of knowledge of material properties is linked to the selected modelling strategy, which, in turn, is based on different assumptions and theories, resulting in diverse complexity levels and costs (and different access for practitioners). One modelling strategy cannot be preferred over the other because different application fields exist for micro- and macro-models. The key aspects to be considered in the decision process are: (1) the relation between the analysis tool and the sought information, (2) the available analysis tools and their understanding by the user and (3) the resources that may be allocated, involving cost and time for analysis. Experimental data play a crucial role to set both micro- and macro-modelling analyses and investigations should include information on the entire material response up to failure (with the possible softening branch). A synthesis of the needed parameters for each approach is shown in Table 4.1.

Table 4.1 Synthesis of the main parameters needed for nonlinear static analysis of masonry structures

Strategy	Elements	Linear parameters	Nonlinear parameters
Detailed micro-modelling	Unit (usually assumed isotropic)	Young's modulus Poisson's ratio	Compressive and tensile strengths Fracture energies in tension and compression Hardening–softening criteria for compression and tension (σ–ε diagrams)
	Mortar (isotropic)[a]		
	Interface for bed and head joint	Dummy normal and shear stiffness (to avoid interpenetration)	Bond tensile strength Cohesion Friction coefficient Dilatancy coefficient Tensile and shear softening criteria (evolution of stresses vs. crack opening and slipping) Mode I fracture energy Mode II fracture energy
Simplified micro-modelling	Expanded units (usually assumed isotropic)	Young's modulus Poisson's ratio	No need as all nonlinearities are lumped at the interfaces (masonry joints and potential cracks in the units)
	Interface for bed and head joints	Normal stiffness Shear stiffness	Equal to detailed micro-modelling but, in addition: Compressive strength of masonry Hardening–softening criterion for compression (evolution of stresses vs. crushing) Fracture energy in compression
	Interface for potential cracks in the unit	Dummy normal and shear stiffness (to avoid interpenetration)	Tensile strength of unit Cohesion Friction coefficient Dilatancy coefficient Tensile and shear softening criteria (evolution of stresses vs. crack opening and slipping) Mode I fracture energy Mode II fracture energy
Macro-modelling	Isotropic	Young's modulus Poisson's ratio	Tensile and compressive strengths of masonry Fracture energies in tension and compression Hardening–softening criteria for tension and compression (σ–ε diagrams)
	Orthotropic	Similar to isotropic but referring to both horizontal and vertical directions as material axes	

[a] The elastic properties of mortar should replicate its actual properties in the composite (which are different from experimental outcome of independent mortar specimens).

Finally, it is worth underlining that, given the wide variability of masonry according to e.g. component materials and workmanships, the following discussion mainly focuses on the behaviour of regular single-leaf masonry, laying the basis for the further understanding that the analyst should develop on a case-by-case basis. Irregular masonry is discussed in Sections 4.1.5 and 4.1.6 only as a composite material.

4.1.3 Mechanical properties of masonry components

Masonry structural response is strongly influenced by the mechanical properties of its components. Whereas, on the one hand, up-to-date manufacturing and construction controls drastically reduce the variability of properties for modern materials, experimental evidence has highlighted large scatters in the structural response of historical units and mortar, even when extracted from the same building. This represents an important issue when dealing with existing structures and makes generalization difficult. For instance, in fired clay bricks the compressive strength is highly dependent on the firing temperature; the same scatter is found in stones, even when extracted from the same quarry, according to the zone of extraction (e.g. the upper part is generally weathered and altered). On top of that, different test conditions, including e.g. specimen dimensions, moisture content or boundary conditions, can affect the results, particularly in the case of postpeak material properties.

In the following, units and mortar are dealt with separately, highlighting experimental evidence and recommended mechanical parameters, as suggested by codes of practice and literature. Regarding the stress–strain relationship, it is widely accepted to consider for units (and masonry, in general) a parabolic shape for the hardening and the subsequent softening in compression, and exponential softening in tension.

4.1.3.1 Unit

Among the innumerable materials adopted for units in masonry constructions, Table 4.2 provides a range of values encountered in practical applications. Significant variations are expected and the analyst needs to gather additional data from the local context or obtain information from testing. As an example, the compressive strength of granite intact rock can be in the range 15–350 MPa, where the low bound is indicative of much weathered stone and the high bound regards geological values likely at great depths, hardly possible in construction materials. The same occurs for basalt in which a hard rock can feature a compressive strength of 400 MPa, while a light porous weathered rock in a volcanic area can provide a compressive strength of 1 MPa. Additionally, dealing with natural materials, a considerable scatter is often observed in real case applications.

Table 4.2 Reference values for mechanical properties of units in masonry
constructions

Material	Compressive strength f_c [MPa]	Tensile strength f_t [MPa]
Natural stone	10–200	1–5
Clay brick	10–70	1
Calcium silicate	10–35	1–2
Concrete	10–35	3
Aerated concrete	3–7.5	0.5–2
Adobe	1–3	0–0.3

Moving to laboratory tests, the uniaxial compressive test is the simplest material characterization and gives a good indication of the mechanical properties. Besides the compressive strength, the test can provide the main elastic properties, namely, Young's modulus and Poisson's ratio, with additional complexity in data acquisition and test procedure. In general, given the easiness in evaluating the compressive strength f_c, simple formulas are available in literature to assess other mechanical data from this value. For instance, in absence of more exhaustive information, Young's modulus may be taken as 200–1,000 f_c, depending on the materials, while the Poisson's ratio ranges from 0.15 to 0.25. The behaviour of units under tensile loading is more complicate to assess in laboratory, and the tensile strength of the units may be taken as 0.03–0.10 f_c, depending on the materials (as discussed below).

Experimental evidence has also showed that units may display an anisotropic response induced by the possible presence of foliation or rift plane in stone units and geometry (e.g. holes), extrusion or vibro-compaction processes in modern units. In this regard, a full material characterization needs to include the analysis along the material axes. This is lacking in literature and units are often considered isotropic. In the case of hollow units, a simplified approach to evaluate the mechanical properties along the perpendicular axis is to consider the material properties according to the net area. For instance, if a 50% hollow unit is characterized by a compressive strength equal to 2 MPa, the net area (thus the material) provides an estimate of the effective strength equal to 4 MPa.

In the absence of laboratory or in situ tests, guidelines (NZSEE, 2017) suggest a simple scratch test for evaluating the clay brick properties according to its hardness. Table 4.3 reports the probable compressive strength and tensile strength (equal to 0.12 f_c). The term "probable" is defined as the *expected or estimated average value*. The previous version of the same guidelines (NZSEE, 2006) proposed additional data, as shown in Table 4.4. In both cases, to ensure that the test is representative of the structural capability of the materials, removing any weathered or remediated surface material prior to assessing the hardness characteristics is recommended.

Table 4.3 Probable strength parameters for clay bricks (NZSEE, 2017)

Brick hardness	Brick description	Compressive strength [MPa]	Tensile strength [MPa]
Soft	Scratches with aluminium pick	14	1.7
Medium	Scratches with 10 cent copper coin	26	3.1
Hard	Does not scratch with above tools	35	4.2

Table 4.4 Strength parameters for preliminary assessment of clay bricks (NZSEE, 2006)

Brick quality	Visual characteristics	Compressive strength [MPa]	Tensile strength [MPa]	Young's modulus [10^3 MPa]	Poisson's ratio [−]
Soft	Weathered, pitted, distinct colour variation with depth (bright orange), probably under-fired	1–5	0.1–0.5	4	0.35
Stiff	Common brick, can be scored with a knife, red. Lower figures if split	10–20	1–2	13	0.2
Hard	Dense (heavy heft), hard surface, well fired, dark reddish brown	20–30	2–3	18	0.2

Moving to the nonlinear response, this can be described according to fracture energy. In this regard, it is possible to refer to Model Code 90 (CEB-FIP, 1993) extending its indications to concrete-like materials. In particular, the code provides equations for the full nonlinear stress–strain relationships for different concrete grades. Their integrations, multiplied by the height of the specimen (assumed as 200 mm), provide the fracture energy in compression G_c. This information has been complemented with the authors' database to compensate for the lack of data regarding low-strength materials (the code deals with a range between 20 and 90 MPa). Accordingly, the recommended fracture energy in compression G_c can be calculated with Eq. (4.1). The dependence of the fracture energy from compressive strength is illustrated in Figure 4.6, where the comparison with the data from Model Code 90 is also shown. It is worth highlighting that high-strength materials are more brittle than low-strength materials and require further investigation (looking at Figure 4.6 the fracture energy stops increasing while the compressive strength does). It is noted that the recent version of the code, namely, Model Code 2010 (fib, 2013), does not provide a mathematical relation for the softening portion of the diagram, as this relation exhibits some size dependency.

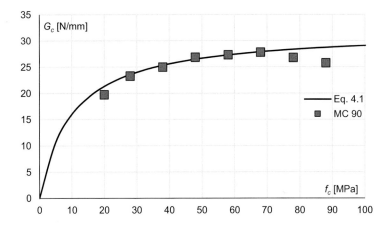

Figure 4.6 Proposed relation between compressive fracture energy and strength for concrete-like materials.

$$G_c = \frac{32 f_c}{10 + f_c} \qquad [\text{N/mm}] \tag{4.1}$$

Regarding the behaviour in tension, the direct tensile test is not easy to perform. Flexural and splitting tensile tests (the latter also called Brazilian test) are often preferred. It must be noted that the mentioned tests are not interchangeable and results must be carefully processed. For instance, looking at Figure 4.7a, when a specimen is subject to bending, a linear distribution of stresses may be assumed (according to Euler–Bernoulli's theory) and the flexural tensile strength can be calculated by dividing the ultimate bending moment M_u by the elastic section modulus of the failing cross-section. This value is coincident with the uniaxial tensile strength if the analysis is limited to linear elastic behaviour.

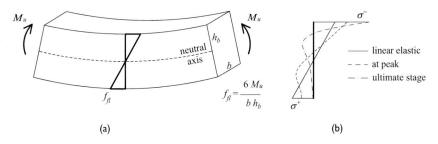

Figure 4.7 Difference between linear and nonlinear analysis in bending. (a) Linear analysis with triangular stress distribution and (b) nonlinear stress distribution in the cross-section for increasing external moment.

This simplification is valid for fully brittle materials but differences are evident between test configurations for concrete-like materials (which exhibit a quasi-brittle behaviour). In the case of bending, an internal stress redistribution occurs, as shown in Figure 4.7b. Once the tensile strength is reached at the extreme fibre (lower side in Figure 4.7b), every fibre of the cross-section behaves according to the uniaxial tensile stress–strain relationship. Consequently, the microcrack propagates upwards and the neutral axis of the cross-section shifts towards the fibres in compression (upwards in Figure 4.7b). Accordingly, the bending moment increases towards a peak value and decreases afterwards, until reaching collapse. This process depends on the height of the specimen (and, necessarily, on the fracture energy in tension). For this reason, flexural strength must be understood as a *structural property* (as it depends on the structural geometry), and it is not an intrinsic property of the material. In particular, the higher is the section, the more brittle is the flexural response, and tensile and flexural strengths tend to be coincident. From the reverse viewpoint, the shorter is the section, the larger is the flexural over the tensile strength.

From the theoretical point of view, the ratio between the uniaxial tensile strength f_t and the flexural tensile strength f_{fl} in rectangular sections can vary between 1 (for fully brittle materials) and 0.33 (for ideal plastic materials). For concrete-like materials, intermediate values are found, which depend on the fracture energy and the size of the specimen. Model Code 2010 (fib, 2013) proposes the expression in Eq. (4.2), where h_b is the height of the specimen expressed in mm (see Figure 4.7a). If the tensile strength is measured according to the splitting tensile strength, Model Code 2010 states that the same ratio varies from 0.67 to 0.95. Synthetic values are proposed for bricks by the New Zealand guidelines (NZSEE, 2017): tensile strength may be taken as 85% of the strength derived from splitting tests or 50% of the strength derived from bending tests. The latter value is in line with Eq. (4.2), if the brick height is about 50 mm.

$$f_t = \frac{0.06 h_b^{0.7}}{1 + 0.06 h_b^{0.7}} f_{fl} \qquad \left(h_b \text{ in mm} \right) \tag{4.2}$$

Model Code 2010 (fib, 2013) also provides the tensile strength as a function of the compressive strength for concrete. The data from Model Code 2010 have been integrated by the authors with test results within a wide range of strengths, especially on the lower strength side. The recommended tensile strength can be calculated with Eq. (4.3). When compared with the concrete values from Model Code 2010 (fib, 2013), Figure 4.8 shows that the proposed relation provides larger values of tensile strength in the low range of compressive strengths. Overall, the tensile strength ranges from 0.16 to 0.04 f_c, for a compressive strength in the range 1–100 MPa. High-strength

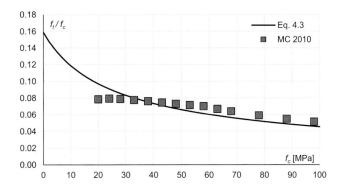

Figure 4.8 Proposed relation between the ratio of tensile and compressive strength and the compressive strength for concrete-like materials.

materials require further investigation as they are more brittle than low-strength materials.

$$f_t = 2.12 \ln(1 + 0.075 f_c) \quad (f_c \text{ in MPa}) \tag{4.3}$$

Model Code 2010 (fib, 2013) provides also tensile fracture energy values. The regression analysis of these values with the authors' database, including low-strength values, led to the formula reported in Eq. (4.4).

$$G_f^I = 0.07 \ln(1 + 0.17 f_c) \quad [\text{Nmm}] \tag{4.4}$$

4.1.3.2 Mortar

Mortar is generally composed of sand and a binder, usually aerial lime, hydraulic lime, cement or a combination of these. The mechanical properties of mortar depend on type and proportions of its main constituents (usually described by the ratio in terms of volume, e.g. 1:2:9 for cement:lime:sand), possible admixtures and amount of water used for the mix (usually indicated by water/binder ratio in terms of volume). The needed amount of water used is related to the workability of fresh mortars but affects in great extent the performance in hardened conditions. On the one hand, water is needed for having a dense and flowable material; on the other hand, a lower ratio usually leads to higher strength and durability. In this regard, the needed amount of water depends significantly on the particle size distribution of the sand, but also on the binder type and admixtures. Regarding the hardening process, the difference between *aerial* (or hydrated) and *hydraulic* lime must be highlighted. Aerial lime needs air for hardening, (a chemical reaction with carbon dioxide called carbonation), while hydraulic limes harden with a chemical reaction with water. This aspect is of great importance when dealing with thick and multileaf walls, as aerial lime takes very long time to harden.

Generally, lime-based mortars are more flexible (lower Young's modulus) while cement-based mortars are stronger but more brittle. With the aim of suggesting indicative values on historical mortars, Kržan et al. (2015) collected the results of experimental tests available in literature, whose synthesis is reported in Table 4.5 in terms of compressive strength and Young's modulus. As it is possible to observe, the scatter is rather high. In addition, the level of degradation, weathering or damage can significantly affect mortar properties.

In this scenario, the New Zealand guidelines (NZSEE, 2017) describe simple tests for a preliminary assessment of the mechanical properties of historical lime/cement mortars. This is based on the material hardness evaluated by means of a simple scratch test, as given in Table 4.6. Again, any weathered or remediated surface material must be removed prior to assessing the hardness characteristics. As a possible mean to evaluate the uniformity of the mortar quality across the wall thickness, the guidelines suggest drilling into the mortar joint and inspect the condition of the extracted mortar dust as the drill bit progresses through the joint. Hard and very hard mortars (typically cement-based) need to be further investigated by means of experimental tests.

Other reference values for preliminary assessment are provided by the previous version of the New Zealand guidelines (NZSEE, 2006) reported in Table 4.7. The mechanical properties of noncohesive mortar are synthetically put equal to zero. About the remaining categories, besides a possible knowledge of components proportion, the guidelines suggest

Table 4.5 Typical values of compressive strength and Young's modulus of mortars, collected by Kržan et al. (2015)

Binder	Earth	Lime	Hydraulic lime	Roman cement	Portland cement
f_c [MPa]	0.5–3	0.5–3	2–10	5–20	10–50
E [10^3 MPa]	<1	1–5	5–15	-	20–30

Table 4.6 Probable strength parameters for lime/cement mortar (NZSEE, 2017)

Mortar hardness	Mortar description	Compressive strength [MPa]
Very soft	Raked out by finger pressure	0–1
Soft	Scratches easily with fingernails	1–2
Medium	Scratches with fingernails	2–5
Hard	Scratches using aluminium pick	To be established from testing
Very hard	Does not scratch with above tools	To be established from testing

Table 4.7 Strength parameters for preliminary assessment of mortar (NZSEE, 2006)

Mortar quality	Visual characteristics and hand tests	Compressive strength [MPa]	Young's modulus [10³ MPa]	Poisson's ratio [−]
Noncohesive	Lime-based mortar that is heavily leached and weathered. Sand-like, easily raked out by hand, aggregate is unbound.	0.0	-	-
Soft	Lime binding, possibly mildly leached, can be raked out of joint, but stays bound. Punch test < 30 mm.	1.0	7	0.05
Firm	Firm lime based (lime:sand = 1:3), but in interior locations (not weathered). Punch test < 20 mm.	4.0	9	0.07
Stiff	Stiff high Portland cement content (typical cement:lime:sand = 1:0.25:3). Punch test < 10 mm.	8.0	12	0.11

the use of a punch test. More specifically, the punch test uses a standard carpenter's nail punch (3 mm diameter at the tip), which is firmly driven with a standard carpenter's hammer for six blows. The total penetration is then used as indicative parameter to assess the mortar mechanical characteristics.

For constructions of the recent past, it is possible to refer to modern design codes. They propose several categories of mortars, typically according to their composition and expected mechanical properties. According to Eurocode 6 (2005), masonry mortars are defined as general purpose (i.e. without special characteristics), thin layer or lightweight. Table 4.8 shows lower bound compressive strengths of selected mix proportions, according to Briceño et al. (2019) and the authors' data collection. It is stressed that results are expected to vary significantly according to the particle size distribution of the sand and amount of water used in the mix. Note that also the Australian standard (AS 3700, 2018) and ASTM C270 (2019) deal with a similar approach, defining mortar grades according to mortar composition.

About nonlinear parameters, it is possible to refer to the previous section in the absence of experimental values.

4.1.4 Mechanical properties of interface

The bond between units and mortar is usually the weakest link in masonry assemblages. The severe nonlinear behaviour of the interface between units and mortar is one of the most relevant features of masonry structural response. In this regard, ASTM C270 (2019) states that bond is the most important single physical property of hardened mortar, being, at the same

Table 4.8 Lower bound compressive strengths of selected mortar compositions

Average mortar compressive strength [MPa]	Proportion by volume		
	Cement	Lime	Sand
0.5		1	5
	1	2	18
1		1	3
	1	2	12
2.5	1	2	8-9
	1		6
5	1	1	5-6
	1		5
10	3	1	12
	1		4
15	5	1	15
	1		3

time, the more variable and less predictable. In this regard, mortar–unit bond mainly depends on:

- **masonry unit properties:** surface absorption, porosity, composition, size, surface finishing, etc.;
- **mortar properties:** composition, water and air content, cohesion, curing conditions, etc.;
- **workmanship:** elapsed time between preparation and spreading of mortar, pressure applied to masonry during placement and tooling, etc.;
- **distribution of mortar over the contact surface:** usually smaller than the gross cross-sectional area following the combined result from shrinkage of the mortar and the process of laying units in the mortar bed.

With reference to the detailed micro-modelling approach of Figure 4.5, two different phenomena occur in the unit–mortar interface, one associated with tensile failure (Mode I), i.e. *crack-opening* behaviour, and the other associated with shear failure (Mode II), i.e. *bond-slip* behaviour. It is worth reminding that the interface element is, to some extent, an abstraction to characterize the material discontinuity between unit and mortar, or the physical discontinuity between units in the case of a dry joint.

In turn, considering the simplified micro-modelling approach, being a zero-thickness element between two expanded units, the interface includes the mortar joint, and its constitutive law may encompass the crushing of masonry as a phenomenological lumped failure mechanism. Consequently, interface elements may be used as potential *crack*, *slip* and *crushing planes*, as originally proposed by Lourenço and Rots (1997). Accordingly, looking at Figure 4.9, the interface model is usually described by a multisurface domain in the normal traction vs. shear (σ–τ) stress space. The three individual failure surfaces represent a straight tension cut-off for Mode I failure, a Mohr–Coulomb friction law for Mode II failure and an elliptical cap mode for shear-compression interaction. The meaning of the symbols was already introduced. The figure schematically represents also the evolution of the three failure surfaces, from the initial (or maximum) to residual envelope. Whereas tensile strength, cohesion and friction angle are assumed to reduce throughout the analysis (softening), compressive strength may increase before decreasing (hardening–softening). Associated flow rules are assumed for tensile and compressive modes, and nonassociated flow is usually adopted for the shear mode. The latter, in particular, is due to the significant difference between friction and dilatancy angles, as shown next in Section 4.1.4.3.

To better understand the evolution of the inelastic domain of Figure 4.9, the diagrams for cracking-slipping-crushing modes shown in Figure 4.10 provide the relationships between tractions and the related displacements at the interface. For instance, the tensile behaviour is usually described with an exponential softening: once the elastic displacement is reached, the tensile strength of Figure 4.9 reduces following the exponential curve of Figure 4.10a. The same trend can be assumed for the Mode II softening diagram of Figure 4.10b that governs the cohesion, i.e. the intercept of the Mohr–Coulomb friction mode with the τ axis. Finally, crushing mode can be defined with a hardening/softening diagram such as the one in Figure 4.10c.

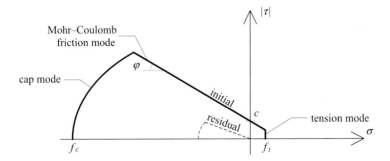

Figure 4.9 Proposed model for interfaces as potential crack (tension), slip (friction) and crushing (cap) planes (after Lourenço and Rots, 1997), not to scale. Failure surface evolution is described through initial (bold) to residual envelope (dashed lines).

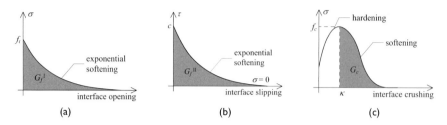

Figure 4.10 Typical postelastic relationships for interface failure modes in terms of stress–displacement. (a) Mode I (cracking), (b) Mode II (slipping) and (c) crushing.

In line with Table 4.1 and according to Figure 4.10, the key parameters needed to define the multisurface failure domain are:

- **Mode I**: softening curve (e.g. exponential, linear or multilinear diagram), tensile strength and Mode I fracture energy;
- **Mode II**: softening curve (as per Mode I), cohesion, Mode II fracture energy, and friction and dilatancy coefficients (the latter to describe the nonassociated plastic flow);
- **Crushing**: hardening/softening curves, compressive strength, compressive fracture energy, normal–shear stress interaction (shape of the cap mode) and the value of interface displacement κ_p at the point where softening starts.

In the following subsections, reference values for these parameters are discussed for masonry joints. It is noted that vertical joints may be weaker than horizontal joints due to the quality of workmanship and execution technique. As a first approximation, the same data can be used for both horizontal and vertical joints.

4.1.4.1 Elastic parameters

Elastic properties of the interface regard normal and tangential stiffnesses, i.e. the relation between (normal and tangential) surface stresses (or tractions) and the interface relative displacements, namely, σ and Δu of Eq. (4.5), respectively, where n and s denote the normal and shear components, respectively. Generally, the interface stiffnesses are considered independent from each other and indicated with k_n and k_s [N/mm³]. It is possible to evaluate their values according to elastic considerations, assuming the mortar joint (with thickness h_m) as series of unit-interface-unit springs (i.e. simplified micro-modelling). The consequent formulas are shown in Eq. (4.6). By assuming a Poisson's ratio $\nu \approx 0.2$ for both unit and mortar, and linear elastic isotropic materials, it is possible to derive k_s from k_n, with a ratio approximately equal to 0.4.

A few considerations on Eq. (4.6) are worthy. This approach, although present in literature, ignores the actual mortar–unit interaction. The interface is a zero-thickness element, which is not rigid but provided with a certain stiffness. Additionally, the use of a mortar Young's modulus calculated according to tests on isolated samples may overestimate the interface properties because the actual mortar stiffness within the masonry composite is unknown. This means that Eq. (4.6) is theoretical and should not be used in practice. Accordingly, the interface stiffness should be established from the masonry and unit properties, as shown below in Eq. (4.7). Lacking additional information, it has been observed that a value close to the half of the theoretical value of Eq. (4.6) may approximate the expected behaviour.

In the case of dry-joint masonry, the simplified micro-modelling represents the true geometry of the bricks or blocks (with height h_b) and the zero-thickness interface element is aimed at describing the joint contact area, which has some stiffness too as contact is not perfect. Also in this case, it is possible to consider a prism made of two units stacked as a series of unit-interface-unit springs and derive the interface stiffness from Young's moduli of a masonry prism and the units, E_M and E_b, respectively. The relative formulas are shown in Eq. (4.7). These are written in the general form that can be used for both mortar and dry joints (for the latter case, $h_m=0$). The respect of Eq. (4.7) guarantees that the structural member, either with micro- or macro-modelling has the same (elastic) stiffness. For Eqs. (4.6) and (4.7), it is worth reminding that the subscripts M, b and m are indicative of masonry, units and mortar, respectively.

$$\sigma = D\Delta u \quad \rightarrow \quad \begin{bmatrix} \sigma \\ \tau \end{bmatrix} = \begin{bmatrix} k_n & 0 \\ 0 & k_s \end{bmatrix} \begin{bmatrix} \Delta u_n \\ \Delta u_s \end{bmatrix} \tag{4.5}$$

$$k_n = \frac{E_b E_m}{h_m(E_b - E_m)}$$

$$k_s = \frac{G_b G_m}{h_m(G_b - G_m)} = \frac{k_n}{2(1+v)} \approx 0.4 k_n \tag{4.6}$$

$$k_n = \frac{E_b E_M}{(h_b + h_m)(E_b - E_M)}$$

$$k_s = \frac{G_b G_M}{(h_b + h_m)(G_b - G_M)} = \frac{k_n}{2(1+v)} \approx 0.4 k_n \tag{4.7}$$

4.1.4.2 Mode I failure

The tensile strength of the interface measures the bond between mortar and unit (null in the case of dry-joint unit assemblage). It generally depends on the surface absorption properties of the unit (porosity, initial rate of absorption, etc.) as well as the properties of the mortar itself. In this regard, it appears evident that the joints play a major role in the tensile response of masonry, being mostly independent from the compressive response of the components.

Regarding tensile behaviour of masonry joints, Van der Pluijm (1992) carried out deformation-controlled tests in small masonry specimens of solid clay and calcium-silicate units. These tests resulted in an exponential tension softening curve with a Mode I fracture energy G_f^I ranging from 0.005 to 0.02 N/mm, and tensile bond strength ranging from 0.3 to 0.9 MPa (according to the unit–mortar combination). Subsequently, Van der Pluijm (1999) found values of fracture energy in tension in the range of 0.005–0.035 N/mm for other combinations of unit and mortar. Figure 4.11 shows the envelope of the stress–crack displacement diagrams obtained by the experimental tests, suitably approximated by the exponential curve.

Rots (1997) proposed a wider range for fracture energy, within 0.001–0.020 N/mm. Lourenço et al. (2002) performed tests to characterize masonry under uniaxial tension considering hollow and solid bricks. For the latter, they found an average value equal to 0.008 N/mm. As the value of Mode I fracture energy is independent of tensile bond strength, in the absence of detailed investigations, the value $G_f^I = 0.01$ N/mm is recommended. It is worth mentioning that different values may be found with different combination of units and mortar, with much lower values for low porosity units or very low-strength mortar. For instance, upon direct tensile tests on regular stone units linked by lime mortar, Bisoffi-Sauve et al. (2019) found Mode I fracture energy equal to 0.0008 N/mm, one order of magnitude smaller than the recommended value.

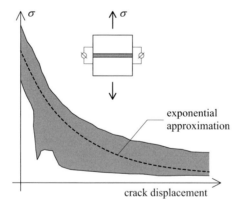

Figure 4.11 Tensile bond behaviour of masonry: direct tension test and typical experimental stress–crack displacement curves for solid clay brick masonry (shaded area is the envelope of four tests). (After Van der Pluijm, 1999.)

As far as tensile strength is considered, as shown for units, given the greater easiness of performing flexural tests, it is possible to derive the tensile bond strength of interface elements from flexural tests (with the plane of failure parallel to the bed joints). The difference between the two parameters, namely, flexural and tensile strengths, was discussed in Section 4.1.3.1 (to which the reader is referred), and Eq. (4.2) can be used to correlate each other. For the sake of completeness, the typical experimental tests performed to assess tensile bond strength of masonry joints are shown in Figure 4.12. For instance, according to bond wrench tests, Kržan et al. (2015) found the bond strength of a weak mortar in the range 0.01–0.13 MPa. However, for mortars with high pozzolanic content, values up to 0.6 MPa may be achievable, depending on the unit porosity and the mortar properties.

Given the vanishing character of the tensile bond strength, building codes often recommend a null value for it. However, in some cases, such as per EN 1996-1-1, 2005, flexural strength is invoked for out-of-plane bending actions. Aiming at evaluating the tensile bond strength, bending with plane of failure parallel to bed joints should be considered. This is calculated as the maximum tensile stress developed along the cross-section (see Figure 4.7a). The proposed average values are shown in Table 4.9, being quite low. Referring to Briceño et al. (2019) for further insight, the UK national annex provides higher values, whose synthesis is shown in Table 4.10. It is worth underlining that Eurocode 6 suggests a null value of flexural strength in the case of permanent mainly lateral loading (e.g. from lateral earth and water pressure).

Australian standard (AS 3700, 2018) provides a maximum average value of flexural strength with plane of failure parallel to bed joints equal to 0.30 MPa for clay, concrete and calcium-silicate masonry, and 0.39 MPa for

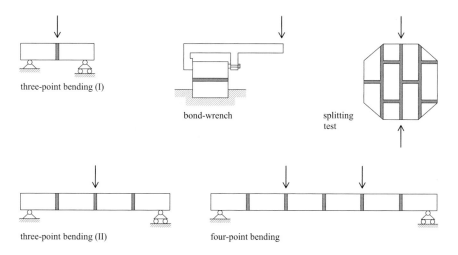

Figure 4.12 Typical experimental tests setups to assess tensile bond strength of masonry.

Table 4.9 Average flexural strength of masonry with general purpose mortar and plane of failure parallel to bed joints

	Mortar compressive strength	
Masonry unit	$f_m < 5\,\text{MPa}$	$f_m \geq 5\,\text{MPa}$
Clay	0.15	0.15
Calcium silicate	0.08	0.15
Aggregate concrete	0.08	0.15
Autoclaved aerated concrete	0.08	0.15
Manufactured stone	0.08	0.15
Dimensioned natural stone	0.08	0.15

Source: Average values according to EN 1996-1-1 (2005).

Table 4.10 Average flexural strength of masonry with general purpose mortar and plane of failure parallel to bed joints

	Mortar compressive strength		
Masonry unit	$f_m = 2\,\text{MPa}$	$f_m = 4\text{-}6\,\text{MPa}$	$f_m = 12\,\text{MPa}$
Clay[a]	0.38–0.60	0.45–0.75	0.60–1.05
Calcium silicate	0.30	0.45	0.45
Aggregate concrete	0.30	0.45	0.45
Aggregate concrete, manufactured stone and AAC in walls of thickness < 100 mm	0.30	0.38	0.38
Aggregate concrete, manufactured stone and AAC in walls of thickness ≥ 250 mm	0.15	0.23	0.23
Aggregate concrete, manufactured stone and AAC in walls of any thickness and $f_{c,u} \geq 10.4\,\text{MPa}$	0.30	0.38	0.38

Source: Average values according to BSI (2005).

[a] According to water absorption of the unit, smaller values are associated with larger water absorption.

Autoclaved Aerated Concrete (AAC) masonry with thin-bed joints, being slightly larger than Eurocode 6 (compare to Table 4.9). The New Zealand guidelines (NZSEE, 2017) suggest that values of tensile strength may be taken as half the cohesion values provided below in Table 4.14.

4.1.4.3 Mode II failure

To investigate the shear bond behaviour at the interface, tests should be able to track the shear stress vs. slip displacement under uniform states of normal stress (i.e. confining stress due to the precompression load). A uniform distribution of shear stresses across the interface is not possible as a null value must be found at the joint extremities, due to equilibrium conditions.

The study of Vasconcelos (2005) highlighted the high dependency of bond strength on the moisture content and porosity of the units, the strength and composition of mortar, and the nature of the interface. This leads to a wide range of shear strength values for various combinations of units and mortar.

Figure 4.13 collects the main test setups for laboratory experiments. In particular, the triplet shear test is recommended by EN 1052-3:2002 to characterize the cohesion and friction coefficient, being thus widely used. A closer inspection to Figure 4.13 shows that, as shear forces are not aligned, they produce bending moment at the interface, with nonhomogeneous distribution of normal confining stresses. Additionally, in triplet tests, joints do not fail simultaneously, and a rotation is likely to occur after the failure initiation of the first joint. Accordingly, the couplet test is normally used to obtain nonlinear parameters. Figure 4.13b shows the experimental test proposed by Van der Pluijm (1993) that avoids the drawback of nonhomogeneous normal stress distribution (the distribution of shear V and bending moment M within the specimen is also displayed).

Moving to mechanical parameters, Mann and Müller (1982) indicated an average friction coefficient of approximately 0.65 on brick-mortar assemblages and a cohesion ranging from 0.15 to 0.25 MPa, depending on the mortar grade. The results of Van der Pluijm (1993) on solid clay and calcium-silicate units are shown in Figure 4.14 about couplets of clay bricks

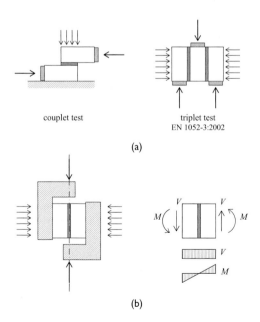

Figure 4.13 Experimental setups for assessing the shear bond behaviour. (a) Couplet and triplet tests and (b) test according to Van der Pluijm (1993).

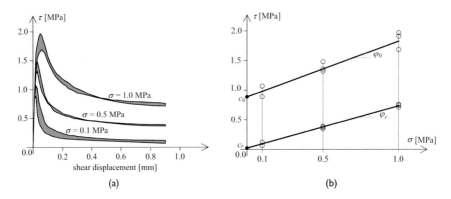

Figure 4.14 Typical shear bond behaviour of the joints for solid clay units (after Van der Pluijm, 1993). (a) Stress–displacement diagram for different confining stress levels σ (the shaded area represents the envelope of three tests) and (b) Coulomb friction law, with initial and residual friction angle and cohesion (φ_0 and φ_r, c_0 and c_r, respectively).

under three different levels of confining stress, namely, 0.1, 0.5 and 1.0 MPa. Looking at the diagrams of Figure 4.14a, after the peak of the shear stress, a gradual softening is observed (in the shape of an exponential diagram), that stabilizes in a residual value. These diagrams can be interpreted in the framework of Mohr–Coulomb criterion, according to which a linear regression in the σ–τ stress space (i.e. normal confining stress vs. tangential stress space) provides the angle of friction φ and cohesion c. Moreover, in the case the conditions at peak are considered, the relative quantities are labelled as *initial* (φ_0 and c_0), and as *residual* if the ultimate conditions are taken (φ_r and c_r). These relations are shown in Figure 4.14b, which is in line with the evolution of the failure surface described in Figure 4.9.

Regarding the friction coefficient, according to Van der Pluijm (1993), the tangent of φ_0 ranges from 0.7 to 1.2 for different unit–mortar combinations. In turn, the tangent of φ_r was found approximately constant and equal to 0.75. As far as cohesion is concerned, Van der Pluijm (1993) found values of c_0 ranging from 0.1 to 1.8 MPa, for different unit–mortar combinations, whereas c_r is null.

Rots (1997), based on the results of Van der Pluijm (1993), proposed to adopt a tangent of φ equal to 0.75, independently of the type of unit and mortar, in the case no additional information is available. From the results of direct shear tests carried out by Van der Pluijm (1999), the friction coefficient ranges between 0.61 and 1.17, whereas cohesion varies from 0.28 up to 4.76 MPa, depending on different types of units and mortar. Table 4.11 summarizes other results published in literature referring to the shear strength properties for different combinations of materials. In the case of dry joints, Vasconcelos (2005) proposed indicative values for the friction coefficient stressing the fact that the same is highly dependent on the roughness characteristics of the bed joint surface.

Table 4.11 Literature review on cohesion and friction coefficient for different unit–mortar assemblages

Reference	Units	Mortar (cement:lime:sand)	c [MPa]	tan φ
Hegemeier et al. (1978)	Concrete block	1:0.5:3	0.25	0.89
Drysdale et al. (1979)	Clay brick	1:0.5:4	0.57	0.90
Stöckl and Hofmann (1986)	Clay brick	1:0.68:15	0.95	0.70
		1:0:9.7	1.45	0.56
Atkinson et al. (1989)	Old clay	1:2:9 (13 mm thick)	0.13	0.70
		1:2:9 (7 mm thick)	0.21	0.64
	New clay	1:1.5:4.5 (7 mm thick)	0.81	0.74
Amadio and Rajgelj (1991)	Solid bricks	Cement mortar	0.65	0.72
		Lime–cement mortar		
Magenes (1992)	Solid bricks	Hydraulic lime mortar	0.21	0.81
		Lime mortar	0.08	0.65
Binda et al. (1994b)	Sandstone	Hydraulic lime mortar	0.33	0.74
	Calcareous stone	Hydraulic lime mortar	0.58	0.58
Roberti et al. (1997)	Bricks	Hydraulic lime mortar	0.23	0.57
Vasconcelos (2005)	Granite block	Lime mortar	0.36	0.63
		Dry joint	0.00	0.67
Augenti and Parisi (2011)	Tuff stone	Hydraulic mortar	0.15	0.29

Source: Augenti and Parisi (2011).

Indicative values of cohesion (indicated with f_{v0}) are also reported in building codes regarding new construction. The recommendations of Eurocode 6 (2005) are reported in Table 4.12, together with the values suggested by

Table 4.12 Average value of cohesion according to Eurocode 6 (EN 1996-1-1, 2005) and the UK national annex (BSI, 2005)

Masonry units	General purpose mortar			
	Eurocode 6		UK national annex	
	f_m [MPa]	f_{v0} [MPa]	f_m [MPa]	f_{v0} [MPa]
Clay	10–20	0.38	12	0.38
	2.5–9	0.25	4 and 6	0.25
	1–2	0.13	2	0.13
Calcium silicate	10–20	0.25	12	0.25
	2.5–9	0.19	4 and 6	0.19
	1–2	0.13	2	0.13
Aggregate concrete	10–20	0.25	12	0.25
Autoclaved aerated concrete	2.5–9	0.19	4 and 6	0.19
Manufactured stone and dimensioned natural stone	1–2	0.13	2	0.13

Table 4.13 Strength parameters for preliminary assessment of mortar (NZSEE, 2006)

Mortar quality	Visual characteristics and hand tests	Cohesion [MPa]	Friction coefficient [–]
Noncohesive	Lime-based mortar that is heavily leached and weathered. Sand-like, easily raked out by hand, aggregate is unbound.	0.0	0.0
Soft	Lime binding possibly mildly leached, can be raked out of joint, but stays bound. Punch test < 30 mm.	0.1	0.4
Firm	Firm lime based (lime:sand = 1:3), but in interior locations (not weathered). Punch test < 20 mm.	0.2	0.6
Stiff	Stiff high Portland cement content (typical cement:lime:sand = 1:0.25:3). Punch test < 10 mm.	0.4	0.8

Table 4.14 Probable strength parameters for lime/cement mortar (NZSEE, 2017)

Mortar hardness	Mortar description	Cohesion [MPa]	Friction coefficient [–]
Very soft	Raked out by finger pressure	0.1	0.3
Soft	Scratches easily with fingernails	0.3	
Medium	Scratches with fingernails	0.5	0.6
Hard	Scratches using aluminium pick	0.7	
Very hard	Does not scratch with above tools	To be established from testing	0.8

UK national annex (BSI, 2005), slightly different in the arrangement of the mortar strength grade. Other values are suggested by the New Zealand guidelines (NZSEE, 2017, 2006), as shown in Tables 4.13 and 4.14, where the description of mortar quality was already introduced in Section 4.1.3.2. More in detail, looking at Table 4.13, the mechanical properties of noncohesive mortar are synthetically put equal to zero.

In the absence of more accurate results, the friction coefficient can be set equal to 0.75 and, according to Lourenço (1996), the cohesion can be calculated as 1.4 times the tensile bond strength.

The nonlinear response is controlled by the fracture energy and, as illustrated in Figure 4.15a, the area defined by the stress–displacement diagram and the residual dry friction shear level is the Mode II fracture energy, G_f^{II}. It should be noted from the stress–displacement diagram of Figure 4.14a that G_f^{II} depends on the confining stress. This is also evident in Figure 4.15b, where the relation between G_f^{II} and the confining stress σ shows a strong

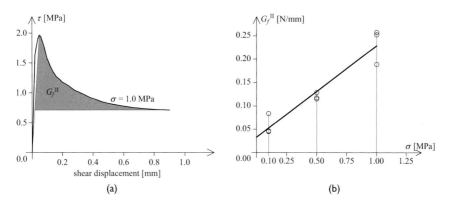

Figure 4.15 Mode II fracture energy G_f^{II} for solid clay units, after Van der Pluijm, 1993. (a) Example of calculation for confining stress level $\sigma = 1.0$ MPa and (b) linear relation between G_f^{II} and confining stress level.

linear correlation. The intercept of the regression line with the vertical axis represents the estimation of the fracture energy at zero confining stress. Lacking further investigations, this value can be assumed as a constant throughout the analysis (independent of the confining stress). Van der Pluijm (1993) found G_f^{II} ranges between 0.01 and 0.25 N/mm for cohesion values between 0.1 and 1.8 MPa. In a subsequent research, Van der Pluijm (1999) stated that the ratio G_f^{II}/c (where c indicates the cohesion) falls in the range 0.062–0.147 mm, for different combinations of units and mortars. In the absence of detailed information, the recommended ratio is the intermediate value G_f^{II}/c equal to 0.1 mm.

As a final remark, a phenomenon encountered dealing with fracture processes is the tendency of solid materials to dilate (i.e. expand in volume) when sheared, the so-called *dilatancy*. A physical interpretation of dilatancy is shown in Figure 4.16a, where a granular material is considered. When compacted, grains are interlocked limiting mutual movements; when stressed, motion occurs, which produces an apparent volume expansion of the material. Intuitively, the level of confining stress plays a capital role in

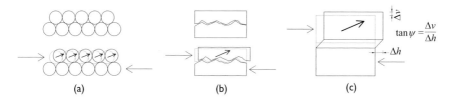

Figure 4.16 Physical interpretation of dilatancy in the case of shear action. (a) Granular material (with a certain level of compaction); (b) masonry dry joint (rough surface) and (c) mortar joint.

the grain re-arrangement, limiting the volume expansion. The same phenomenon occurs for masonry, with both dry joint (and rough surface) or mortar joint, as shown in Figure 4.16b and Figure 4.16c, respectively. The volume expansion can be measured at the macroscopic level by means of the dilatancy angle ψ, whose tangent is equal to the ratio between normal and shear displacement (Figure 4.16c). The angle of dilatancy basically measures the uplift of one unit over the other upon shearing.

Regarding dilatancy, Figure 4.17a shows the results of a cyclic shearing test of dry stone masonry joints with rough surface (Lourenço and Ramos, 2004). The diagram shows the relation between normal and shear displacements, which in this case indicates volume loss (i.e. negative dilatancy) for each load reversal due to compaction of the joints and regularization of the rough surface. This phenomenon contributes to interlocking loss in the case of irregular masonry units and likely out-of-plane masonry disintegration in the case of an earthquake. For the sake of clarity, the axis scales of the diagram are not the same and the ordinal numbers of cycles are indicated on the right. In the bottom, the diagram is drawn with the same scale for both axes. Negative dilatancy is not usually the case for mortar joints and, in the fashion of what was shown for fracture energy (Figure 4.15b), the dependence of tangent of ψ with confining pressure can be estimated thanks to multiple shear tests and subsequent regression analyses (Figure 4.17b). For low confining pressures, the average value of tangent of ψ falls in the range 0.2–0.7, depending on mortar and unit surfaces at the interface. For high confining pressures, dilatancy decreases to zero. An extrapolation of the results from Van der Pluijm (1993) indicates that a normal confining pressure ranging from 1.0 to 2.0 MPa is enough to yield zero dilatancy.

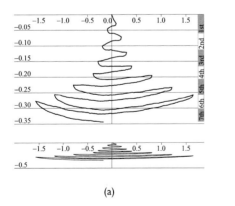

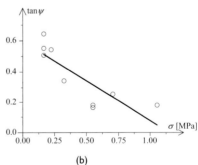

(a)   (b)

Figure 4.17 Dilatancy in masonry joints. (a) Dry stone masonry joints with rough surfaces under cyclic shearing (after Lourenço and Ramos, 2004) – horizontal and vertical displacements of the joints in the respective axes, in mm; (b) tangent of the dilatancy angle ψ as a function of the confining stress level σ for solid clay units after Van der Pluijm, 1993.

With increasing slip, dilatancy also decreases to zero due to the smoothing of the sheared surfaces.

In general, for practical applications and in the absence of further investigations, the dilatancy angle can be assumed equal to zero throughout the analysis.

4.1.4.4 Shear–tension and shear–compression interaction

Regarding tension and shear interaction, softening phenomena are expected to be coupled because they are both related to the bond between unit and mortar. This assumption is supported by the fact that, at the microscopic scale, the two phenomena are related to the breakage of the same material bridges. From the mechanical perspective, cohesion and tensile strength softening are linked in the degradation process. For simplicity, according to Figure 4.18a, the failure surfaces of Mode I and Mode II may be assumed to contemporarily shrink in the stress space.

Regarding shear and compression interaction, the evolution of the failure surfaces may be assumed uncoupled because they describe different phenomena: unit–mortar interface on the one side and masonry crushing on the other side. In the stress space their interaction may be represented by means of a linear cut-off (as done for tensile mode) or an elliptical cap. The latter represents a good compromise between the ultimate Mohr circle in

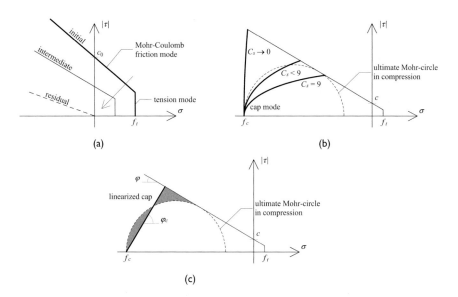

Figure 4.18 Shear–tension and shear–compression failure surface interaction. (a) Equal softening for Mode I and Mode II; (b) compressive elliptical cap with different values of the parameter C_s controlling the shear contribution to failure and (c) linearized cap in compression.

compression and a further limit to shear strength. A generic ellipse centred in the origin is governed by the equation $\sigma^2 + C_s\,\tau^2 = f_c^2$ (with f_c the masonry compressive strength). Figure 4.18b shows the cases with different values of C_s that controls the shape of the cap. If $C_s \to 0$, the cap degenerates into a straight vertical line; if $C_s = 9$, which is recommended, the maximum tangential stress is lower than one-third of the compressive strength. Other values can be chosen to fit the ultimate Mohr circle in compression (such as $C_s < 9$ in Figure 4.18b). Alternatively, a linearized cap mode through the masonry compressive value has been used in literature. This is depicted in Figure 4.18c, where the slope for compression is indicated with φ_c. As a general recommendation, Eq. (4.8) guarantees the best match with the ultimate Mohr circle in compression.

$$\varphi_c = 67° - 0.3\varphi \quad \text{[angles in degrees]} \tag{4.8}$$

4.1.4.5 Compressive failure

Compressive failure of masonry is addressed in Section 4.1.5. For the sake of completeness related to the interface, with reference to Figure 4.10c, the last parameter needed to define the crushing mode is the relative displacement κ_p. This can be calculated so that the total masonry strain at the end of the hardening phase is equal to 0.2%, which is a value typically found for concrete-like materials. In this regard, Eq. (4.9) can be used, where the underlined parenthesis represents the inverse of the equivalent Young's modulus in simplified micro-modelling. In the case dry joints are considered, $h_m = 0$.

$$\kappa_p = \left(h_b + h_m\right)\left(0.002 - f_c\left(\underline{\frac{1}{E_b} + \frac{1}{k_n\left(h_b + h_m\right)}}\right)\right) \tag{4.9}$$

4.1.5 Mechanical properties of masonry as a composite material

Macro-modelling is aimed at describing masonry as an equivalent homogenous material, with no distinction between units and mortar (and interfaces). A complete macro-model must reproduce the orthotropic material behaviour with different tensile and compressive strengths along the material axes (namely, vertical and horizontal), as well as different inelastic behaviour. Contrarily to modern masonry structures, given the complexity of this goal in existing structures, a simplified isotropic model is usually preferable in engineering applications.

To describe the overall material response, two main strategies are possible. The first approach regards laboratory experiments or in situ investigations performed directly on masonry elements, aimed at getting homogenous properties of the composite material. The dimensions of the specimen should be representative of the masonry as a whole, thus relatively large and costly to execute if compared to the standard cube or cylinder tests for concrete. The second approach stems from the analysis of the constituents from which, mathematically, the homogenized properties are derived. Whereas the former is rather intuitive, the latter is still actively pursued in the research community and is not discussed here.

From the computational point of view, if an isotropic model is considered, the most relevant mechanical parameters are limited to the compressive and tensile responses for uniaxial description (additional parameters are needed to describe the bi- and tri-axial material behaviour, e.g. the compressive strength increases with the confining action). The uniaxial description can be done in terms of hardening or softening criteria, fracture energies and strengths (in the fashion of what was shown in Section 4.1.3 about masonry components), usually considering the material behaviour in the vertical direction as this is assumed as the primary loading direction of masonry. In the framework of smeared crack models, cracks can be associated with tension in the principal direction at each integration point, and similarly crushing can be related with compression in the principal direction. As far as shear is concerned, by selecting the rotating smeared crack model, softening occurs implicitly as a result of the principal stress and strain conditions, with no need of further parameters.

4.1.5.1 *Axial behaviour (compression)*

Since the pioneering work of Hilsdorf (1969) it has been accepted by the masonry community that the difference of the elastic properties for unit and mortar is the precursor of failure. For illustrative purpose, the stress distribution following a uniaxial compression of a masonry prism is depicted in Figure 4.19. Given the typical low Young's modulus of mortar, and due to Poisson's effect, the compressed mortar tends to have a more pronounced lateral expansion (upper part of Figure 4.19b). Due to friction at the interface, this phenomenon leads to the interface stress distribution shown in the lower part of Figure 4.19b, leading to a twofold result. On the one hand, the lateral expansion of mortar produces a state of tension in the units, with the possible occurrence of cracks in the middle line; on the other hand, equilibrium of the joint leads to compressive forces for mortar, which becomes laterally confined and is able to reach a compressive stress much larger than the (unconfined) uniaxial compressive strength. Figure 4.19c illustrates the three-dimensional states of stress for the two masonry components, compression/biaxial tension for units and triaxial compression for mortar. Still, it is important to stress that a masonry prism under compression fails in a similar way to any concrete-like material, even if dry-stacked (i.e. without mortar).

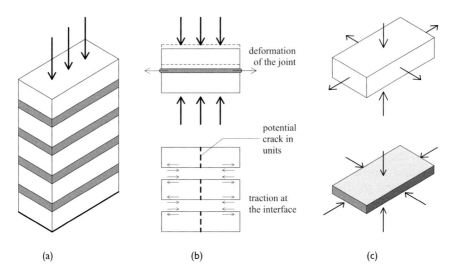

Figure 4.19 Stress distribution in a compressed masonry element. (a) Prism with units and mortar (the latter grey-shaded); (b) details of mortar deformation and relative traction at the interface (front view of the prism); (c) compression/biaxial tension in the unit and triaxial compression in the mortar (three-dimensional view).

The typical crack pattern of a masonry wall subjected to uniaxial compression state is shown in Figure 4.20a. Despite the influence of the boundary conditions (evident when the crack direction departs from verticality), the cracks on the right part of the specimen follow the reasoning described in Figure 4.19. Additionally, as vertical joints represent a natural discontinuity within the masonry elements, a certain continuity between unit cracks and vertical joints is noticeable, even at an early stage of loading. This phenomenon is schematized in Figure 4.20b by means of the simplified micro-modelling.

As an example, Figure 4.21 shows the stress–displacement diagrams of masonry prisms made with the same solid soft-mud brick (formed by moulding and not by extrusion) for different mortar compressive strengths, after Binda et al. (1988). As shown, for a given unit, the elastic properties of mortar may accommodate or exacerbate the Poisson's effect, leading to relatively high strength/brittle failure, or relatively low strength/ductile failure.

In the case of loading parallel to the direction of the bed joints, the compressive behaviour is quite similar to the previous case. In general, for solid units and irregular masonry, the two strengths are not significantly different but, for units with high or unfavourable perforation, large differences may be evident and further investigation is needed. In this regard, Hoffmann and Schubert (1994) found that the ratio between the uniaxial

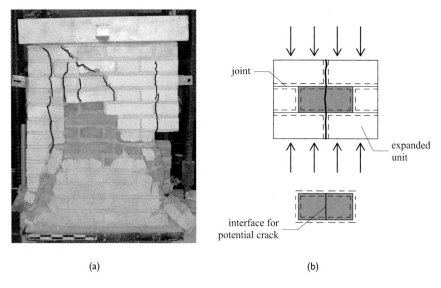

(a)   (b)

Figure 4.20 Masonry wall subjected to uniaxial compression. (a) Crack pattern and (b) simplified micro-modelling with interface for potential crack in the unit. (Photo courtesy of Giorgos Karanikoloudis.)

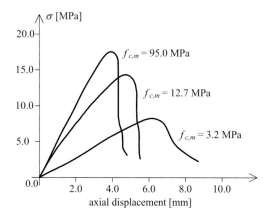

Figure 4.21 Typical uniaxial behaviour of masonry upon compressive loading normal to bed joints (Binda et al., 1988).

compressive strength parallel and normal to the bed joints ranges from 0.2 to 0.8. These ratios were obtained for masonry samples of solid and perforated clay units, calcium-silicate units, lightweight concrete units and aerated concrete units. These ratios cannot be used for horizontally perforated units, in which the ratio is typically larger than 1.0.

Moving to code recommendations, the compressive strength perpendicular to bed joints can be estimated from analytical formulas or tables proposed in building standards. Regarding formulas, they are generally in the form of Eq. (4.10). Eurocode 6 (EN 1996-1-1, 2005) proposes several values for the parameter K according to unit and mortar types, where α and β are equal to 0.7 and 0.3, respectively. For instance, for dimensioned natural stone and general purpose mortar $K=0.54$. The result is the average value of masonry strength, with unit and mortar strengths also considered with their average values. The main limitations about its applicability regard the compressive strength of the masonry unit, $f_{c,b}<75$ MPa, and the compressive strength of the mortar, $f_{c,m}<20$ MPa (in any case, lower than 2 $f_{c,b}$). In the case of thin-layer mortar (0.5–3 mm thick), Eurocode 6 provides a different formula where the compressive strength of masonry is only related to the compressive strength of the unit.

The New Zealand guidelines (NZSEE, 2017) proposes K, α and β equal to 0.75, 0.75 and 0.3, respectively, for mortar strength larger than 1 MPa, and 0.75, 0.75 and 0 (as mortar is uninfluential), respectively, for the case mortar strength is smaller than 1 MPa. The formula is referred to clay brick masonry and all the strengths involved (masonry, brick and mortar) are assumed with their probable values.

$$f_{c,M} = K f_{c,b}{}^{\alpha} f_{c,m}{}^{\beta} \qquad [\text{strengths in MPa}] \qquad (4.10)$$

Regarding recent masonry, Italian (NTC, 2018) and Australian (AS 3700, 2018) codes propose a tabular approach where the expected masonry strength is interpolated according to the strength of units and the strength of mortar. NTC (2018), in particular, deals with the strength of solid or hollow artificial units (voids $\leq$ 45%, bed and head joints 5–15 mm thick) and regular stone units. The recommended average values are synthesized in Tables 4.15 and 4.16, respectively. In turn, Table 4.17 reports the values proposed by AS 3700 (2018). These values should be multiplied by the joint thickness factor, whose formula is shown in the last line. For AAC masonry with thin-bed mortar, masonry compressive strength shall be taken as the unit one. The reference values proposed by TMS (2016) for clay and concrete masonry units in combination with different mortar types are shown in Table 4.18. Note that these values are based on the net surface area of a cross-section of masonry, given the common practice in the United States to grout specimens (only in this case is the net area equal to the gross area). The proposed values are applicable if the thickness of bed joints is less than 16 mm.

Regarding Young's modulus, many researchers and standards propose a linear proportion with the average or characteristic value of masonry strength. For instance, Eurocode 6 (2005) and Italian code (NTC, 2018) suggest a value equal to 800 times larger than the average masonry compressive strength. TMS (2016) suggests constants of proportion equal to

Table 4.15 Average values of masonry strength [MPa] normal to bed joints for solid and hollow artificial units

Average strength of units [MPa]	Mortar compressive strength [MPa]			
	2.5	5	10	15
2.4	1.4	1.4	1.4	1.4
3.6	2.4	2.6	2.6	2.6
6.0	3.6	4.0	4.1	4.2
9.0	4.2	4.9	5.4	6.0
12.0	4.9	5.6	6.4	7.4
18.0	6.1	7.2	8.0	9.8
24.0	7.3	8.4	9.6	11.6
36.0	8.6	10.3	12.0	14.4
≥ 48.0	-	12.5	14.4	17.2

Source: NTC (2018).

Table 4.16 Average values of masonry strength [MPa] normal to bed joints for regular natural units

Average strength of units[a] [MPa]	Mortar compressive strength [MPa]			
	2.5	5	10	15
2.7	1.2	1.2	1.2	1.2
4.0	2.4	2.6	2.6	2.6
6.7	3.6	4.1	4.1	4.2
10.0	4.2	4.9	5.4	6.0
13.3	4.9	5.6	6.4	7.4
20.0	6.1	7.2	8.0	9.8
26.6	7.3	8.4	9.6	11.6
39.9	8.6	10.3	12.0	14.4
≥ 53.2	-	12.5	14.4	17.2

Source: NTC (2018).

[a] As explicitly assumed by the code, the average compressive strength of natural units is conventionally set equal to 1.33 times the characteristic value.

700 and 900 for clay and concrete masonry, respectively. Other constants have been proposed in the last decades to fit experimental outcomes. According to Tomaževič (1999), Young's modulus is about 200–1,000 times larger than the compressive strength. A coefficient equal to 500, in the mid-range of the interval of Tomaževič (1999), is possibly more adequate for traditional masonry. The New Zealand guidelines (NZSEE, 2017) recommends for clay brick masonry a constant equal to 300, indicative of

Table 4.17 Average values of masonry strength [MPa] normal to bed joints. Mortar classes are based on mortar composition

Type of unit	Bedding type	Mortar class	Average compressive strength of unit [MPa]							
			6.0	12.0	18.0	24.0	30.0	36.0	48.0	≥60.0
Clay	Full	M2	3.0	4.2	5.2	5.9	6.6	7.2	8.4	9.4
Clay	Full	M3	3.7	5.3	6.5	7.6	8.4	9.2	10.6	11.9
Clay	Full	M4	5.4	7.6	9.2	10.7	12.0	13.1	15.2	16.9
Clay	Face shell	M3	4.3	6.1	7.4	8.6	9.6	10.6	12.1	13.6
Concrete	Full	M3	3.7	5.3	6.5	7.6	8.4	9.2	10.6	11.9
Concrete	Face shell	M3	4.3	6.1	7.4	8.6	9.6	10.6	12.1	13.6
Calcium silicate	Full	M3	3.7	5.3	6.5	7.6	8.4	9.2	10.6	11.9
Calcium silicate	Full	M4	5.4	7.6	9.2	10.7	12.0	13.1	15.2	16.9

Source: AS 3700 (2018).

The tabular values should be subsequently multiplied by the joint thickness factor calculated as follows:

$$\text{Joint thickness factor} = 1.3\left(\frac{h_b}{19h_m}\right)^{0.29} \leq 1.3.$$

Table 4.18 Compressive strength of masonry based on the compressive strength of clay and concrete masonry units vs. type of mortar. Types N, S and M have a minimum compressive strength equal to 5.2, 12.4 and 17.2 MPa, respectively

Net area compressive strength of clay masonry units [MPa]		Net area compressive strength of masonry [MPa]	Net area compressive strength of concrete masonry units[a] [MPa]		Net area compressive strength of masonry [MPa]
Type M or S mortar	Type N mortar		Type M or S mortar	Type N mortar	
11.72	14.48	6.90	-	13.79	12.07
23.10	28.61	10.34	13.79	18.27	13.79
34.13	42.75	13.79	17.93	23.44	15.51
45.51	56.88	17.24	22.41	28.96	17.24
56.88	71.02	20.69	26.89	-	18.96
68.26	-	24.13	31.03	-	20.69
79.29	-	27.58			

Source: TMS (2016).

[a] For units of less than 102 mm height, 85% of the values listed.

the secant modulus of elasticity between 5% and 70% of the maximum strength. Australian standard (AS 3700, 2018) proposes a set of relations according to unit types and mortar grades, as shown in Table 4.19.

Table 4.19 Relation between Young's modulus and average value of masonry strength normal to bed joints according to type of units and mortar grade

Type of unit	Mortar compressive strength [MPa]	Young's modulus
Clay unit $f_{c,u} = 5$–30 MPa	2–3	$580\ f_{c,M}$
Clay unit $f_{c,u} > 30$ MPa	3–4	$830\ f_{c,M}$
Concrete unit with mass density $\geq 1{,}800$ kg/m^3 and calcium silicate units	3–4	$830\ f_{c,M}$
Concrete units with mass density $< 1{,}800$ kg/m^3	2–3	$630\ f_{c,M}$
Grouted concrete or clay masonry	Any	$830\ f_{c,M}$
AAC	Thin bed	$420\ f_{c,M}$

Source: AS 3700 (2018).

Finally, for what concerns shear modulus in modern masonry, it is recommended to set it equal to 0.4 times Young's modulus E, i.e. Poisson's ratio equal to 0.2. For existing masonry, the Italian code (Circolare 2019) recommends a ratio equal to one-third. Regarding fracture energy in compression, this can be estimated according to Eq. (4.1), as discussed in Section 4.1.3.1. However, critical reasoning may lead to different values to be assumed for structural analysis, depending on the brittleness or ductility expected.

4.1.5.2 Axial behaviour (tension)

In the case of tensile loading perpendicular to the bed joints, failure is generally triggered by the relatively low tensile bond strength between bed joint and unit, see Section 4.1.4.2. The different nature of compressive and tensile strengths is evident, the former related to the composite behaviour of units and mortar, the latter related to the joint behaviour. An attempt to relate masonry tensile and compressive strengths is thus unrealistic. If tensile behaviour is considered along the direction perpendicular to bed joints, the elastic properties can be assumed equal to the case in compression.

In turn, Figure 4.22 describes the typical behaviour of masonry under tension in the direction parallel to the bed joints. According to mortar–unit combinations, two main failure possibilities are evident: (1) a stepped crack through head and bed joints (smoother curve, with residual strength) and (2) a straight crack through head joints and units (stepped curve due to the sudden breakage of units in tension). A stepped crack is more frequent and is related to the cohesion c at the horizontal interface, units overlapping length l_o and unit height h_b, as shown in Eq. (4.11). In the formula, the thickness of mortar is considered negligible and the contribution of the tensile bond strength at the vertical joints is neglected, as well as the beneficial effect of normal stress in the bed joints. Differently, a straight crack is indicative of masonry composed of low-strength units or greater bond strength between bed joints and units, e.g. high-strength

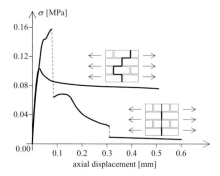

Figure 4.22 Typical uniaxial behaviour of masonry upon tensile loading parallel to bed joints (Backes, 1985).

mortar and units with numerous small perforations, which produce a dowel effect. In this case, tensile strength of masonry as a composite $f_{t,M}$ can be assumed equal to half of the unit tensile strength $f_{t,b}$, as shown in Eq. (4.12), again neglecting the contribution of the tensile bond strength at the vertical joints.

$$f_{t,M} = \frac{cl_o}{h_b} \tag{4.11}$$

$$f_{t,M} = \frac{f_{t,b}}{2} \tag{4.12}$$

4.1.5.3 Biaxial behaviour

Due to the intrinsic anisotropy of the material, the biaxial characterization of masonry is complex and requires advanced experimental procedures. Nonetheless, for the sake of completeness, a brief introduction follows.

The most complete set of experimental data on masonry subjected to proportional biaxial loading is shown in Figure 4.23 (Page, 1981, 1983). As it is possible to notice, the principal stresses are indicated according to the rotation angle θ with respect to the material axes (i.e. bed and head directions). The tests were carried out with half-scale solid clay units. Both the orientation of the principal stresses and the principal stress ratio considerably influence the failure mode and strength. As expected, no matter the inclination of the principal stresses, the tensile strength is much lower than the compressive strength. The different modes of failure are illustrated in Figure 4.24, essentially a combination of cracking and sliding through head and bed joints, except for biaxial compression, in which masonry fails out of plane.

A plane stress continuum model, which can capture different strengths and softening characteristics in orthogonal directions, was formulated

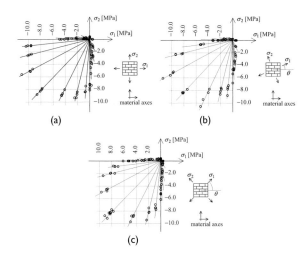

Figure 4.23 Biaxial strength of solid clay unit masonry (after Page, 1981, 1983). Principal direction rotated of (a) 0°, (b) 22.5° and (c) 45° with respect to the material axes.

by Lourenço (1996), based on multisurface plasticity. More in detail, it comprises of an anisotropic-type Rankine's failure criterion combined with an anisotropic Hill's criterion for compression, accounting for different strength values for tension and compression (namely, f_t and f_c) along each material axis x and y. The model is illustrated in Figure 4.25a in the plane of the normal stress components along the material axes (described in Figure 4.25b). As it is possible to observe, the failure surface is three-dimensional and shrinks with an increasing level of shear stress.

The study of masonry components in the previous subsections made clear the difficulties for the complete mechanical characterization of historical masonries. This gets more complicated when the masonry assemblage is considered, together with its evident orthotropy. On top of that, it is noted that the diagrams shown in Figures 4.23 and 4.24 are of limited applicability for other types of masonry as different strength envelopes are likely to be found for different masonry materials, unit shapes and geometry.

4.1.6 Structural behaviour of masonry as a composite

Whereas the previous sections shed light on the material properties for both micro- and macro-modelling, the present section aims at giving an overview on the structural behaviour of masonry elements. Basically, as shown in Section 4.1.1, a good-quality masonry guarantees a significant unit interlocking and, consequently, masonry elements behave monolithically, displaying an appreciable capacity for in-plane loading. In turn, a poor-quality masonry is usually characterized by low-resistance mortar, irregular and

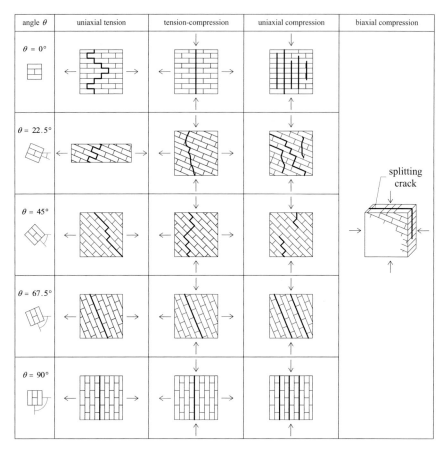

Figure 4.24 Modes of failure of solid clay unit masonry under biaxial loading, after Dhanasekar et al., 1985.

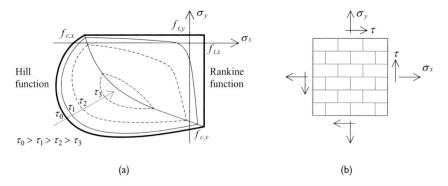

Figure 4.25 Rankine–Hill anisotropic model for masonry. (a) Multisurface elastic domain with iso-shear stress lines and (b) description of the three stress components involved in plane stress.

small-sized units, with an overall insufficient unit interlocking along height and thickness. In this case, the behaviour is not monolithic, and masonry walls may disintegrate abruptly, leading to the collapse of the entire building.

Although the insight of the present subsection may be not directly exploitable in FE analyses, it is still considered crucial for the overall understanding of the material and for verifying a numerical model. In this regard, two aspects are discussed next. The former regards the experimental activity on different types of walls subjected to combined compressive and shear actions (i.e. shear walls). The latter illustrates the Masonry Quality Index (MQI), based on the Italian code (Circolare, 2019). This is a method originally developed by Borri and co-workers (2015, 2019) according to which the mechanical characteristics of masonry can be assessed through visual inspections.

4.1.6.1 Shear walls

The response of masonry subjected to shear actions is strongly dependent on wall morphologies. Vasconcelos (2005) carried out an experimental campaign performing shear–compression tests on three types of single-leaf walls. Figure 4.26a–c shows the elevation of the three walls made of granite, with regular (dry assembled), irregular and rubble stones, respectively; Figure 4.26d illustrates the scheme of the tests, i.e. uniform constant vertical stress and horizontal force monotonically increasing. The three masonries exhibit differences in terms of capacity, ductility, energy dissipation and failure mechanisms. Figure 4.27 shows the shear failure modes for the three types of walls with increasing normal stress. As it is possible to notice, the arrangement of masonry units plays a crucial role in the crack formation, while units break only under high level of stresses. This is paradigmatic of the lower tensile and shear strengths of joints (in terms of both mortar and interface) when compared to the strength of units. Additionally, with particular regard to the two last walls, although the components are the same with equal individual

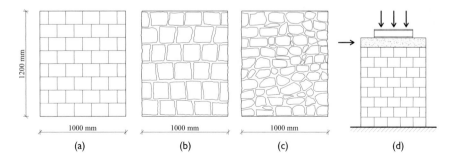

Figure 4.26 Shear–compression tests on three types of walls (after Vasconcelos, 2005). (a) Regular (dry assembled), (b) irregular and (c) rubble stone configuration. (d) Typical test configuration.

(a)   (b)   (c)

Figure 4.27 Shear failure mechanisms of the tests described in Figure 4.26, with normal
stress equal to 0.50, 0.88 and 1.25 MPa (from top to bottom). (a) Regular
(dry assembled), (b) irregular and (c) rubble stone configuration. (Photos
courtesy of Graça Vasconcelos.)

mechanical properties, unit shape and arrangement revealed themselves crucial for the response. Therefore, visual inspection and a critical evaluation of masonry elements are of fundamental importance for numerical simulations.

Limiting the discussion to regular masonry, the behaviour of walls under in-plane loading largely depends on the aspect ratio (i.e. height over base of the structural element) and the vertical compression. For slender walls and walls with relatively low axial stress, the behaviour is usually dominated by flexure. The strength is limited by rocking, possibly followed by toe crushing. For stocky walls and walls with moderate to heavy axial stress, shear usually dominates through bed joint sliding or diagonal tension modes of failure. For increasing values of normal stress, diagonal cracking involving units and diagonal-compression cracking–crushing are possible. The mentioned cases are depicted in Figure 4.28, together with an indicative subdivision of the σ–τ stress domain (with elliptical cap in compression). The reader should not be misled by the similitude of the graph in Figure 4.28 with the one in Figure 4.9 (and Figure 4.18). Although both are based on Mohr–Coulomb's criterion, Figure 4.28 refers to masonry shear walls, not to the interface.

4.1.6.2 Masonry Quality Index

According to the analysis of masonry morphologies, Borri and co-workers (2015, 2019) proposed the so-called Masonry Quality Index, shortly MQI, according to which the quality of masonry is evaluated in accordance with seven parameters. In other words, the respect of the rules of the art (that ensure compactness and monolithic behaviour) in the given conservation state is "measured" by means of a qualitative appraisal, ending up with values of MQI in the range of 0–10. Based on MQI outcomes, according to correlation curves, it is possible to obtain the expected mechanical properties. This approach is appealing because the material properties are assessed thanks to visual inspection only. According to the Italian code (Circolare, 2019), this must include plaster removal to identify the nature and dimensions of masonry components (units and mortar), unit arrangement and, possibly, must also include the inspection of the cross-section (see Section 4.5.1.2).

Referring the reader to the mentioned works for a comprehensive description, the methodology assumes the evaluation of distinct MQIs according to three load directions, namely, vertical (MQI_v), horizontal in-plane (MQI_{ip}) and horizontal out-of-plane (MQI_{op}). Regarding FE analysis, only MQI_v is of interest but, for the sake of completeness and to provide more insight on the structural behaviour of masonry, the discussion encompasses all indexes. In the following, rubble, rounded or pebble stones are referred to as irregular stones and, by analogy, irregular is the masonry that comes thereof.

The seven parameters are presented in Tables 4.20 and 4.21, whereas the possible outcomes are *Not Fulfilled* (NF), *Partially Fulfilled* (PF) and *Fulfilled* (F). As it is possible to notice, the two tables are mainly based on

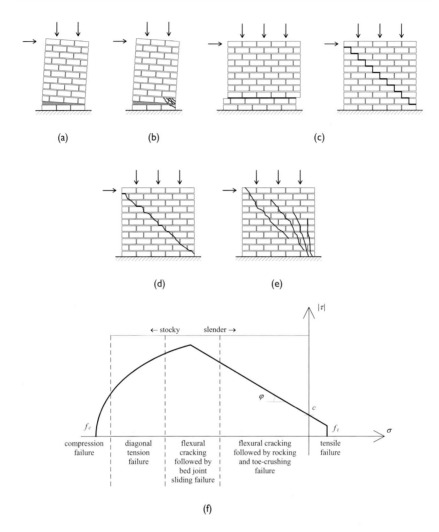

Figure 4.28 Different failure modes for masonry wall subjected to shear and normal stress. (a) Flexural cracking; (b) rocking and toe crushing; (c) shear sliding along one course or with stepped crack; (d) diagonal cracking involving unit failure; (e) diagonal-compression cracking–crushing and (f) indicative subdivision of σ–τ stress domain of masonry shear walls according to failure type.

qualitative criteria. However, for the two parameters in Table 4.21, namely, WC and VJ (related to wall leaf connections and vertical joint, respectively), a quantitative approach is also proposed. This is based on the interlocking between units through the wall thickness and on the leaf surface, respectively. The quantitative procedure is detailed in Figures 4.29 and 4.30 and it is based on the evaluation of M_I, which is the ratio between the minimum distance to connect two points on the wall (passing only through

Table 4.20 Criteria for evaluating SM, SD, SS, HJ, MM parameters and numerical values according to load direction

Parameter	Outcome	Criteria	Vertical (v)	In-plane (ip)	Out-of-plane (op)
SM Conservation state and mechanical properties of units	NF	More than 50% of units are deteriorated/damaged Hollow bricks (void area >70%) Adobe bricks or unfired clay bricks ($f_{c,u}$=0.5–5 MPa)	0.3	0.3	0.5
	PF	Deteriorated/damaged elements in the range 10%–50% Hollow bricks (void area in the range 70%–55%) Sandstone or tuff elements ($f_{c,u}$=5–20 MPa)	0.7	0.7	0.7
	F	Less than 10% of deteriorated/damaged units Hollow bricks (void area <55%) Solid fired bricks Concrete blocks and hardstone	1	1	1
SD Unit dimension	NF	More than 50% of units having largest dimension <0.20 m Header bond (single leaf with all units placed as headers on the faces of the walls, see Figure 4.1b)	0	0	0
	PF	More than 50% of units having largest dimension in the range 0.20–0.40 m Presence of units of different dimensions	0.5	0.5	0.5
	F	More than 50% of units having largest dimension >0.40 m	1	1	1
SS Unit shape	NF	Predominance of rubble, rounded or pebble stonework on both leaves	0	0	0
	PF	Presence on both leaves of irregular stones, together with regular stones and bricks Two-leaf walls, of which one is made of regular stones or bricks and the other of irregular stones Irregular stones (rubble, rounded, pebble) with presence of effective wedge stones[a]	1.5	1	1
	F	Predominance of regular stones or bricks on both leaves Brickwork	3	2	2

(Continued)

Table 4.20 (Continued) Criteria for evaluating SM, SD, SS, HJ, MM parameters and numerical values according to load direction

Parameter	Outcome	Criteria	Vertical (v)	In-plane (ip)	Out-of-plane (op)
HJ Type of bed joints	NF	Lack of continuous bed joints	0	0	0
	PF	Intermediate situation between NF and F For double-leaf walls, only one leaf with continuous bed joints	1	0.5	1
	F	Continuous bed joints (for at least 1.0 cm) on both leaves Stone masonry wall with brick courses (distance between courses < 1.00 m)	2	1	2
MM Mortar mechanical properties and bonding	NF	Very weak or dusty mortar with no cohesion Lack of mortar in irregular masonry Bed joints made of weak mortar whose thickness is comparable to the unit height Porous stones (e.g. tuff) or bricks with a weak bond to mortar	0	0	0
	PF	Medium-quality mortar, with bed joints not largely notched or degraded Irregular masonry and weak mortar, but with effective wedge stones[a]	0.5	1	0.5
	F	Good-quality and nondegraded mortar in bed joint, whose thickness does not exceed unit height Large bed joint thickness made of very good-quality mortar Large ashlar stones with no mortar or thin layer mortar	2	2	1

Source: Borri and co-workers (2015, 2019).

a Small stones or brick pieces used to fill an excessive gap between two units.

Table 4.21 Criteria for evaluating WC and VJ parameters in the MQI assessment

Parameter	Outcome	Quantitative analysis (visible section)	Criteria — Quantitative analysis (nonvisible section)	Vertical (v)	In-plane (ip)	Out-of-plane (op)
WC Wall leaf connections (for M_i, see Figure 4.29)	NF	$M_i < 1.25$ Or small stones (regardless of M_i)	Small units compared to wall thickness No headers (i.e. diatones) or less than 2/m²	0	0	0
	PF	$1.25 < M_i < 1.55$	For double-leaf wall, at least one with a good unit arrangement Presence of some headers ($\approx$ 2–5/m²) Wall thickness lightly larger than unit largest dimension	1	1	1.5
	F	$M_i > 1.55$	Wall thickness similar to stone largest dimension Systematic presence of headers (>4–5/m²)	1	2	3
VJ Vertical joint (for M_i, see Figure 4.30)	NF	For single-leaf wall, $M_i < 1.4$ For double-leaf wall, $M_i < 1.4$ for one leaf and $M_i < 1.6$ for the second one Wall made of very small stones	Aligned vertical joints (e.g. stack bong of Figure 4.2a) Aligned vertical joints for at least two large units in large portions of the wall Header bond (Figure 4.2b), despite the minimum interlocking Clear and visible deficiency of interlocking along one or more vertical lines	0	0	0
	PF	For single-leaf wall, $1.4 < M_i < 1.6$ For double-leaf wall: (a) both leaves $1.4 < M_i < 1.6$ (b) for at least one leaf $M_i > 1.6$ (c) first leaf $M_i > 1.6$ (d) second leaf $1.4 < M_i < 1.6$	Partially staggered vertical joints (vertical joint between two units is not placed in the middle of adjacent upper and lower units)	0.5	1	0.5
	F	For single-leaf wall, $M_i > 1.6$ For double-leaf wall, both leaves $M_i > 1.6$	Properly staggered vertical joints (vertical joint between two units is placed in the middle of adjacent upper and lower unit)	1	2	1

Source: Borri and co-workers (2015, 2019).

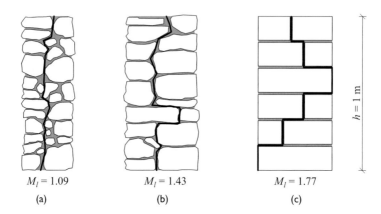

$M_l = 1.09$ $M_l = 1.43$ $M_l = 1.77$

(a)   (b)   (c)

Figure 4.29 Quantitative analysis of wall leaf connection parameter (WC) through the wall thickness (after Borri and co-workers, 2015, 2019). (a) NF, (b) PF and (c) F.

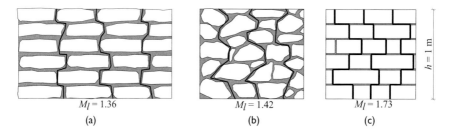

$M_l = 1.36$ $M_l = 1.42$ $M_l = 1.73$

(a)   (b)   (c)

Figure 4.30 Quantitative analysis of vertical joint parameter (VJ) (after Borri and co-workers, 2015, 2019). (a) NF, (b) PF and (c) F.

mortar joints) and the straight distance between the same two points. The straight distance is usually assumed to be equal to 1 m, but smaller values up to 0.5 m may be used. As noticeable, the quantitative evaluation of WC (Figure 4.29) is possible only when the cross-section is visible (or carefully inspected).

Tables 4.20 and 4.21 show also the numerical outcomes in terms of MQI, according to the three load directions (MQI_v, MQI_{ip} and MQI_{op}). On closer inspection, it is possible to notice that the numerical values are given to each parameter based on its fulfilment to the rule of the art. However, based on Figures 4.2 and 4.3, the respect of the rules may be more or less influential for a given load direction. For instance, according to these figures, the parameters WC and VJ (inherent to wall connection and vertical joint, respectively) are more decisive for out-of-plane and in-plane actions, respectively.

All the numerical values are included in Eq. (4.13) to calculate the three MQIs, one per load direction. The formulas are differentiated according to two typologies of masonry, namely, made with solid regular or hollow / rubble units. Based on the recommendations of Italian code (Circolare, 2019), three parameters are also introduced (intermediate values may be reasonable):

- m is a corrective coefficient related to the compressive strength of mortar: $m=0.7$ for poor-quality ($f_c<0.7\,\text{MPa}$) and $m=1$ for good-quality mortar.
- g is a corrective coefficient about joint thickness (only for solid unit masonry): $g=0.7$ for joints larger than 13 mm and $g=1$ for smaller values. It is worth reminding that Eurocode 6 (EN 1996-1-1, 2005) defines normal thickness between 6 and 15 mm.
- r is a corrective coefficient about the effect of mortar quality on compressive and shear strength (only for solid unit masonry). It depends on the parameter MM as shown in Table 4.22.

Solid regular unit masonry

$$\text{MQI}_v = m\ g\ r_v\ \text{SM}_v\left(\text{SD}_v+\text{SS}_v+\text{HJ}_v+\text{MM}_v+\text{WC}_v+\text{VJ}_v\right)$$

$$\text{MQI}_{ip} = m\ g\ r_{ip}\ \text{SM}_{ip}\left(\text{SD}_{ip}+\text{SS}_{ip}+\text{HJ}_{ip}+\text{MM}_{ip}+\text{WC}_{ip}+\text{VJ}_{ip}\right)$$

$$\text{MQI}_{op} = m\ g\ r_{op}\ \text{SM}_{op}\left(\text{SD}_{op}+\text{SS}_{op}+\text{HJ}_{op}+\text{MM}_{op}+\text{WC}_{op}+\text{VJ}_{op}\right)$$

$$(4.13)$$

Other cases (e.g. rubble or hollow units)

$$\text{MQI}_v = m\ \text{SM}_v\left(\text{SD}_v+\text{SS}_v+\text{HJ}_v+\text{MM}_v+\text{WC}_v+\text{VJ}_v\right)$$

$$\text{MQI}_{ip} = m\ \text{SM}_{ip}\left(\text{SD}_{ip}+\text{SS}_{ip}+\text{HJ}_{ip}+\text{MM}_{ip}+\text{WC}_{ip}+\text{VJ}_{ip}\right)$$

$$\text{MQI}_{op} = m\ \text{SM}_{op}\left(\text{SD}_{op}+\text{SS}_{op}+\text{HJ}_{op}+\text{MM}_{op}+\text{WC}_{op}+\text{VJ}_{op}\right)$$

Table 4.22 Corrective coefficient r_v, r_{ip} and r_{op}, according to MM parameter

Parameter MM	r_v	r_{ip}	r_{op}
NF	0.2	1	0.1
PF	0.6	1	0.85
F	1	1	1

Source: Borri et al. (2019).

Table 4.23 Masonry categories and qualitative behaviour according to MQIs

MQIs	Masonry categories (provided the respect of the rule of the art)		
	A	B	C
MQI_v Vertical actions	5–10 Cracking is very unlikely	2.5–5 Possible cracks but collapse is unlikely	0–2.5 Evident cracks Out-of-plumb Possible collapse for high-stress states
MQI_{ip} In-plane lateral actions	7–10 Cracking is unlikely	4–7 Limited cracking	0–4 High probability of major and critical cracks
MQI_{op} Out-of- plane actions	5–10 Monolithic behaviour Cracking is unlikely	3–5 Possible cracks and bowing Macroblock analysis conservatively with two separate leaves Collapse is detained by transversal elements (floors and walls) if efficiently connected	0–3 No monolithic behaviour Ineffective connections with other structural elements Macroblock analysis unrealistic High probability of collapse

Source: Borri et al. (2019).

The indexes MQIs provide a qualitative understanding of the behaviour of masonry elements when subjected to vertical and lateral in-plane and out-of-plane actions. Table 4.23 illustrates three categories of masonries, namely, A, B and C and their expected performances. Of the three loading directions, the horizontal out-of-plane actions (last line of the table) deserve particular attention. Even though the related MQI_{op} is not meant to provide mechanical parameters (as shown ahead), it gives precious information on the possibility of adopting simplified analyses. Finally, it is worth highlighting that poor-quality masonries are likely to lead to a structural behaviour that is not monolithic with material disintegration, making FE analysis ineffective and the results meaningless.

Additionally, instability under compressive actions is a delicate issue that must be verified before proceeding with any analysis. In this regard, the Italian code explicitly warns the analyst about the possibility of leaves not adequately connected or with rubble infill, that may trigger premature instability failure. Three-leaf walls represent a typical case. These may be composed of two outer leaves with ashlars or brickwork filled with irregular stones or rubble with poor cohesion and without any interlocking. Due to the inadequate mechanical properties of the infill and the lack of through stones, instability may occur, likely emphasized by the eccentricity of vertical loads (e.g. on one leaf only). In the case instability is not prevented, the norm suggests considering the two leaves as independent.

With the aim of evaluating the expected mechanical parameters of a given masonry, Borri et al. (2019) performed correlation analyses between

Table 4.24 Values of compressive strength, shear strengths and elastic moduli of masonry

MQIs	f_c [MPa] min average max	$f_{v0}{}^a$ [MPa] min average max	E [MPa] min average max	G [MPa] min average max
$x = MQI_v$	$1.0360e^{0.1961x}$ $1.4211e^{0.1844x}$ $1.8021e^{0.1775x}$	-	$599.03e^{0.1567x}$ $731.5e^{0.1548x}$ $863.74e^{0.1535x}$	-
$x = MQI_{ip}$	-	$0.0300x^{0.9093}$ $0.0475x^{0.8536}$ $0.0654x^{0.8219}$	-	$204.50e^{0.1464x}$ $247.62e^{0.1457x}$ $290.56e^{0.1452x}$

Source: Borri et al. (2019).

[a]   Only for regular masonry.

the three MQIs and reference values per the Italian code (Circolare, 2019). Once MQIs are known, the mechanical parameters of masonry can be estimated according to Table 4.24. The formulas already include multiplication factors needed to evaluate the condition detailed in Section 4.1.7.3 (namely, good-quality mortar, normal thickness joints and transverse connections). As it is possible to notice, the quantities related to axial behaviour depend on MQI_v, whereas the ones related to shear depend on MQI_{ip}. Regarding shear, f_{v0} is the shear strength under zero compressive stress, provided only for regular masonry, i.e. masonry that can develop a shear stepped failure mechanism such as the one presented in Figure 4.28. Consequently, f_{v0} can be used as the cohesion for bed joints in the micro-modelling approach.

4.1.7 Synthesis of the mechanical properties

The goal of the present section is to provide an overview about estimates of mechanical properties to perform advanced analysis of masonry constructions. The reader is referred to the previous sections where a description of the mechanical properties is presented. Additional information may be available locally for a specific geography, may be gathered from laboratory testing or obtained from the building itself.

The following two subsections deal with micro- and macro-modelling. The data given may be of help to assess missing data due to incomplete investigations or to set up values for preliminary analyses, in the absence of more information. The subsequent subsections regard recommended mechanical properties for macro-modelling as proposed by selected building codes.

4.1.7.1 Individual masonry components

The essential mechanical properties for detailed and simplified micro-modelling of masonry structures are given in Tables 4.25 and 4.26, respectively.

4.1.7.2 Masonry as a composite material

Table 4.27 reports a synthesis of the mechanical properties for macro-modelling analysis of masonry, considered as an isotropic material.

Table 4.25 Mechanical properties of masonry for detailed micro-modelling approach

Elements	Recommended mechanical properties			
Unit (isotropic)	Young's modulus	E	MPa	See Tables 4.2–4.4
	Poisson's ratio	$\nu = 0.2$	-	
	Compressive strength	f_c	MPa	
	Fracture energy in compression	$G_c = \dfrac{32 f_c}{10 + f_c}$	N/mm	Eq. (4.1)
	Flexural to tensile strength	$f_t = \dfrac{0.06 h^{0.7}}{1 + 0.06 h^{0.7}} f_{fl}$	MPa	Eq. (4.2)
	Tensile strength	$f_t = 2.12 \ln(1 + 0.075 f_c)$	MPa	Eq. (4.3)
	Fracture energy in tension	$G_f^I = 0.07 \ln(1 + 0.17 f_c)$	N/mm	Eq. (4.4)
	Hardening–softening criteria for both compression and tension (σ–ε diagrams)	Parabolic for compression and exponential for tension		
Mortar (isotropic)	Young's modulus	E	MPa	Tables 4.5–4.8
	Poisson's ratio	ν	-	
	Compressive strength	f_c	MPa	
	Fracture energy in compression	$G_c = \dfrac{32 f_c}{10 + f_c}$	N/mm	Eq. (4.1)
	Flexural to tensile strength	$f_t = \dfrac{0.06 h^{0.7}}{1 + 0.06 h^{0.7}} f_{fl}$	MPa	Eq. (4.2)
	Tensile strength	$f_t = 2.12 \ln(1 + 0.075 f_c)$	MPa	Eq. (4.3)
	Fracture energy in tension	$G_f^I = 0.07 \ln(1 + 0.17 f_c)$	N/mm	Eq. (4.4)
	Hardening–softening criteria for both compression and tension (σ–ε diagrams)	Parabolic for compression and exponential for tension		

(Continued)

Table 4.25 (Continued) Mechanical properties of masonry for detailed micro-modelling approach

Elements	Recommended mechanical properties			
Interface	Dummy normal and shear stiffness (to avoid interpenetration)	$> 10^4 – 10^6$	N/mm³	
	Bond tensile strength	$f_t = 0.1$	MPa	Weak mortar (Null for dry joint)
	Flexural to tensile strength	$f_t = \dfrac{0.06h^{0.7}}{1+0.06h^{0.7}} f_{fl}$	MPa	Eq. (4.2)
	Mode I fracture energy	$G_f^I = 0.01$	N/mm	
	Cohesion	$c = 1.4 f_t$	MPa	See also Tables 4.11, 4.13, 4.14
	Friction coefficient	$\tan \varphi = 0.75$	-	For cohesion, see also Section 4.1.6.2 and the following ones
	Dilatancy coefficient	$\tan \psi = 0$	-	
	Tensile and shear softening criteria (σ–ε diagram)	Exponential		
	Mode II fracture energy	$G_f^{II} = 0.1\ c$	N/mm	

Note: h is the height of the specimen adopted in flexural testing; f_{fl} is the flexural strength.

Note that strength and cohesion must be given in MPa, whereas height h must be given in mm.

Based on literary review and on-site testing, Lourenço and Pereira (2018) proposed the mechanical parameters for historic adobe masonry structures as shown in Table 4.28.

4.1.7.3 Data from Italian code

According to the Italian code (Circolare, 2019), the recommended mechanical properties for masonry are reported in Table 4.29, where f_{v0} is the shear strength under zero compressive stress. Except for the last line, all the masonry typologies refer to precise conditions: low-strength lime mortar ($0.7 < f_c < 1.5$ MPa), nonconnected or weakly connected masonry leaves, as-built unreinforced masonry and masonry bond pattern according to the rules of art. With the aim of assessing the mechanical properties of better-quality masonries (e.g. good-quality mortar, normal thickness joints and

Table 4.26 Mechanical properties of masonry for simplified micro-modelling approach

Elements	Recommended mechanical properties		
Expanded units (isotropic)	Young's modulus E	MPa	As per detailed micro-modelling, Table 4.25
	Poisson's ratio ν	-	
Interface for bed joints	Equal to the interface properties of Table 4.25, except for		
	Normal and shear stiffness (Mortar joint)[a] $k_n = \dfrac{E_b E_m}{h_m (E_b - E_m)}$ $k_s \approx 0.4 k_n$	N/mm³	See Eq. (4.6)
	Normal and shear stiffness $k_n = \dfrac{E_b E_M}{(h_b + h_m)(E_b - E_M)}$ $k_s \approx 0.4 k_n$	N/mm³	See Eq. (4.7)
	Compressive strength of masonry $f_{c,M} = K f_{c,b}{}^{\alpha} f_{c,m}{}^{\beta}$ $\alpha = 0.7$ $\beta = 0.3$ $K = 0.54$ (general purpose mortar)	MPa	Eq. (4.10) See also Section 4.1.6.2 and the following ones
	Hardening–softening criterion for compression (σ–ε diagram)	Parabolic	
	Fracture energy in compression $G_c = \dfrac{32 f_{c,M}}{10 + f_{c,M}}$	N/mm	Eq. (4.1)

Note: subscripts b, m and M indicate unit, mortar and masonry, respectively (e.g. h_m is the thickness of the mortar).

Note that strength, Young's modulus and shear modulus must be given in MPa, whereas the height h must be given in mm.

[a]   In numerical applications, if mortar Young's modulus is estimated from experimental tests on individual mortar specimens, it is recommended to divide by two the theoretical values obtained.

transverse connections), the code provides multiplication factors as reported in Table 4.30. These values should be considered as indicative, and further insight can be gathered on a case-by-case basis. When the contemporary presence of several beneficial effects is observed, no more than two coefficients (the most effective ones) can be adopted per time. The mentioned coefficients are distinguished according to the following list:

- **Good mortar** affects strengths, Young's modulus and shear modulus.
- **Presence of regularizing courses** affects only strengths, and it can be used only for masonry typologies where this technique is observed.
- **Transversal connection** (e.g. through stone) affects only strengths.

Table 4.27 Mechanical properties for macro-modelling approach of masonry as an isotropic material

Young's modulus	$E = 500\text{-}800\ f_c$	MPa	
Poisson's ratio	$\nu = 0.2$	-	
Compressive strength	$f_{c,M} = 0.54 f_{c,b}^{0.7} f_{c,m}^{0.3}$	MPa	See Eq. (4.10) and the following sections
Fracture energy in compression	$G_c = \dfrac{32 f_{c,M}}{10 + f_{c,M}}$	N/mm	Eq. (4.1)
Flexural to tensile strength	$f_t = \dfrac{0.06 h^{0.7}}{1 + 0.06 h^{0.7}} f_{fl}$ (h in mm)	MPa	Eq. (4.2)
Tensile strength	f_t	MPa	Equal to interface bond tensile strength of Table 4.25, if available
Fracture energy in tension	$G_f^I = 0.02$	N/mm	
Hardening–softening criteria for both compression and tension (σ–ε diagrams)	Parabolic for compression and exponential for tension		

Note: h is the height of the specimen adopted in flexural testing; f_{fl} is the flexural strength; subscripts b, m and M indicate unit, mortar and masonry, respectively.

Note that the compressive strength f_c must be given in MPa and the height h must be given in mm.

Table 4.28 Mechanical properties of historic adobe masonry as an isotropic material

Young's modulus	$E = 100$	MPa
Poisson's ratio	$\nu = 0.2$	-
Compressive strength	$f_c = 0.5$	MPa
Fracture energy in compression	$G_c = 1.5$	N/mm
Tensile strength	$f_t = 0.05$	MPa
Fracture energy in tension	$G_f^I = 0.02$	N/mm

Source: Lourenço and Pereira (2018).

As far as strengthening measures are concerned, the same reasoning can be adopted, provided that the maximum coefficient does not exceed the one in the last column. More in detail:

- **Grout injection** affects strengths and elastic moduli. It is worth reminding that grouting is beneficial when it effectively fills inner gaps with a good bond with existing materials. Generally, the poorer is the existing mortar, the more beneficial is the effect of grouting.

Table 4.29 Values of compressive strength, shear strength, Young's modulus, shear modulus and mass density of masonry

Masonry typology	f_c [MPa] min–max	f_{v0} [MPa] min–max	E [MPa] min–max	G [MPa] min–max	ρ [kN/m³] min–max
Irregular stone masonry (pebbles, erratic, irregular stones)	1.0–2.0	-	690–1,050	230–350	19
Uncut stone masonry with inhomogeneous thickness of leaves	2.0	-	1,020–1,440	340–480	20
Regular stone masonry with good bonding	2.6–3.8	-	1,500–1,980	500–660	21
Soft stone masonry (tuff, limestone, etc.)	1.4–2.2	-	900–1,260	300–420	13–16
Regular soft stone masonry (tuff, limestone, etc.)	2.0–3.2	0.10–0.19	1,200–1,620	400–500	
Dressed rectangular (ashlar) stone masonry	5.8–8.2	0.18–0.28	2,400–3,300	800–1,100	22
Solid brick masonry with lime mortar	2.6–4.3	0.13–0.27	1,200–1,800	400–600	18
Hollow bricks (voids<40%) with cement mortar	5.0–8.0	0.20–0.36	3,500–5,600	875–1,400	15

Source: Circolare (2019).

- **Reinforced structural render** affects strengths and elastic moduli. Besides the values recommended, its effect may be estimated thanks to considerations on render and wall thicknesses, as well as on the properties of mortar and reinforcement. It is worth reminding that the reinforced structural render becomes ineffective in absence of transversal connections with the wall, being generally less effective if realized only on one side. In the case of internal loss of cohesion in the masonry, grouting should precede the render application.
- **Artificial transversal connections** affect strengths and moduli. It is noted that they are effective only if adequately distributed and provided with sufficient shear stiffness. The relative coefficient is the same as reported in the "transversal connection" column. In the case of low shear stiffness (i.e. ties), only compressive strength f_c should be increased.
- **Mortar joint repointing with reinforcing cords and transversal connection** affects strengths and moduli, the latter in the amount of 50%. In particular case, the cords on one side can be substituted with reinforced structural render, provided with adequate transversal connections.

The first set of multiplication factors reported in Table 4.30 regarding the current condition is already included in the MQI (see Section 4.1.6.2) and should not be used in combination with this method.

Table 4.30 Maximum multiplication factors to evaluate the mechanical properties of masonry in the current condition and after strengthening measures

Masonry typology	Current condition			Strengthening measures			Maximum allowable coefficient
	Good mortar	Presence of regularizing courses	Transversal connection (e.g. through stone)	Grout injection[a]	Reinforced structural render[b]	Mortar joint repointing with reinforcing cords and leaf transversal connection[b]	
Irregular stone masonry (pebbles, erratic, irregular stones)	1.5	1.3	1.5	2	2.5	1.6	3.5
Uncut stone masonry with inhomogeneous thickness of leaves	1.4	1.2	1.5	1.7	2.0	1.5	3.0
Cut stone with good bonding	1.3	1.1	1.3	1.5	1.5	1.4	2.4
Soft stone masonry (tuff, limestone, etc.)	1.5	1.2	1.3	1.4	1.7	1.1	2.0
Regular soft stone masonry (tuff, limestone, etc.)	1.6	-	1.2	1.2	1.5	1.2	1.8
Dressed rectangular (ashlar) stone masonry	1.2	-	1.2	1.2	1.2	-	1.4
Solid brick masonry with lime mortar	[c]	-	1.3[d]	1.2	1.5	1.2	1.8
Hollow bricks (voids <40%) with cement mortar	1.2	-	-	-	1.3	-	1.3

Source: Circolare (2019).

[a] The coefficient should be correlated to the effectiveness of the grouting according to the volume injected (during the strengthening) and sonic test outcomes (after the strengthening).

[b] Smaller values should be adopted in the case of thick walls (e.g. larger than 0.7 m).

[c] For brickwork, good mortar is with $f_{cm} > 2$ MPa. In this case, the coefficient may be estimated as $f_{cm}^{0.35}$ (f_{cm} in MPa).

[d] For brickwork, transversal connection means masonry bond following the rules of good practice.

4.1.7.4 Data from American code

The past American standard ASCE 41-06 (2007) estimated the mechanical properties of masonry according to the condition assessment made. In particular, three cases are considered:

- **good condition:** mortar and units intact with no visible cracking;
- **fair condition:** mortar and units intact but with minor cracking;
- **poor condition:** degraded mortar, degraded masonry units, or significant cracking.

The default lower bound mechanical properties are shown in Table 4.31. The standard suggested a multiplication factor equal to 1.3 to get the expected values (Young's modulus is 550 times larger than the expected compressive strength). The updated version of the standard, namely, ASCE 41-17 (2017), still maintaining the multiplication factor equal to 1.3, includes masonry with lime mortar. The default lower bound compressive strength is 1.97 MPa, whereas the flexural strength is 0.03 MPa. In the case of poor conditions, a careful investigation is required.

4.2 TIMBER

Wood is a heterogeneous material with intrinsic singularities such as knots, grain deviation and fissures. It is an orthotropic material that has independent mechanical properties in the directions of three mutually perpendicular axes. According to Figure 4.31 it is possible to define: (1) the longitudinal axis (parallel to the grain), (2) the radial axis (perpendicular to the grain in the radial direction) and (3) the tangential axis (tangent to the growth rings).

Table 4.31 Default lower bound masonry properties according to the current condition

Mechanical properties [MPa]	Good condition	Fair condition	Poor condition
Compressive strength	6.21	4.14	2.07
Young's modulus		550 f_c	
Flexural tensile strength	0.14	0.07	0.00
Shear strength			
Masonry with a running bond lay-up	0.19	0.14	0.09
Fully grouted masonry with a lay-up other than running bond	0.19	0.14	0.09
Partially grouted or ungrouted masonry with a lay-up other than running bond	0.08	0.06	0.03

Source: ASCE 41-06 (2007).

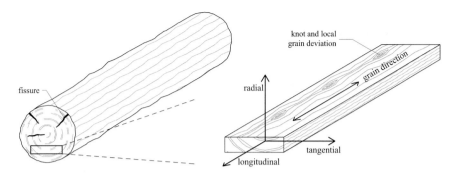

Figure 4.31 Wood principal directions with respect to grain direction and growth rings.

The behaviour of timber is strongly direction-dependent with large ratios of mechanical properties such as Young's modulus or strength between the respective values parallel and perpendicular to the grain directions. In tension, timber exhibits a nearly linear elastic-brittle failure behaviour and, under compression, the hardening part of the stress–strain diagram is nonlinear with limited ductility up to the ultimate stress, followed by mild softening (Glos, 1978), see Figure 4.32a and Figure 4.32b. In terms of professional applications, timber structures in existing buildings are mostly made of one-dimensional elements, such as beams or ties. Accordingly, the behaviour of timber under tension is usually considered linear elastic up to failure, while, for compression, a trilinear diagram is usually adopted, see Figure 4.32c. The European Standard (EN 408, 2012) describes experimental tests for the determination of ultimate strength and Young's modulus in the elastic range for structural timber. In turn, the numerical simulation of the two- and three-dimensional behaviour of timber is complex and not discussed in the following. The reader is referred to the works of Oudjene and Khelifa (2009a, b) for further details. As expected, several material constants must be determined, which require testing a large number of timber specimens.

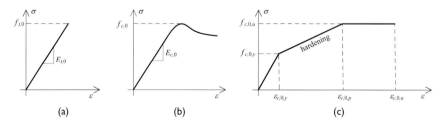

Figure 4.32 Stress–strain diagrams for timber (not to scale). (a) Tension and (b) compression behaviour according to the constitutive model developed by Glos (1978); (c) simplified stress–strain diagram in compression.

Moving to codes of practice, timber structures can be assumed with a linear elastic constitutive behaviour, with the same stiffness in tension and compression (ANSI/AWC, 2018; EN 1995-1-1, 2014). Moreover, the material is considered transversally isotropic with respect to the parallel-to-grain axis and its perpendicular plane (radial and tangential axes). The mechanical properties of timber depend on the species and provenance, this is why international codes define groups of structural qualities depending on the strength, stiffness and mass density (hereinafter, simply density). In the European code (EN) these groups are called strength classes (EN 338, 2016). The state-of-the-art in characterization of structural sawn timber (according to EN) is described in Ridley-Ellis et al. (2016).

It is stressed that in the modelling of timber structures, two aspects are crucial. The first modelling aspect is related to the definition of the internal forces in the structure as a function of the modelling of the connections of the one-dimensional elements, which may be considered hinged, semi-rigid or rigid, depending on the actual joints used. A careful visual inspection may contribute to a sound choice on the stiffness of the connections and the definition of the ends of each structural member, which must be replicated in the numerical model. In existing traditional buildings, hinged joints are likely to be the best representation of the actual condition of connections. The second modelling aspect is related to the capacity of the joints themselves, which must be assessed and usually control the failure of traditional timber structures. This topic is outside the scope of the present book.

Regarding buildings of the recent past, besides sawn timber, other timber products may be encountered in engineering practice, such as Laminated Veneer Lumber (LVL), Glued Laminated Timber (Glulam), Cross-Laminated Timber (CLT) and Oriented Strand Board (OSB). It is noted that LVL is a wood product that uses multiple layers of thin wood assembled with adhesives; Glulam is constituted by layers of dimensional lumber bonded together with durable, moisture-resistant structural adhesives; CLT is a wood panel product made from gluing together layers of solid-sawn lumber in alternating directions; OSB is similar to particle board: it is formed by adding adhesives and then compressing layers of wood strands (flakes) in specific orientations. In general, they all exhibit an anisotropic behaviour (orthotropic or two-dimensionally isotropic) with different strengths in tension and compression.

4.2.1 Strength grading

Strength grading of existing timber members is made by indirect prediction of the mechanical properties using nondestructive techniques and visual inspection. The former are generally correlated to mechanical parameters following destructive tests, and include stress wave, ultrasound and other technologies (Bodig, 2000; Thomas, 1995). For strength prediction, a major drawback of the nondestructive techniques is the relatively poor correlation between the measured nondestructive quantity and material strength. In addition, large coefficients of variation on solid wood are found, which make the final result less reliable, see Table 4.32.

In turn, visual inspection is subjective and may be even inaccurate as evidenced by Huber et al. (1985) for grading red oak lumber, when inspectors achieved only 68% of the expected result with respect to recognizing, locating and identifying defects. To overcome these obstacles and to unify criteria in Europe, general guidelines for grading are issued in the European standard EN 14081-1 (2005), by specifying the minimum requirements for structural timber with rectangular cross-section. Subsequent national standards were developed in order to establish strength classes for local wood species, whereas assignment of species and visual grades from national standards to strength classes is provided by EN 1912 (2012). For a comparison between visual strength grading standards for different countries, the reader is referred to Almazán et al. (2008), Muñoz et al. (2011), among others.

The need to relate the presence of defects to the stress distribution in a member has been pointed out by the Italian standard UNI 11119 (2004). Overall, the result of visual strength grading is the attribution of nominal values or allowable stress levels for indirect prediction of key properties which, due to its correlation to other properties, allow for the mechanical characterization of the material. One key property for the prediction of strength of timber elements is the bending Modulus Of Elasticity (MOE), or stiffness. The relation between strength and MOE can, however, be affected by the presence of significant local defects, and a decrease of the MOE is also expected when decay is present.

Table 4.32 Average coefficients of variation (CoV) of the main mechanical properties of solid wood based on small-clear specimen tests

Property	CoV [%]
Bending strength	7–20
Modulus of elasticity in bending	9–23
Impact bending	25
Compression parallel to grain	8–20
Compression perpendicular to grain	28
Shear parallel to grain, maximum shearing strength	14–22
Tension parallel to grain	25
Side hardness	20
Toughness	34
Density	5–13

Source: Kasal and Anthony (2004).

4.2.2 Mechanical properties

In the absence of accurate inspections, the Italian standards UNI 11035-2 (2010) (applicable to timber elements that do not show signs of deterioration) and UNI 11119 (2004) (applicable to timber elements, regardless of their state of conservation) allow a classification of the resistance classes based on the evaluation of the visual characteristics of the timber elements, namely, knots, sagging, grain direction, warps, growth rate and ring cracks. Depending on the visual classification, values of the mechanical properties are assigned. Table 4.33 shows the suggested values for mechanical properties of pine for the visual resistance classes S3 and III in the standards UNI 11035-2 (2010) (providing characteristic values) and UNI 11119 (2004) (providing allowable design values), respectively. It is worth highlighting that S3 and III are lower bound classes. In particular, S3 class regards the classification of coniferous wood elements (in a range that goes from S1 to S3), while III class is based on the in situ diagnosis (in the range I–III).

The analogous data for oak are shown in Table 4.34. According to UNI 11035-2 (2010), a unique class S is present for this type of structural timber.

Although based on characteristic values of certain properties, it is worth mentioning the relations proposed by EN 384 (2018) to evaluate the mechanical parameters of timber. Often, in literature the letter "C" is referred to Coniferous whilst "D" is referred to Deciduous (i.e. the tree that loses its foliage for part of the year). These references are not accurate and are becoming less popular. In turn, the letter "T" is referred to classes intended

Table 4.33 Mechanical parameters for pine according to UNI 11035-2 (2010) and UNI 11119 (2004) for visual classification S3 and III, respectively

Parameters	UNI 11035-2 S3	UNI 11119 III
Bending, f_m	22	8
Tension // to grain, $f_{t,0}$	13	6
Tension ⊥ to grain, $f_{t,90}$	0.4	0
Compression // to grain, $f_{c,0}$	20	7
Compression ⊥ to grain, $f_{c,90}$	3.7	2
Shear, f_v	3.8	0.8
Average modulus of elasticity // to grain, $E_{0,average}$	10,500	11,000
Characteristic modulus of elasticity // to grain, $E_{0,k}$	7,000	-
Average modulus of elasticity ⊥ to grain, $E_{90,average}$	350	-
Average shear modulus, $G_{average}$	660	-
Characteristic density, ρ_k	530	-
Average density, $\rho_{average}$	575	-

Note that strength and stiffness properties must be given in MPa, whereas density must be given in kg/m³.

Table 4.34 Mechanical parameters for oak according to UNI 11035-2 (2010) and UNI 11119 (2004) for visual classification S and III, respectively

Parameters	UNI 11035-2 S	UNI 11119 III
Bending, f_m	42	8.5
Tension // to grain, $f_{t,0}$	25	7
Tension $\perp$ to grain, $f_{t,90}$	0.6	0
Compression // to grain, $f_{c,0}$	27	7.5
Compression $\perp$ to grain, $f_{c,90}$	11	2.2
Shear, f_v	4	0.9
Average modulus of elasticity // to grain, $E_{0,average}$	12,000	11,500
Characteristic modulus of elasticity // to grain, $E_{0,k}$	10,100	-
Average modulus of elasticity $\perp$ to grain, $E_{90,average}$	800	-
Average shear modulus, $G_{average}$	750	-
Characteristic density, ρ_k	760	-
Average density, $\rho_{average}$	825	-

Note that strength and stiffness properties must be given in MPa, whereas density must be given in kg/m³.

for use in making laminated timber products (mainly glulam). Table 4.35 reports the relative equations according to the characteristic value for bending strength $f_{m,k}$ (for C or D classes), or tensile strength $f_{t,0,k}$ (for T classes), average modulus of elasticity $E_{0,average}$ and density ρ. Basically, for C and D classes, tensile strength is a secondary property, conservatively estimated from bending strength (the same is true for shear strength and compression strength parallel to grain). For T classes, bending strength is a secondary property, conservatively estimated from tensile strength (the same is true for shear strength and compression strength parallel to grain). With reference to Table 4.35, the term "given" refers to known data. In absence of more information, the average values may be obtained by multiplying the characteristic values by a coefficient equal to 1.2.

4.3 REINFORCED CONCRETE

The research on concrete has been ample if compared to the research dedicated to masonry and timber. Much literature exists, including numerous guidelines, so that nonlinear finite element analyses are swiftly moving from research to the professional field. Regarding mechanical properties and approaches, the following discussion is based on Model Code 2010 (fib, 2013), globally accepted as an international reference. Guidelines from fib (fib Bulletin 45, 2008) and Dutch recommendations (Hendriks et al., 2017) are available based on this code. These guidelines apply to structural

Table 4.35 Equations to characterize timber according to EN 384 (2018)

Parameters	C classes	T classes	D classes
Species	Softwood	Softwood	Hardwood
Based on	Edgewise bending	Tension	Edgewise bending
Bending, $f_{m,k}$	Given	$3.66 + 1.213 f_{t,0,k}$	Given
Tension // to grain, $f_{t,0,k}$	$-3.07 + 0.73 f_{m,k}$	Given	$0.6 f_{m,k}$
Tension $\perp$ to grain, $f_{t,90,k}$	0.4	0.4	0.6
Compression // to grain, $f_{c,0,k}$	$4.3 (f_{m,k})^{0.5}$	$5.5 (f_{t,0,k})^{0.5}$	$4.3 (f_{m,k})^{0.5}$
Compression $\perp$ to grain, $f_{c,90,k}$	$0.007 \rho_k$	$0.007 \rho_k$	$0.010 \rho_k$ or $0.0015 \rho_k$ if $\rho_k \geq 700 \, \text{kg/m}^3$
Shear, $f_{v,k}$	$f_{m,k} \leq 24 \, \text{MPa}$ $1.6 + 0.1 f_{m,k}$ $f_{m,k} > 24 \, \text{MPa}$ 4.0	$f_{t,0,k} \leq 14 \, \text{MPa}$ $1.2 + 0.2 f_{t,0,k}$ $f_{t,0,k} > 14 \, \text{MPa}$ 4.0	$f_{m,k} \leq 60 \, \text{MPa}$ $3.0 + 0.03 f_{m,k}$ $f_{m,k} > 60 \, \text{MPa}$ 5.0
Average modulus of elasticity // to grain, $E_{0,average}$	Given	Given	Given
Characteristic modulus of elasticity // to grain, $E_{0,k}$	$0.67 E_{0,average}$	$0.67 E_{0,average}$	$0.84 E_{0,average}$
Average modulus of elasticity $\perp$ to grain, $E_{90,average}$	$E_{0,average}/30$	$E_{0,average}/30$	$E_{0,average}/15$
Average shear modulus, $G_{average}$	$E_{0,average}/16$	$E_{0,average}/16$	$E_{0,average}/16$
Characteristic density, ρ_k	Given	Given	Given
Average density, $\rho_{average}$	$1.2 \rho_k$	$1.2 \rho_k$	$1.2 \rho_k$

Note that strength and stiffness properties must be given in MPa, whereas density must be given in kg/m³.

concrete with normal and lightweight aggregates, composed and compacted so as to retain no appreciable amount of entrapped air.

4.3.1 Mechanical properties of concrete and reinforcement

Like masonry, concrete behaviour is inherently nonlinear, even when subjected to modest loads under service conditions. The main nonlinear effects observed in structural elements are synthesized in Table 4.36. These are organized in terms of effects that are primarily associated to concrete, reinforcing steel and the combination of the two. In practice, not all nonlinear effects must be considered, but the analyst should isolate the ones more pertinent for the problem at hand and judge whether or not a particular set of analysis assumptions are appropriate. As done for masonry, cyclic and dynamic loading (e.g. loading rate, earthquake or wind loads), thermal and

Table 4.36 Main nonlinear effects for concrete components and reinforced concrete as a whole

	Plain concrete	*Reinforcement*	*Combined concrete and reinforcement*
Tension	Cracking Tension softening Crack closing	Yielding Strain hardening	Bond[a] Tension stiffening Tension splitting[a]
Compression	Crushing Postpeak unloading Confinement Volume change	Buckling[a]	Compression softening
Shear	-	Dowel action[a]	Aggregate interlocking[a]

Source: fib Bulletin 45 (2008).

[a] Not easy to model in commercial finite element codes. Often neglected in practice.

transient effects (e.g. creep and shrinkage) fall outside of the scope of the book and are not discussed further.

Regarding the nonlinear behaviour of concrete, the importance of mesh sensitivity to avoid strain localization phenomena (see Chapter 3) must be stressed. Mesh sensitivity may be particularly problematic in the case of punching shear, shear in members without transverse reinforcement, over-reinforced columns and beams, and large unreinforced concrete structures such as dams. In these cases, the behaviour at failure is more controlled by the softening response of plain concrete than by the ductile response of the reinforcement.

4.3.1.1 Elastic parameters

According to Model Code 2010 (fib, 2013), concrete is classified on the basis of its compressive strength. Concrete grades for normal-weight concrete are indicated with a "C" and a number that denotes the specified characteristic cylinder compressive strength f_{ck} (in MPa), e.g. C30 indicates a normal-weight concrete with f_{ck} equal to 30 MPa. Unless specified otherwise, the compressive strength of concrete (and the other mechanical characteristics) is understood at an age of 28 days. As per Model Code 2010, the most important material properties of concrete can be related to f_{ck} or the average compressive strength, here indicated with $f_c = f_{ck} + 8$ MPa.

The linear elastic material properties are Young's modulus and the Poisson's ratio. The former is calculated according to Eq. (4.14), assuming typical quartzite aggregates. Regarding the Poisson's ratio, ν usually ranges between 0.14 and 0.26, and the average value $\nu = 0.20$ is suggested. However, as underlined in Section 4.3.1.3, in presence of cracks, the Poisson's effect ceases and a value ν equal to zero is often used; in presence of crushing, the equivalent nonlinear Poisson's effect due to dilatancy is normally larger.

$$E_c = 21,500 (f_c / 10)^{1/3} \qquad \text{[MPa]} \qquad (4.14)$$

Finally, as far as density is concerned, concrete is classified into three different categories, namely, lightweight (800–2,000 kg/m³), normal (2,000–2,600 kg/m³) and heavyweight (> 2,600 kg/m³). In absence of further insight, a density equal to 2,400 kg/m³ for plain concrete and 2,500 kg/m³ for reinforced concrete may be assumed.

4.3.1.2 Compressive behaviour

The axial behaviour of plain concrete is similar to the one of masonry (and its components), see Sections 4.1.3 and 4.1.5.1. A typical stress–strain diagram of plain concrete in compression is shown in Figure 4.33a, where a parabolic hardening–softening behaviour is noticeable. Young's modulus was already introduced in Eq. (4.14). The diagram, in line with smeared crack models, is based on the compressive fracture energy G_c and the finite

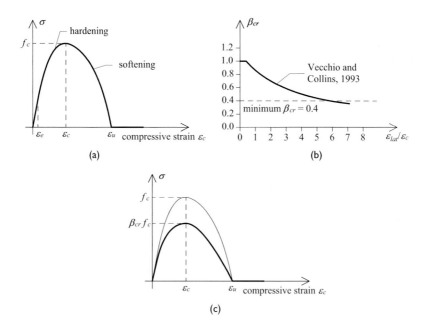

Figure 4.33 Compressive behaviour of concrete. (a) Stress–strain diagram with parabolic hardening–softening; (b) compressive strength reduction factor due to lateral cracking (after Vecchio and Collins, 1993) and (c) combined effect of compression and lateral (tension) loading.

element equivalent size h_{eq}. The word "equivalent" means it is related to the overall dimension of the finite element (such as area or volume), possibly supplemented with information about crack directions with respect to the element alignment. Automatic computation according to the element area or volume may be standard in the adopted software code. If this is not the case, the element size perpendicular to the crushing direction (or cracking in the case of tension, discussed ahead in Section 4.3.1.3), or the average element size, should be assumed for h_{eq}.

The compressive fracture energy has been found experimentally in the range 25–50 N/mm, about 50–100 times as large as the tensile fracture energy. As discussed in Section 4.1.3.1, in order to have quantitative indications, the full nonlinear stress–strain relationships proposed by Model Code 90 (CEB-FIP, 1993) have been integrated. The fracture energy in compression G_c can be obtained by multiplying this area by the height of the specimen (conventionally made equal to 200 mm). The consequent regression analysis is shown in Eq. (4.15), where f_c is the average compressive strength. This can be used as a rough estimation in the absence of more detailed investigations. It is noted that the recent version of the Model Code (fib, 2013) does not provide a mathematical relation for the softening portion of the stress–strain diagram, as this relation exhibits some size dependency.

$$G_c = -0.004f_c^2 + 0.496f_c + 11.786 \quad \text{[N/mm]} \quad 20 \le f_c \le 88 \text{ MPa} \quad (4.15)$$

When compressive stress is accompanied by cracks due to lateral tensile actions, which is usual for reinforced concrete, an overall performance deterioration occurs, whose neglection is nonconservative. According to literature, the deterioration affects both the compressive strength and the relative strain at peak (still maintaining the same initial stiffness). However, it is possible to focus only on the compressive strength and assume a reduction factor $\beta_{cr} < 1$ (due to lateral cracking). In this regard, Vecchio and Collins (1993) proposed to evaluate β_{cr} as shown in Figure 4.33b. More in detail, ε_{lat} is the lateral tensile strain and ε_c is the compressive peak strain. The consequent stress–strain diagram is displayed in Figure 4.33c, where the peaks of the two curves are in correspondence of the same strain ε_c. Basically, the diagram shrinks only along the vertical direction. The diagram of the reduction factor in Figure 4.33b can be further manipulated to avoid excessive reduction that would lead to a nonrealistic response of the structure. It is suggested to adopt a minimum value $\beta_{cr} = 0.4$.

Finally, in the case of lateral confining stress, its effect is beneficial for concrete in terms of both strength and ductility, the so-called *pressure-dependent behaviour*. The increase of the strength with confining stress can be modelled with specific failure surfaces or by simplified uniaxial stress–strain relationships, marginally discussed in Sections 4.3.1.4 and 4.3.2, respectively.

4.3.1.3 Tensile behaviour and tension stiffening

Regarding pure tensile action, Model Code 2010 (fib, 2013) suggests the diagrams in Figure 4.34a, where the difference between uncracked and cracked states is marked. Two FE-oriented approaches are shown in Figure 4.34b and Figure 4.34c in terms of stress–strain diagrams with exponential and linear softening branches, respectively. It must be noted that the more pronounced is the stress drop (after reaching the tensile strength), the more localized is the crack (because it avoids large areas of diffuse cracking). The softening diagrams are defined by tensile strength (f_t), fracture energy (G_f^I) and the equivalent length of the finite element (h_{eq}, already introduced). The first two quantities may be computed according to Eqs. (4.16) and (4.17), respectively.

In the case the tensile strength is evaluated experimentally by means of splitting or flexural tests, the approach discussed in Section 4.1.3.1 is recommended. As far as fracture energy is concerned, this depends primarily

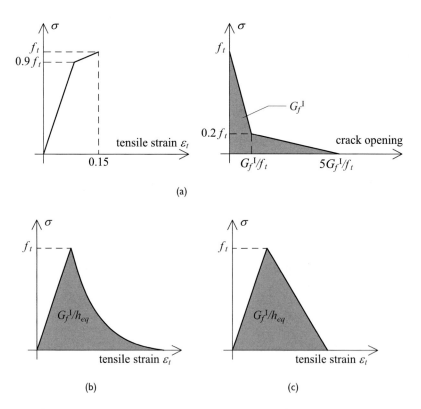

Figure 4.34 Schematic constitutive relations for the tensile behaviour of concrete. (a) Stress–strain and stress–crack diagrams according to Model Code 2010 (fib, 2013). Stress–strain diagrams with (b) exponential and (c) linear softening.

on the water/cement ratio, the maximum aggregate size and the age of concrete. Furthermore, G_f^I is affected by the size of a structural member and by the depth of the ligament above a crack or notch. About aggregates, in particular, type and content seem to have a more pronounced effect than size.

$$f_t = 0.3(f_c - 8)^{2/3} \qquad \text{with } f_c \le 58 \text{ MPa}$$

$$f_t = 2.12\ln(1 + 0.1f_c) \qquad \text{with } f_c > 58 \text{ MPa}$$

(4.16)

$$G_f^I = 0.073 f_c^{0.18} \qquad [\text{N/mm}]$$

(4.17)

Dealing with the cracked state, it is important to consider the reduction of the apparent Poisson's ratio ν. In a cracked state, the Poisson's effect of a material ceases to exist because the stretching of a cracked direction does no longer contribute to the contraction of the perpendicular directions. To model this phenomenon, software codes often apply orthotropic formulations.

The cracking phenomenon is more complex if reinforced concrete is considered. With reference to Figure 4.35a, when a concrete structural member (or part of it, e.g. under flexure) is subjected to tension, microcracks develop

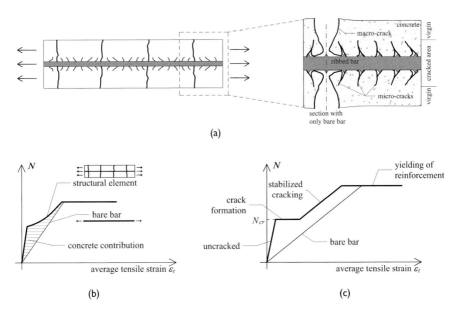

(a)

(b)

(c)

Figure 4.35 Schematic description of tension stiffening. (a) Concrete structural element subjected to tension; (b) comparison between structural response of the element vs. bare bar and (c) simplified load-deformation behaviour of a centrally reinforced member subjected to tension, as proposed by Model Code 2010 (fib, 2013).

around the bars (i.e. bond cracking) and macrocracks extend to the external surface. Although both processes follow the tensile softening behaviour of plain concrete, their integration within the entire cross-section can be represented with the load-deformation diagram of Figure 4.35b. As it is possible to notice, the comparison between the behaviour of the structural element and the one of the bare bar shows that concrete, although cracked, still provides a positive contribution. This contribution is named *tension stiffening* because the stiffness of the structural element increases (but not its ultimate strength). This is clear if Figure 4.35a is considered. Simple equilibrium considerations require that the force resultant N of internal stresses must be the same for every section, including the one with no surrounding concrete, whose response is governed exclusively by the bare bar.

Figure 4.35c shows the simplified representation of the load-deformation behaviour of a centrally reinforced member subjected to tension (or imposed deformation) as proposed by Model Code 2010 (fib, 2013). According to the simplification, after a linear behaviour in the uncracked stage, the crack formation stage is characterized by a constant axial tensile force N_{cr}. This corresponds to the tensile cracking force, which depends on the concrete tensile strength. The assumption of constant N_{cr} for the crack formation stage is regarded as sufficiently accurate for practical applications. When microcracks are diffused throughout the element, no additional cracks appear, marking the onset of the stabilized cracking stage where existing cracks widen. Finally, the reinforcement starts yielding at ultimate stage.

From the practical point of view, tension stiffening is the difference in behaviour between the response of a bare bar and the observed response of the composite reinforced element (thanks to the shear transfer from the reinforcement to the concrete between cracks). Although the interaction between concrete and reinforcement is complex at the microscopic scale (Figure 4.35a), it is possible to model the tension stiffening macroscopically. With this aim, the simplest approach is the one shown in Figure 4.35c where tension stiffening is described by the change of the overall stiffness of the element. While such an approach is suitable for simple one-dimensional calculation where structural elements are subjected to pure tension, this is generally less applicable to two- or three-dimensional problems with more complex stress states.

A more versatile approach, in line with smeared crack models, modifies the concrete constitutive relation to account for the averaged behaviour of a cracked volume. This is shown in Figure 4.36a with a beam subjected to bending, highlighting the measure of the crack spacing. Figure 4.36b illustrates the presence of multiple smeared cracks in the concrete volume defined by two subsequent macrocracks, in which the fracture energy is exhausted. In a smeared approach, it is possible to account for the concrete–bar interaction by providing the concrete between the two macrocracks with an extra equivalent fracture energy. By assuming a perfect

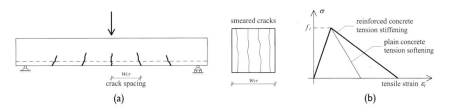

Figure 4.36 Smeared crack model for tension stiffening. (a) Beam subjected to bending and finite element with smeared crack; (b) comparison between plain concrete and reinforced concrete (tension softening and tension stiffening, respectively).

bond between concrete and reinforcement, the consequent constitutive relationship for concrete is shown in Figure 4.36b. Although the diagram describes de facto a tension softening (i.e. reduction of stress with increasing deformation), the reader should not misinterpret the reasoning behind tension stiffening.

The increment of fracture energy of Figure 4.36b depends on the relation between the finite element equivalent size (h_{eq}) and the estimated average crack spacing w_{cr}. The amount of fracture energy that can be dissipated by the element depends on the average number of cracks n_{cr} expected within the finite element. The calculations are shown in Eq. (4.18), where G^I_{fRC} is the fracture energy when tension stiffening is considered. The crack spacing can be computed according to e.g. Eurocode 2 (2004). From Eq. (4.18), it is clear that, for smaller elements, tension stiffening is not influential. Additionally, this condition is particularly favourable because smaller elements provide less stiffness to the structural member and they simulate better individual discrete cracks. Quite the opposite, for finite elements with dimensions much larger than the crack spacing, it is practical to assign an ultimate strain in the tension-stiffening diagram equal to the yield strain of the reinforcement ε_y.

$$ G^I_{fRC} = n_{cr} G^I_f \qquad \text{with} \qquad n_{cr} = \max\left(1, \frac{h_{eq}}{w_{cr}}\right) \qquad (4.18) $$

It is worth mentioning that tension stiffening is valid only in the effective concrete area around the longitudinal reinforcement, while plain-concrete tension-softening relationship is assumed elsewhere. The effective concrete area can be calculated according to e.g. Model Code 2010 (fib, 2013) or other standards. As a first approximation, for elements subjected to bending (Figure 4.37), it is the rectangular region with the height equal to the minimum value amongst one-third of the section in the state of tension (according to the position of the neutral axis) and 2.5 times the distance between the level of the steel centroid of the reinforcement and the most

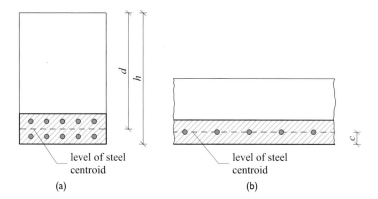

Figure 4.37 Effective concrete area in tension (not to scale) for (a) beam and (b) slab subjected to bending.

tensioned side of the structural element (namely, $h - d$ and c for Figure 4.37a and Figure 4.37b, respectively).

As long as smeared crack models are considered, the fixed crack approach is recommended to be used with a variable shear retention factor (see Chapter 3). Conservatively, it is possible to assume the same rate for both secant shear and secant tensile stiffness degradation (the latter due to cracking). Alternatively, a variable shear retention model may consider the progressive reduction of shear stiffness until reaching a null value in correspondence of a crack width of half the average aggregate size. Constant shear retention models are not advisable or should at least be accompanied with thorough postanalysis checks of spurious principal tensile stresses. In turn, compared to the fixed model, the rotating model usually results in a lower limit failure load because it does not suffer as much from spurious stress-locking. Additionally, it has been seen that for large elements subjected to shear, the angle of principal strains, averaged over a sufficient gauge length to include several cracks, can rotate significantly between first cracking and ultimate limit states. In these cases, the rotating crack model may be a suitable choice.

4.3.1.4 Biaxial behaviour

There is extensive information about the biaxial and triaxial behaviour of concrete, which is less complex than masonry, given the isotropic behaviour. For instance, it is assumed that the biaxial compressive strength (ratio of $\sigma_{II}/\sigma_I=1.0$) is 1.16 f_c whereas, the multiplicative coefficient reaches 1.25 with a stress ratio of $\sigma_{II}/\sigma_I=0.5$. There is also considerable information on the strength and volumetric behaviour under triaxial loading.

Upon these considerations, Model Code 2010 (fib, 2013) suggests a failure criterion whose plane case is displayed in Figure 4.38. It is worth stressing

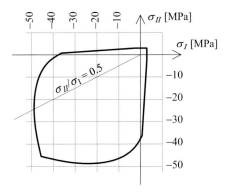

Figure 4.38 Plane failure surface for concrete C30 (i.e. $f_c = 38\,\text{MPa}$) according to Model Code 90 (CEB-FIP, 1993).

that this failure criterion is one among several acceptable formulations. An elastoplastic formulation, damage or different combinations may be used to describe the triaxial nonlinear concrete behaviour at the macroscopic scale.

4.3.1.5 Reinforcement behaviour and modelling

Reinforcing bars in structural concrete, either passive or active (pre-stressed), are generally assumed to be one-dimensional line elements without transverse shear stiffness and flexural rigidity. Material data, namely, tensile strength, Young's modulus, possible hardening modulus and ultimate strain, should be determined according to available documents or experimental tests. Mechanical properties can also be derived from Model Code 2010 (fib, 2013).

Regarding reinforcing bars and stirrups, embedded finite elements are usually recommended, either as bars or grids (i.e. localized and distributed, respectively). Accordingly, reinforcement is directly related to the "mother" elements (i.e. those crossed by the bars) without adding new displacement variables, adding their stiffness and strength contribution in the mesh nodes of the continuum. In turn, by considering perfect bond with the surrounding, the strains in the reinforcements are computed from the displacement field of the mother elements. In more advanced "embedded bond-slip models", reinforcements and interface models are combined, such that slip can be modelled explicitly, an alternative not discussed in the present text.

Regarding steel reinforcement, an elastoplastic material model with (multi-)linear hardening is usually considered, such as the bilinear stress–strain diagram shown in Figure 4.39a. The diagram regards only the tension response, which is the same in compression due to the tension–compression symmetry of steel. According to Model Code 2010 (fib, 2013), depending on the ductility of the steel, the ultimate strain ranges between

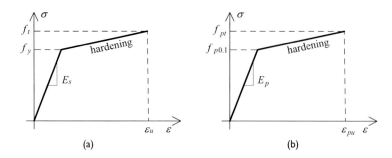

Figure 4.39 Main mechanical parameters for (a) steel reinforcement and (b) prestressing
steel according to Model Code 2010 (fib, 2013).

2.5% and 8% whereas the ultimate strength vs. yield strength ratio typically varies between 1.05 and 1.25. Young's modulus E_s can be set equal to
200,000 MPa, the Poisson's ratio equal to 0.3 and the mass density equal
to 7,850 kg/m³. In the case no hardening information is available, perfect
plasticity can be assumed with the minimum ultimate strain ($\varepsilon_u=2.5\%$). To
improve the stability of the analysis, a vanishing hardening modulus equal
to e.g. 0.02 E_s can be adopted.

   In the case of prestressing, the diagram of Figure 4.39b is analogous to
what was already described. In particular, neglecting thermal or fatigue
aspects, it is possible to distinguish ultimate tensile strength, 0.1% proof
stress, total elongation at ultimate tensile strength and Young's modulus.
The meaning of the symbols is self-explaining. In literature, the 0.1%
proof stress is also addressed as "yield stress". According to Model Code
2010 (fib, 2013), Young's modulus can be assumed, in this case, equal to
205,000 MPa for wires and 195,000 MPa for strands, whereas the value for
bars can be estimated from manufacturer documentation or by means of
further investigation. The analyst should also consider the prestress losses.
Short-term prestressing losses regard wobble, friction and anchor retraction.
Long-term prestressing losses are influenced by relaxation, shrinkage and
creep of the structure. The actual level of prestressing should be assessed
as accurately as possible. If no data are available, the design prestressing
level should be reduced to 70% for both serviceability and ultimate limit
state. For simulation of construction stages, the prestressing levels should
be increased to 110%.

4.3.2 Nonlinear models of frame elements

The approach discussed in the previous subsection is less adequate in the
case of large frame structures, which are a usual typology for concrete
structures. In this regard, it is possible to use line finite elements with *distributed* or *concentrated plasticity*. In this case, the corresponding cross-
sections are discretized with fibres with uniaxial behaviour, each one with

its own constitutive characteristics (*fibre model*), then integrated to obtain the cross-sectional response. This approach considers the through-depth and through-thickness integration points (see Chapter 1). As this option may become computationally expensive in the case of three-dimensional frames, a possible simplification may be to assign a pre-integrated moment–curvature (or moment–rotation) law at each integration point of the line element.

Focusing on the fibre model shown in Figure 4.40a, the typical fibre discretization of a reinforced concrete cross-section is shown in Figure 4.40b, where steel reinforcement, and unconfined and confined concretes are isolated. Of those, the constitutive relationship of confined concrete is essential for the correct analysis of the structural elements. This can be evaluated thanks to simplified models, such as the one proposed by Mander et al. (1988) or the model proposed by Selby and Vecchio (1993). Basically, according to experimental tests, the complex triaxial stress state is included in the analysis by means of a unified uniaxial stress–strain diagram. Thanks to the beneficial effect provided by stirrups, in fact, confined concrete may achieve higher strength and larger ultimate strain along the main load direction, as shown in Figure 4.41. The confining effect depends on the efficacy of the stirrups (e.g. effectiveness of hooks), stirrups diameter and spacing, and location of longitudinal reinforcing bars. For new structures, these aspects are implicitly accounted for by means of constructive details imposed by the codes. The reader is referred to e.g. Eurocode 2 and Eurocode 8 for further insight.

One of the advantages of fibre models is that they allow an easy description of the element response for any arbitrary-shaped cross-section. Additionally, the interaction between axial force and bending moment, plus the coupling effect of biaxial bending moments, are implicitly accounted for with the fibre approach. The only mechanical parameters needed are the uniaxial constitutive laws of the section components (Figure 4.40b).

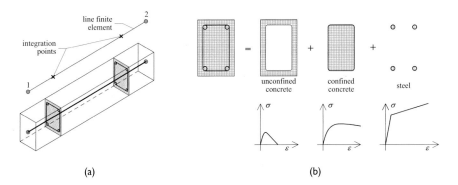

(a)   (b)

Figure 4.40 Schematic representation of a fibre model. (a) Integration points along the element and (b) discretization of the reinforced concrete section (diagrams not to scale).

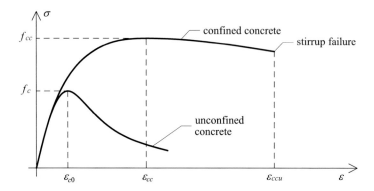

Figure 4.41 Comparison in terms of stress–strain relationships between confined and unconfined concrete.

A reasonably large number of integration points through the height and width must be provided for nonlinear analysis and softening behaviour. These are needed to integrate the internal forces for a given curvature and obtain rotation and displacement of the beam (conversely, in the case of plane linear analysis, two integration points are sufficient). For a three-dimensional nonlinear analysis, the recommended value is 100–400, i.e. a subdivision of each side in 10–20 fibres, to account for possible biaxial bending moments. In the case of plane analysis, fibres are represented by horizontal layers (i.e. discretization only through the height). In both cases, a smaller number of fibres (or integration points) may lead to a poor discretization of the element response, acceptable only if the global behaviour is sought and when too many integration points result computationally inefficient.

As stated, the second approach for the nonlinear analysis of frame structures is the use of concentrated plasticity (also called *lumped plasticity*). This is illustrated in Figure 4.42 and assumes that plasticity occurs at the end regions of beam or column elements, keeping elastic the rest of the structural member. In this regard, plasticity is lumped into zero-length

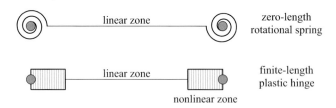

Figure 4.42 Schematic representation of the concentrated approach for nonlinear analysis of a beam element by means of zero-length rotational spring and finite-length plastic hinge.

sections for which the mechanical properties are idealized by means of nonlinear rotational springs. These are usually referred to as "plastic hinges" to underline the nonlinear rotation of the section subjected to plastic bending. The basic advantage of the lumped model is the intrinsic simplicity and the low computational effort, with an overall numerical stability. The introduction of an artificial element, where all nonlinearities are lumped, allows the inclusion and the control of additional mechanical properties. For instance, the axial force-bending moment interaction is usually described by a yield surface for the stress resultants and an associated flow rule according to the principles of classical plasticity theory. Multilinear constitutive representations that include cracking and cyclic stiffness degradation are also possible. Along this line, by means of opportune constitutive laws, to be tuned on a case-by-case basis, it is possible to model other features, such as shear deformability, reinforcement bond, concrete frame-masonry infill iteration and beam-column node flexibility.

Two main limitations must be underlined when using this approach. On the one hand, the position of the plastic hinge is assumed a priori, usually at the element extremities. Although this normally occurs, it automatically excludes the possibility of inelastic behaviour somewhere else along the element (e.g. mid-span when gravitational loads are predominant). Conventionally, this case is addressed by adding internal nodes within the finite element to divide it into several subelements so that the presence of internal hinges is also accounted for. On the other hand, the zero-length inelastic zone may lead to an overestimation of the ultimate strength of structures. An alternative solution is to extend the zero-length plastic hinge to a finite inelastic region defined by a so-called *equivalent plastic hinge length*. This is the length of the structural element (not the physical length of the real plastic hinge region) where the flexural moment exceeds the yielding capacity, and the curvature is assumed constant. The product between curvature and equivalent plastic hinge length defines the rotation.

As a final remark, a thorough analysis of concrete elements may require addressing issues not discussed in the present text. For compressed elements, these include spalling of concrete cover and buckling of the reinforcement, whose neglection may overestimate the response of the structural element (in the case of large inelastic deformations) in terms of strength and ductility. The same applies for beam-to-column joints, commonly considered critical regions for reinforced concrete frames (usually modelled with panel zones and rigid offsets). The need of further investigating these issues remains at the discretion of the analyst.

4.4 STEEL AND IRON

Pure iron is a soft material, whereas alloys of iron and carbon cover a wide range of strength, from low to very high. These characteristics make metals one of the most versatile and strong engineering materials. Modern steel currently used in construction (hot rolled structural steel) presents a yield strength that varies from 235 up to 355 MPa and an ultimate strength between 360 and 510 MPa, respectively, although higher strength steels are becoming more common. This section discusses aspects related to the elastic deformation, plastic deformation and ductility of metals and alloys, subjected to monotonic loading.

4.4.1 General behaviour and material properties

The basic behaviour of a ductile metal is obtained under monotonic loading. The stress–strain response of metals and alloys at low strain is almost always reversible and linear until the yield limit. Above the yield limit, permanent or inelastic deformations occur. In the plastic range, the stress usually increases with the total amount of accumulated plastic strain and becomes the new yield stress if the material is unloaded. This behaviour has been extensively discussed in Chapter 3, to which the reader is referred. For the sake of completeness, Figure 4.43a shows a typical stress–strain curve of hot rolled carbon steel. In the elastic range, the slope is linear and is defined by

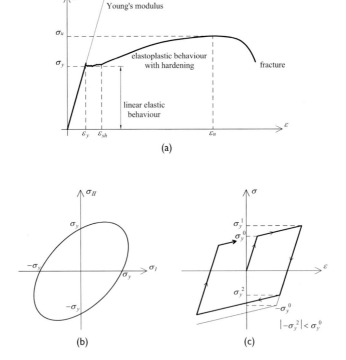

Figure 4.43 Main features of ductile metals. (a) Stress–strain relation in uniaxial tension, showing yielding, hardening and ductile fracture; (b) yield surface in biaxial stress space and (c) Bauschinger effect.

the Young's modulus E. The linear path is limited by the yield stress σ_y and the corresponding yield strain ε_y, followed by a region of plastic flow at an approximately constant stress until the strain hardening strain ε_{sh} is reached. At this point, the plastic yield plateau ends and strain hardening initiates. Beyond this point, stress accumulation recommences at a reducing rate up to the ultimate tensile stress σ_u and the corresponding ultimate tensile strain ε_u.

Plastic yielding occurs approximately at the same magnitude of stress in tension as in compression. In general, plastic deformation occurs without volume change and hydrostatic pressure has no influence on yielding. Figure 4.43b shows the typical yield surface in biaxial stress, which is approximately elliptical for a polycrystalline metal (e.g. von Mises yield surface). Finally, Figure 4.43c shows the so-called Bauschinger effect, evident when the material is deformed up to a given strain, then it is unloaded and, subsequently, it is loaded in the reverse direction. The yield stress after strain reversal is lower than the stress level before unloading from the first deformation step. As an example, an increase in tensile yield stress from σ_y^0 to σ_y^1 occurs at the expense of compressive yield stress σ_y^2 that results smaller than σ_y^0 in absolute terms (Figure 4.43c). The Bauschinger effect has important consequences for the inelastic deformation when materials are exposed to complex strain paths and loading changes. These conditions apply to practically all operations of metal forming, which are less important in the context of civil engineering.

4.4.1.1 Tensile strength

According to European codes for structural carbon steels, the material is classified on the basis of the yield strength. In the case of hardening, the ultimate tensile strength is also an important property to some aspects of design. In ductile materials, the ultimate tensile strength can be significantly higher than the yield strength. A range of values for ultimate tensile strength is specified in the product standards. According to Eurocode 3 (2014), the nominal values of the yield strength f_y and the ultimate strength f_u of structural steel should be obtained either by experimental testing according to EN ISO 6892-1 (2019) or by using the values from the corresponding steel grade proposed in EN 1993-1-1. Typical values of structural steel encountered in civil engineering are shown in Table 4.37.

4.4.1.2 Ductility

Ductility is the ability of a material to sustain large plastic deformation after yielding and up to its fracture. It can be measured by the amount of permanent deformation obtained from the tensile stress–strain diagram. The types and percentage of alloying elements present in the material have a major influence on ductility. Fabrication processes such as bending and straightening are influenced by the ductility of the material used. The same occurs with structural design, where ductility gives the possibility of stress redistribution or bolt group design at the ultimate limit state.

Table 4.37 Nominal values (in MPa) of yield strength f_y and ultimate tensile strength f_u for hot rolled structural steel

Steel grade according to EN 10025-2	Nominal thickness t of the element			
	$t \leq 40$ mm		40 mm $< t \leq 80$ mm	
	f_y	f_u	f_y	f_u
S 235	235	360	215	360
S 275	275	430	255	410
S 355	355	510	335	470
S 450	440	550	410	550

Source: EN 1993-1-1 (2014).

The various standards that define the grades of steel give recommendations on a minimum value for ductility. If this minimum value is attained, the designer can assume that the material is able to deform plastically up to a large extent and, therefore, use the corresponding design assumptions. In EN 1993-1-1 (2014), the minimum ductility required for steels used in buildings is expressed by the following limits: a ratio $f_u/f_y \geq 1.10$; elongation at failure on a gauge length of $5.65 \sqrt{A_0}$ not less than 15% (where A_0 is the original cross-sectional area); ultimate strain $\varepsilon_u \geq 15\varepsilon_y$, where ε_y is the yield strain ($\varepsilon_y = f_y/E$) ε_u corresponds to the ultimate strength f_u. Considering that the yield strain is typically close to 0.2%, and values of ultimate strain should be higher than 15%, an increase of around 100 times in strain is expected. Ductility tends to decrease with higher values of yield strength, but this reduction is usually not relevant for engineering practice.

4.4.1.3 Code properties

Section 3.2.6 of EN 1993-1-1 (2014) defines values for significant properties of steel, as Young's modulus, shear modulus, Poisson's ratio, coefficient of linear thermal expansion and density (Table 4.38). Young's modulus E does not vary appreciably for the different steel grades.

4.4.2 Mechanical properties of various types of iron and steel

The mechanical properties of iron and steel are the result of choices made regarding chemical composition, mechanical working and heat treatments. According to Bussell (1997), the most relevant properties of structural iron and steel are strength, mode of failure and stiffness. Table 4.39 presents a qualitative evaluation of these properties, while Table 4.40 presents typical ranges of mechanical properties for iron-based products.

Table 4.38 Main mechanical parameters for structural steel

Young's modulus	$E=210,000\,\mathrm{MPa}$
Shear modulus	$G=81,000\,\mathrm{MPa}$
Poisson's ratio	$\nu=0.3$
Coefficient of linear thermal expansion, when $T\leq100°C$	$\alpha=12\times10^{-6}/°C$
Density	$\rho=7,850\,\mathrm{kg/m^3}$

Source: EN 1993-1-1 (2014).

Table 4.39 Qualitative evaluation of important properties of cast iron, wrought iron and steel

	Cast iron	Wrought iron	Steel
Tensile strength	Poor	Good	Good–excellent
Compressive strength	Good	Good	Good–excellent
Ductility	Poor–brittle	Good	Good

Table 4.40 Typical mechanical properties of structural grey cast iron, wrought iron, and early steel [MPa]

	Grey cast iron	Wrought iron	Steel	
Year	Any	Any	Pre-1906	1906–1968
Young's modulus [$\times 10^3$]	66–94 (Tension) 84–91 (Comp.)	154–220	200–205	205
Elastic limit (yield strength)	-	154–408	278–309	225–235
Ultimate tensile strength (UTS)	65–280	278–593	386–494	Mild steel: 386–509
Ultimate compressive strength	587–772	247–309	= UTS	= UTS
Ultimate shear strength	$\geq$ UTS	$\geq$ 2/3(UTS)	3/4(UTS)	$1/\sqrt{3}$ (practically taken as 0.6) UTS

Source: Bussell (1997).

As recommended by ASCE 41-17 (2017), mechanical properties for structural and cold-formed steel materials, components and assemblies shall be based on available construction documents, test reports, manufacturers' data and as-built conditions for the particular structure. Where such documentation fails to provide adequate information to quantify material properties and the capacities of assemblies, or the condition of the structure changed significantly from the original, material tests, mock-up tests of assemblies and careful assessment of existing conditions are needed. However, the code furnishes default lower bound material properties, as

Table 4.41 Default lower bound material properties for cast iron, wrought iron and steel [MPa]

	Cast iron[a]	Wrought iron	Steel	
Year	Any	Any	Pre-1900	1900–1960
Young's modulus [× 10^3]	103.4	172.4	200	200
Lower bound yield strength	Null	124	165	207
Lower bound tensile strength	Null	-	248	345
Lower bound compression strength	117 (for slenderness ratio < 108)	-	-	-

Source: ASCE 41-17 (2017).

[a]   The use of cast iron components to resist tensile stresses is not permitted.

shown in Table 4.41. Default expected strength material properties can be determined by multiplying lower bound values by a factor equal to 1.10. For further details on aspects related to mechanical properties and condition assessment, the reader is referred to the mentioned code.

Cast iron strength is mainly affected by carbon content, rate of cooling in the mould, size and shape of the member, as well as the quality of the foundry. The microstructure of cast iron consists of a distribution of graphite flakes in a steel matrix. The graphite flakes have no strength and act as voids or cracks within the iron. Therefore, the overall behaviour of the material is brittle and inhomogeneous. A typical stress–strain diagram is presented in Figure 4.44, with nonlinear behaviour and plastic deformation, exhibiting the following macroscopic properties: (1) different yield strengths in tension and compression, with the yield strength in compression being a factor of three or more higher than the yield strength in tension; (2) inelastic volume change in tension, but little or no inelastic volume change in compression and (3) different hardening behaviour in tension and compression.

Wrought iron is an iron-based alloy with a very low carbon content (less than 0.1%), with good compressive and tensile strength. Wrought iron is tough, malleable, ductile, corrosion resistant and easily weldable. Its strength is mostly influenced by the orientation of slag strands and the degree of working after manufacture, which give wrought iron its unique fibrous structure, but it is also dependent on the quality of manufacture. The working process results in an anisotropic laminar structure with a low tensile strength perpendicularly to the grain. Wrought iron is ductile and can undergo substantial deformation before fracture.

The strength of steel is principally influenced by the carbon content, other trace elements, heat treatment and degree of working. In addition, the strength of earlier steels is more variable than that of modern steels due to the lower level of quality control in the steel-making process. It may be assumed that the properties of steel in compression are the same as those in

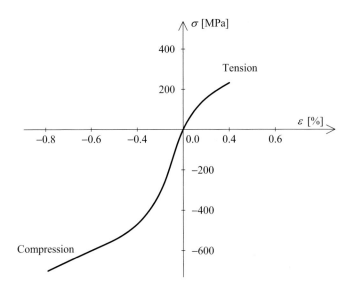

Figure 4.44 Typical stress–strain diagrams for cast iron.

tension. Hot rolled is the type of steel most widely used in structural members and joints. Its carbon content varies between 0.06% and 1%. Yield strength reduces with increasing plate or section thickness, as thinner material is worked more than thick material and working increases the strength. EN 1993-1-1 (2014) and EN 10025-2 (2019) define various grades and corresponding yield and ultimate strength values for hot rolled carbon steel (see Table 4.37). EN 10210-1 (EN 10210-1, 2006) does the same for steel used in hollow sections.

Looking at the experimental behaviour of hot rolled carbon steels under uniaxial tension (Figure 4.43a), various simplified models have been proposed to represent the material response of this material. EN1993-1-1 recommends the use of a bilinear stress–strain relationship for the grades of steel considered, if a nonlinear material behaviour is enforced. EN1993-1-5 provides additional guidance regarding FE analysis. According to this specific standard, in FEM, material properties should be taken as characteristic values, choosing between the four stress–strain relationships of Figure 4.45. It is worth reminding that the true stress–strain diagrams are modified from the test results according to Eqs. (4.19) and (4.20). This is due to the fact that, in the case of large strain, the deformed configuration is sensibly different from the original one. In a simplistic way, when metal necks, a strong variation of the cross-sectional area occurs and, to account for it, nonlinear strain measures need to be introduced (so-called *true*). However, as explained in Chapter 5, this aspect is of marginal interest in civil engineering. If this is not the case, the assumption of large strain deserves a careful investigation.

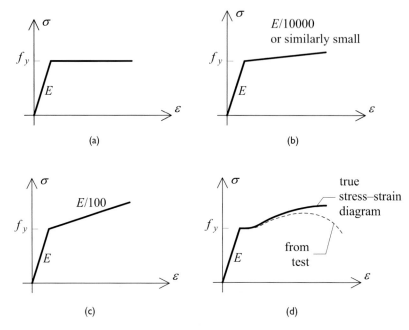

Figure 4.45 Stress–strain relationships for carbon steel modelled with FE software codes. (a) Elastoplastic without strain hardening; (b) elastoplastic with a nominal plateau slope; (c) elastoplastic with linear strain-hardening and (d) true stress–strain diagram.

$$\sigma_{true} = \sigma\left(1+\varepsilon\right) \tag{4.19}$$

$$\varepsilon_{true} = \ln\left(1+\varepsilon\right) \tag{4.20}$$

Several authors consider that due to the existence of a yield plateau, the linear hardening model proposed by EN1993-1-5 is not suitable for hot rolled carbon steels. Therefore, they propose the use of the trilinear model that considers a yield plateau and strain hardening, as shown in Figure 4.46.

Stainless steel is a group of iron-based alloys whose chemical composition prevents the iron from rusting and provides heat-resistant properties. Stainless steels can be classified into five groups, according to their chemical composition and thermomechanical treatment (SCI, 2006). Each group has different properties, regarding strength, corrosion resistance and ease of fabrication. In these steels, the stress–strain relationship does not have the clear distinction of a yield point and, therefore, yield strength corresponds to a proof strength defined for a particular permanent strain, which is conventionally set equal to 0.2%. EN 1993-1-4 also defines the values of the material parameters that may be assumed (Table 4.42).

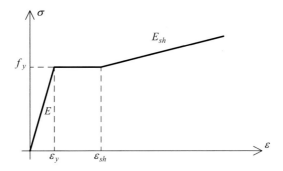

Figure 4.46 Trilinear stress–strain relationship for hot rolled carbon steel.

Table 4.42 Main mechanical parameters for stainless steel

Young's modulus	$E=200 \times 10^3$ MPa for austenitic–ferritic grades
	$E=195–200 \times 10^3$ MPa for austenitic grades
	$E=200–220 \times 10^3$ MPa for ferritic grades
Poisson's ratio in elastic stage	$\nu=0.3$

Annex C of EN1993-1-4 (CEN, 2006) gives guidance for modelling the material behaviour of stainless steel according to the two-stage Ramberg–Osgood model, as given in Figure 4.47. More in detail, E is the Young's modulus and $\sigma_{0.2}$ is the 0.2% proof strength. The parameters $\varepsilon_{0.2}=0.2\%$ and $\varepsilon_{0.2p}$ are the total and plastic strains at yielding, respectively; analogously, ε_{up} and ε_u are the ultimate total and plastic strains (in correspondence of the ultimate tensile strength σ_u); ε_f is the strain at fracture.

Finally, strip steels are steel plates of small thickness that are used in cold forming processes. The most common form of strip steel used in

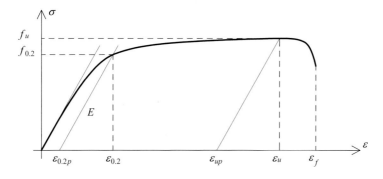

Figure 4.47 Stress–strain diagrams for stainless steel.

construction is hot-dip galvanized. Different types of elements and cross-sections result from the cold forming process, which are then mostly applied in residential buildings and secondary elements. Different levels of cold work (plastic deformation) are generated during the manufacturing process, which results in changes to the stress–strain characteristics of the material. Cold-formed steels are generally characterized by a stress–strain response with no sharply defined yield point. According to Gardner and Yun (2018), cold work results in an increased yield strength and, to a lesser extent, an increased ultimate strength, but lower ductility. In addition, the corner regions of cold-formed cross-sections undergo large plastic deformation from cold bending due to their tight corner radii, hence resulting in a higher degree of strength enhancement but a corresponding loss in ductility.

A wide range of steel grades is available for cold forming. It may be assumed that the properties of steel in compression are the same as those in tension. Gardner and Yun (2018) recommend the adoption of the two-stage Ramberg–Osgood model also in this case, as described above.

4.5 SAFETY ASSESSMENT

The safety assessment of an existing structure is a quantitative procedure, aimed at determining the extent of the actions that the structure is able to support, according to applicable regulations. The focus assumed here is on ultimate conditions, also considering that serviceability conditions are often not complied with by existing buildings. It is up to the analyst to decide whether other performance verifications are needed. The safety assessment must make possible to establish whether:

- the use of the building is allowed to remain as is, without any intervention;
- the use must be adapted (downgrading, changing the use and/or imposing limitations and/or precautions in use);
- it is necessary to increase structural safety through interventions (by acting on the global structural performance, but also by addressing possible local deficiencies).

According to Italian code (NTC, 2018), the safety level of a construction can be expressed by two ratios regarding seismic and nonseismic actions. In both cases, the capacity (assumed as the maximum action the structure can sustain) is divided by the analogous demand adopted for new buildings, based on the applicable limit states. In particular, seismic actions are usually evaluated in terms of peak ground acceleration (for the given site), whereas nonseismic actions are usually evaluated in terms of live loads for every subcomponent of the structure. It is clear that a deficiency in terms of gravitational loads must be tackled immediately, even with partial or temporary measures (e.g. by downgrading, imposing limitations or prop-ping). In turn, given the aleatoric nature of the seismic action, the measures must consider factors such as the level of deficiency, the possible conse-quences for public safety and the available economical resources.

Besides the recommendations presented in the present section, the reader is referred to codes of practice in force in the building location, if applicable. Additionally, it is noted that no information is provided in the present text about the geotechnical investigation or soil–structure interaction effects, which may be required for gravitational loading (regarding the building capacity) and seismic loading (regarding the building capacity and the defi-nition of the seismic demand).

4.5.1 Eurocode 8 and Italian code

As detailed in Section 4.5.1.1, both Eurocode 8 and the Italian code (Circolare, 2019) consider the concepts of the *knowledge level* KL, to account for different levels of inspection (related to geometry, structural organization, construction details and material classification), and of the *confidence factor* CF (which affects the material strength according to the knowledge level attained). The reliability of the results of a struc-tural analysis is strictly related to the knowledge level, thus less detail of the investigation leads to precautionary values with lower mechanical properties.

An adequate knowledge of the building at hand is essential for under-standing its structural behaviour and to highlight possible local defects. These are particularly frequent in masonry constructions and regards defi-ciencies that affect the structural capacity of the structural element itself or the connections between elements. Typical cases are the presence of shafts, niches, flues and openings with no lintels, or rubble masonry and multi-leaf walls whose leaves are not interconnected. Only when local defects are addressed, it is possible to discuss about local and global mechanisms, which involve selected parts of the building or the building as a whole, respectively.

Whereas, on the one hand, local mechanisms can be tackled by means of limit analysis or simplified models (not considered in the present text), the definition of a global and accurate finite element model represents one of the most complex phases in the analysis procedure. This is particularly relevant in the case of masonry constructions where the uncertainties related to the current internal force distribution, material properties, degree of element interconnection and internal morphology play a significant role. Besides, the analyst should ponder about the effects of subsequent structural trans-formations, damage due to past actions, deterioration phenomena, material defects, inadequate detailing, past repair measures, as well as foundation settlements or extreme events (such as earthquakes, floods or fires), and how these may have influenced the actual structural behaviour. These are at the basis of the so-called *historical-critical analysis*, which is aimed at collecting enough insight to interpret the current building condition as a consequence of a series of events and transformations that have led the hypothetical original structure to the actual configuration.

4.5.1.1 Knowledge levels KL and confidence factors CF

The knowledge level **KL** must be understood as an indicator of the completeness of the investigation achieved, which defines the confidence factor **CF**. The factors that determine the appropriate knowledge level are: (1) *geometry* of structural and nonstructural elements (as they may affect the structural response); (2) *details*, which include morphology of masonry, amount and detailing of reinforcement in reinforced concrete, and connections between steel members; (3) *material characterization*, i.e. the mechanical properties of the constituent materials. The quantity and type of information required to achieve a certain knowledge level are briefly introduced here and further specified in the next subsections. It is worth reminding that values or procedures discussed are recommendations, but the analyst can pursue a tailored strategy on a case-by-case basis.

> **KL1** (limited knowledge) includes historical-critical analysis (commensurate with the aimed KL), complete geometric survey, *limited* inspections on the construction details and *limited* testing on the mechanical characteristics of the materials. The corresponding confidence factor is **CF = 1.35**;
>
> **KL2** (normal knowledge) includes historical-critical analysis (commensurate with the aimed KL), complete geometric survey, *extended* inspections on the construction details and *extended* testing on the mechanical characteristics of the materials. The corresponding confidence factor is **CF = 1.20**;
>
> **KL3** (full knowledge) includes historical-critical analysis (commensurate with the aimed KL), complete and accurate geometric survey, *comprehensive* inspections on the construction details and *comprehensive* testing on the mechanical properties of the materials. The corresponding confidence factor is **CF = 1.00**. To reach the KL3 level, the availability of original construction documents can be considered as good as the complete geometric survey with exhaustive knowledge of the construction details (to be checked appropriately in their completeness and correspondence to the real situation).

It is noted that CF values may be differentiated for different materials or for different structural elements, according to the relative knowledge level achieved.

4.5.1.2 Masonry

Regarding the *geometric survey*, three levels of investigation are identified:

Limited: Generally based on visual inspections of masonry elements and, *at least locally*, of the construction technique, as well as the characteristics of the components behind the plaster and across the thickness. The aim is to define the wall cross-section, the degree of connection between orthogonal walls, the characteristics of floor supports and possible ties. The entire building should be investigated to detect cracks and degradation phenomena, and adjacent buildings potentially interacting with the building under consideration should be carefully examined.

Extended: In addition to the requirements of the previous point, inspections need to be more extensive and widespread, for an adequate characterization of materials and construction details.

Comprehensive: In addition to the requirements of the previous point, inspections are systematically extended. The aim is for the analyst to form a clear idea about masonry morphology and quality (for both surface and cross-section), efficacy of connection between orthogonal walls, characteristics of floor supports and possible ties.

For the *mechanical characterization* of the material, three levels of inspection can be distinguished:

Limited: Generally nondetailed and nonextensive, mainly based on visual assessment of masonry constituents (with local removal of plaster in selected locations). Additionally, also making use of historical-critical analysis, the investigation should make possible the classification of the structural elements into areas that can be considered homogeneous. The objective is to identify the masonry typologies in order to assign the mechanical properties. As explained in Section 4.1.6, the visual assessment should consider the masonry bond and the inner composition of walls.

Extended: Visual, extensive and systematic, accompanied by local deeper inspections. These should interest both the exterior and the thickness of masonry elements (also by means of endoscopic inspections) to evaluate constitutive materials, internal morphology, possible homogeneous areas, transverse connections (i.e. through stones, timber lacing or other systems) and existing damage or deterioration (if any). From the mechanical point of view, the analysis of mortars (and of units, if relevant), accompanied by nondestructive diagnostic techniques (penetrometric, sclerometric, sonic, thermographic, radar, etc.), possibly integrated by moderately destructive techniques (e.g. flat-jacks), should be able to classify more accurately the typology and the quality of the structural elements.

Comprehensive: In addition to the requirements of the previous point, tests on the constitutive materials are needed to evaluate their mechanical properties. The analyst is asked to define test typologies and quantities according to the knowledge level to be reached and with respect to the masonry morphology of the building at hand that, for instance, may influence the choice of how investigating the shear strength. The tests must be performed either in situ or in laboratory on undisturbed samples. These may include, if applicable, compression tests (e.g. on wallettes or using double flat-jacks) and shear tests (e.g. compression and shear, diagonal compression or direct shear on joints). The tests should consider the homogeneous typologies defined by the visual inspection, or at least those related to the elements that, from a preliminary sensitivity analysis, are significant for the safety assessment. The material properties to be used for the structural analysis should combine the average/expected values and the tests' results (according to the reliability of the test method). In place of direct tests on the building at hand, it is possible to refer to the results of tests performed on other buildings in the same area, provided the typological correspondence of materials and morphology.

In masonry constructions, the likely presence of elements made with other materials (timber floors and roofs, reinforced concrete floors, iron/steel ties, etc.) should be investigated according to the following subsections.

According to the knowledge level, Circolare (2019) provides information on how to calculate the mechanical properties to be used in the analysis. In the case of nonlinear analysis (with particular reference to horizontal actions), these values are equal to the mean values divided by CF. With reference to Table 4.29 (and Table 4.24 if MQI is considered), the mean values are defined as follows:

KL1: minimum strength and average Young's modulus

KL2: average strength and average Young's modulus

KL3: strength and Young's modulus should be combined from the code recommended values and the test results. For the generic material property X, assuming μ' is the average value provided by the code and X' is the average value of n tests, the mean value μ'' is computed according to Eq. (4.21). The formula describes a weighted average where κ is the weight of the literature values, whose goal is to compensate for possible uncertainties intrinsically present in the test measurements. In the absence of a more exhaustive analysis about the statistical distribution of the mechanical parameter at hand, values of κ are shown in Table 4.43 according to the most used in situ

Table 4.43 Values of κ for the most used in situ and laboratory tests on masonry

Test typology	Mechanical parameter	κ
Compressive test on masonry	Young's modulus	1.5
	Compressive strength	1.0
Double flat-jack test	Young's modulus	1.5
	Compressive strength[a]	2.0[a]
Compression–shear test on an isolated portion of the wall	Shear modulus	1.5
	Shear strength	1.0
Diagonal compression test	Shear modulus	1.5
	Shear strength	1.0
Bed-joint shear test	Shear strength	2.0
Laboratory test on masonry components[b]	Compressive strength	2.0

Source: Circolare (2019).

[a] This test typology does not usually lead to the compressive strength. If the compressive strength is estimated from Young's modulus (by means of an appropriate reducing factor deduced e.g. from Section 4.1.5.1), κ should be taken equal to 3.
[b] For regular masonry, these results can be used to estimate the masonry strength (according to the formulas of Section 4.1.5.1), also noting that the ratio between average and characteristic values may be assumed equal to 1.25.

tests. Particular care should be taken on laboratory tests, due to the intrinsic difficulties in extracting undisturbed samples. Finally, the analyst should examine in depth and justify values of μ'' significantly different from μ' (namely, when tests provide results X' not in line with literature).

$$\mu'' = \frac{nX' + \kappa\mu'}{n + \kappa} \tag{4.21}$$

4.5.1.3 Timber

Either in full timber constructions or isolated timber structural parts (e.g. floors or roof truss in masonry constructions), it is important to determine the wood species. The geometric survey should investigate the cross-section of the elements, their arrangement and their connections (either joinery or mechanical). Additionally, element morphology, such as possible change of shape or section, defects or damage should be also carefully evaluated. The morphology can also play an important role for the structural stability of the element. In this case, the analysis should assess whether the actual deformation state is the consequence of applied loads or due to anatomical imperfection, cutting or manufacturing process.

Regarding the state of conservation, an existing timber structure is likely to exhibit deterioration that may affect the capacity of the structure. Timber deterioration and damage are usually caused by fungal decay, insect infestations, structural overload or mechanical damage. The condition assessment of a timber structure begins with a visual examination to appraise the extent of deterioration and to identify signs of structural distress. Particular attention must be paid in areas prone to rot. For instance, where persistent roof leaks are evident, where timber elements rest in pockets into a masonry wall, in the case of elements hidden by false ceilings or elements that rest on the foundation. In this regard, an awl is a simple and indispensable tool for probing the surface of a timber element to identify the depth and extent of deterioration, from either rot or insect damage. Any portion of the wood that can be penetrated with modest pressure should be assumed to be deteriorated. A hammer may help identifying inner deterioration not visible from the outside. The sound that the timber makes when struck with a hammer helps to reveal the soundness of timber. A dull sound is an indication that there may be internal deterioration. Furthermore, the state of conservation should be associated to the microclimate around the structural element.

Three levels of investigation can be identified, according to their degree of accuracy:

Limited: Mainly based on the visual inspections of the surfaces (at least three faces and one end face of each element of the structural system) supported by limited investigations on the construction elements and connections. Local removal of the protecting coating may be required for assessing the state of conservation, such as proposed by building standards.

Extended: Based on widespread visual investigations on the element surfaces (supported by some instrumental controls), as well as on the conditions of the connections. In this case too, local removal of the protecting coating may be requested for the assessment of the state of conservation. Direct tests should include the estimation of humidity conditions in areas of the element considered particularly sensitive.

Comprehensive: Based on widespread and systematic visual investigations, accompanied by instrumental tests, possibly by means of a resistograph (controlled perforation). The investigations are meant to identify the species, the measurement of the humidity in the material and at the interface with different materials, the analysis of the connections and possible degradation phenomena of the connection elements. Such analyses may also require laboratory activities. It is advisable to use nondestructive or partially invasive techniques to evaluate the mechanical characteristics of the material or to identify degraded areas within the elements.

Regarding the mechanical properties of timber, given their large scatter even within the same building, it is recommended to extend the investigations to a wide number of elements (in accordance with the aimed knowledge level) to assess possible damage or decay (either biotic or abiotic) along the entire development of the structural element and at the extremities where connections are usually placed.

4.5.1.4 Concrete and steel

Structural geometry and member size must always be evaluated as completely as possible to achieve a reliable numerical model. Regarding construction details (e.g. reinforcement location and amount), often hidden from the view, they can be detected with local checks using pacometer tests, inspection windows (opening up the structure) and extending the assessments to the other elements by analogy (also in accordance with current regulations, relevant practice and products on the market at the time of construction). For nonstructural elements (e.g. partition walls, false ceilings or building services), they must be inspected to identify excessive risk for users, such as slender infill walls weakly connected to the structural system or inadequately connected masonry leaves.

The geometric survey must contribute to create a clear idea of structural organization, dimensions of structural and nonstructural elements, floors typology and cross-section. For steel buildings, additional information regarding original steel profiles and relative dimensions, as well as typology and morphology of the connections, are needed. In the case original drawings are available, local checks must be carried out to verify the correspondence with what is actually built, including construction details. In the case construction details are not inspectable, they can be hypothesized from a simulated design (according to technical standards and construction practice in force when the construction was designed and built) to be checked with spot investigations in situ.

Three levels of *geometric survey* can be identified, according to their degree of accuracy:

Limited: By means of local inspections, which allow evaluating the correspondence between original drawings (or simulated design) and what is actually built.

Extended: Carried out when the original drawings are not available (also as an alternative to the simulated design followed by limited investigations) or when the original documents are incomplete.

Comprehensive: Carried out when an accurate knowledge level is desired and the original design documents are not available.

For illustrative purpose (thus not limited to) and when possible, the investigation on construction details should include the quantitative evaluation of the following information:

Reinforced concrete structures:
- Amount and arrangement of longitudinal reinforcement in beams, columns, walls and slabs;
- Amount and location of possible shear-resistance bent-up bars in beams;
- Amount and details of transverse reinforcement in critical areas and in beam-column nodes;
- Amount of longitudinal reinforcement in the upper part of T-shaped beams (such as reinforcement within the floor section);
- Dimension of the supports for horizontal elements and relative constraint conditions;
- Cover thickness;
- Overlapping length of longitudinal reinforcement and possible anchorage technique (e.g. hooks);
- Floor slab details.

Steel structures:
- Type and location of connections;
- Details of floor supports;
- Details of column-to-foundation connections.

For all materials, possible aging or degradation phenomena should be investigated and considered for the material characterization. These include at least carbonation depth of concrete cover, and corrosion of reinforcement and steel members. The *mechanical characterization* of the material is based on three levels of inspections:

Limited: Including a limited number of in situ or laboratory tests on samples aimed at completing the information on the material properties, obtained either from the regulations in force at the time of construction, or from the nominal characteristics shown on the construction drawings or in the original test certificates.

Extended: Including a larger number of tests (with respect to the previous point) aimed at providing information to respond to the lack of both construction drawings and original test certificates, or when the values obtained with the limited tests are lower than those reported in the original drawings or certificates.

Comprehensive: Including a larger number of tests (with respect to the previous point) aimed at obtaining information to respond to the lack of both construction drawings and original test certificates, or when the values obtained from the tests, limited or extended, are lower than those shown on the original drawings or certificates, or in those cases where a particularly accurate knowledge is desired.

For all the cases, the tests should investigate strength and Young's modulus for concrete, and yield strength, ultimate strength and strain for steel. Regarding steel structures, the tests should be extended to connections also. Additionally, for what concerns both geometric and mechanical investigations, it is not unusual that, based on a preliminary structural analysis of the construction, further investigations (in terms of number, type and location) are needed for a better description of the most critical structural elements. The same applies when significant scatter is found in the material properties.

For illustrative purpose, Table 4.44 shows the indications to define the knowledge level and the relative confidence factor. From the table, it is also clear that a poor knowledge of the building does not allow a nonlinear analysis. The subsequent Tables 4.45 and 4.46 give quantitative insight on the levels of material investigation for reinforced concrete and steel constructions, respectively. These are not stringent, and the analyst may adopt a different approach, also according to possible repetitive situations or homogeneous areas. It is noted that the percentages relative to the construction details (second column of Tables 4.45 and 4.46) are more demanding

Table 4.44 Knowledge level according to available information, and relative confidence factor for concrete and steel constructions

Knowledge level	Geometry	Construction details	Material properties	Type of analysis	Confidence factor CF
KL1		Simulated design and limited in situ inspection	According to literature (e.g. guidelines in force at the time of construction) and limited in situ testing	Linear	1.35
KL2	From original construction drawings (with sample visual survey) or from full survey (for all KLs)	Incomplete original drawings and limited in situ inspection, or extended in situ inspection	From original documents and limited in situ testing; or extended in situ testing	Linear and nonlinear	1.20
KL3		Complete original drawings and limited in situ inspection, or comprehensive in situ inspection	From original test reports and extended in situ testing, or comprehensive in situ testing	Linear and nonlinear	1.00

Source: Circolare (2019).

Table 4.45 Indications for evaluating the level of investigation on reinforced concrete
            structures

| Level of investigation | For each type of primary element (beam, column, wall): | |
	Construction details	Tests
Limited	Amount and arrangement of reinforcement is verified for more than 15% of elements	One concrete sample every 300 m^2 per floor, one steel sample per floor
Extended	Amount and arrangement of reinforcement is verified for more than 35% of elements	Two concrete samples every 300 m^2 per floor, two steel samples per floor
Comprehensive	Amount and arrangement of reinforcement is verified for more than 50% of elements	Three concrete samples every 300 m^2 per floor, three steel samples per floor

Source: Circolare (2019).

Table 4.46 Indications for evaluating the level of investigation on steel structures

| Level of investigation | For each type of primary element (beam, column, wall): | |
	Construction details	Tests
Limited	Connection characteristics are verified for more than 15% of the elements	One steel sample and one bolt sample per floor
Extended	Connection characteristics are verified for more than 35% of the elements	Two steel samples and two bolt samples per floor
Comprehensive	Connection characteristics are verified for more than 50% of the elements	Three steel samples and three bolt samples per floor

Source: Circolare (2019).

in Eurocode 8 (2005), namely, 20%, 50% and 80% for the three levels of investigation, respectively. Additionally, it is allowed to replace destructive tests, no more than 50%, with at least three times as numerous nondestructive tests, provided their reliability is documented and they are calibrated on the destructive tests.

For nonlinear analysis of reinforced concrete and steel structures, the demand is obtained using the average values of mechanical properties. In turn, the capacity is evaluated considering the average values divided by CF for ductile failure mechanisms, and average values divided by both CF and partial safety factor (PSF) for brittle failure mechanisms. The structural assessment of ductile mechanisms is made in terms of displacement, whereas, for brittle mechanisms, the assessment is made in terms of strength. According to the Italian code (NTC, 2018), the partial factors can be assumed 1.5 for concrete, 1.15 for steel reinforcement and 1.05 for steel structural members. For the sake of completeness, in seismic analysis, ductile behaviour is associated to beams, columns and walls subjected to bending moment, without compression or small compressive stresses.

In turn, brittle behaviour is associated to shear in beams, columns, walls, and, in general, joints. In the case of large compressive stresses, the structural behaviour of the elements can be considered as brittle.

4.5.2 Italian guidelines for built heritage

Given the importance and vulnerability to earthquakes of cultural heritage buildings, in 2011 the Italian Presidency of the Council of Ministers approved the guidelines for their analysis and conservation (DPCM, 2011). These guidelines give indications about the process of knowledge, seismic assessment and design of interventions according to the principles of conservation. These are aimed at protecting the cultural value and safeguarding the distinctive features of the building at hand. Whereas the technical aspects are conceptually similar to what is stated in Circolare (2019), the approach and criteria to follow are those proposed by the ISCARSAH recommendations (ICOMOS/ISCARSAH, 2003). In this regard, the reader is referred to Roca et al. (2019) where the conservation of historic construction is treated extensively. As most of the existing built heritage in Italy is made in masonry, the mentioned guidelines focus exclusively on this material. In the following, seismic demand and strengthening measures are not examined, limiting the discussion to the structural modelling.

Without entering the merit of the definition of serviceability and ultimate limit states (addressed in national codes), plus a possible damage limit state for artworks (e.g. frescoes, decorations, balusters, altarpieces or statues), it is important to highlight similarities and differences in the analysis of ordinary and heritage buildings. In both cases, dealing with historic constructions, the intrinsic heterogeneity of material properties (even within the same building), possible alterations of the original structural system, existing damage, effective morphology of structural elements and their interactions are only a few of the aspects to be considered in the analysis. These make structural modelling of historic constructions a challenging task based not only on quantitative data (e.g. geometry and mechanical parameters), but also on qualitative insights based on a preliminary historical-critical analysis.

In turn, invasiveness is a key issue when dealing with built cultural heritage. Besides the nature of the tests for the material characterization, the safety levels usually adopted for new buildings may require a set of interventions too drastic (sometimes irreversible) that would compromise the authenticity of the building. In this case, according to the conservation principles, the analyst may consciously accept the idea of a lower safety level than for an ordinary structure and select only unavoidable interventions. This approach is accepted, provided that an accurate monitoring and maintenance plan is enforced, and a safety analysis is planned at a later stage. In probabilistic terms, the building is considered safe only for a defined period of time (i.e. *design working life*), and the required measures are postponed to the end of the period, when new advances in technology, strengthening techniques, and seismic risk assessment may be available. According to

Section 4.5.3.1, the design working life defines the period during which the structure has to meet the performance criteria agreed.

As shown in the previous section, the structural analysis stems from the knowledge level of the building. Given the intrinsic cultural values of heritage buildings, this process should balance the least possible number of invasive tests with the most complete historical-critical analysis. The latter should focus on existing damage, original structural system and use, and their changes through history. This information assists in understanding the causes of alterations occurring through history, and in depicting a clearer picture of the current state. Differently from Circolare (2019), the confidence factor is now calculated as a summation of four partial factors (related to the aspects of the investigation) and not in tabular form, offering greater flexibility to the analyst. The related formula is shown in Eq. (4.22), whereas the partial confidence factors are described in Table 4.47. As it is possible to notice, CF falls in the range 1.00–1.35, which is the same range considered in Circolare (2019).

$$CF = 1 + \sum_{i=1}^{4} CF_i \qquad\qquad (4.22)$$

The guidelines provide also suggestions about structural modelling of the most recurrent construction typologies. Palaces, villas and alike belong to the structural typology of buildings with shear walls and intermediate floors. Although the plan may be complex, with walls oriented in different directions, the analogy with ordinary masonry buildings is evident. In this regard, the same approach can be pursued, sometimes enhanced by the better quality of materials and construction details, higher care in the design and during the building phases. The presence of vaulted structures may be modelled by means of equivalent floors. The relative stiffness should be calculated according to the type of vault, material properties, thickness and level of connection with vertical walls. To obtain equivalent data, a linear approach may be adopted, provided an ultimate in-plane shear deformation is defined. Lateral thrust should be taken into consideration. Additionally, when large divisions are present (e.g. large atriums, loggias or cloisters), the global model may consider them in an approximated way and proceed with local verifications of the substructures at a later stage. Finally, the seismic analysis of the global model does not exclude the need of analysing local mechanisms of failure.

The second typology regards churches, places of worship and, generally, buildings with great halls without intermediate floors. The systematic analysis of damage following strong earthquakes in the past highlighted that their seismic behaviour is governed by the behaviour of macro-elements (i.e. a portion of the building with a structural behaviour substantially autonomous, such as a façade, triumphal arch or bell tower). Accordingly, the

Table 4.47 Definition of partial confidence factors related to the four aspects of investigations

Geometric survey	Complete	$CF_1=0.05$
	Complete, with cracking and deformation pattern restitution	$CF_1=0$
Historical information and construction techniques	Hypothetical identification of constructive phases based on a limited survey on materials and construction elements, associated with the understanding of the transformation events (by means of documental and thematic investigations)	$CF_2=0.12$
	Partial identification of construction phases and structural behaviour based on: (1) limited survey on materials and construction elements, associated with the understanding and verification of the transformation events (by means of historical documentary and thematic investigations, plus diagnostic verification of historiographic hypotheses) and (2) extended survey on materials and construction elements associated with the understanding of the transformation events (by means of documental and thematic investigations)	$CF_2=0.06$
	Complete identification of construction phases and of structural behaviour based on comprehensive survey on materials and construction elements, associated with the understanding of transformation events (by means of documental, and possible diagnostic investigations)	$CF_2=0$
Mechanical parameters of materials	Based on available data (e.g. literature or building codes)	$CF_3=0.12$
	Limited testing	$CF_3=0.06$
	Extended testing	$CF_3=0$
Foundations and soil (related to seismic input and structure–soil interaction)	Limited investigations on soil and foundations, without geotechnical data and without foundation information	$CF_4=0.12$
	Limited investigations on soil and foundations, with geotechnical data and with foundation information	$CF_4=0.06$
	Extended or comprehensive investigations on soil and foundations	$CF_4=0$

Source: DPCM (2011).

analysis of local mechanisms or portions may be preferred (e.g. nave section or triumphal arch). A global model is suggested for central-plan buildings, with one or more symmetry axes, and good-quality materials and connections. The same applies for the modelling of portions of the building, such as for the plane analysis of a single bay. The effect of vaults, when relevant, should be considered in terms of both stiffness and thrust.

Another typology regards towers and, generally, predominantly vertical configurations. The seismic behaviour of this typology depends on specific factors, such as slenderness of the structure, degree of connection between walls defining the cross-section, possible presence of lower adjacent structures (capable of providing a horizontal constraint), presence of slender architectural elements (e.g. spires, bell-gable or parapets) or vulnerable elements (e.g. belfry) at its top. Vulnerability is also influenced by the presence of existing damage due to, for instance, vibrations induced by bells, problems at foundations or lightning. Of the mentioned factors, the wall connections play a significant role because they make the building behave as a cantilever with full cross-section working together, not as a group of distinct walls. This behaviour is guaranteed by the presence of ties, iron rings and good connection with intermediate floors. In general, given the lower complexity with respect to the two previous building typologies, all the peculiarities can be modelled confidently with a global model.

The last typologies mentioned in the guidelines are bridges, triumphal arches and other arched structures. The seismic response of this type of structures is complex, not supported by a systematic observation of damage during postseismic surveys. In this regard, the structural behaviour of arched structures is usually connected to that of a bare masonry arch, whose main mechanisms under gravitational and seismic loads, as well as settlements and scouring, are discussed in Roca et al. (2019). About FE analysis, it is important to dedicate particular care also to components other than the arch, such as the infill, backing and spandrel. This is particularly evident for masonry bridges, whose strategic interest is highlighted by their likely current use. Given the large amount of volume occupied, the structural response of masonry bridges and vaults may be significantly affected by the backing presence and properties. In many cases, the backing reduces the effective span and neglecting its contribution generally leads to rather conservative results. Additionally, the out-of-plane behaviour of the spandrel wall in bridges must be carefully investigated, resembling a retaining wall and being often the most vulnerable bridge component.

4.5.3 Safety formats for nonlinear analysis according to Model Code 2010

In the distant past, design was made according to prescriptive criteria on geometry, materials, configuration, detailing, strength and stiffness. With the advent of the elasticity theory, safety started to be checked according to the concept of admissible stresses. In the last decades, this approach has been progressively abandoned in favour of the so-called *performance-based design*, adopted by modern design specifications. This method is founded on the premise that structural systems (and the nonstructural systems they support) must meet specific performance objectives when subjected to defined loading. The basic steps are as follows: (1) establish the performance objectives, (2) conduct initial design (for new constructions) and (3) verify performance through analytical simulation, prototype testing or a combination thereof. An adequate performance is reached when a structure or a structural component has demonstrated satisfactory behaviour to meet the performance requirements. If not, the performance of a structure or a structural component is considered to be inadequate. This condition is referred to as *failure*, indicating the missed fulfilment of one or more performance criteria rather than the collapse of the building.

In the following, a brief synthesis of the performance-based approach is proposed with a particular focus on nonlinear analysis of existing buildings, being, however, the principles valid for the design of new constructions too. In this regard, Model Code 2010 (fib, 2013) is used here as the main reference. The reader is referred to other building codes, e.g. EN 1990 (2002), for further clarifications.

4.5.3.1 Basic principles

With no loss of generality, it is possible to state that *structures and structural members must be designed, constructed and maintained in such a way that they perform adequately and in an economically reasonable way during construction, working life and dismantlement*. For both new or existing constructions, the three *performance categories* to be considered are *serviceability*, *structural safety* and *sustainability*. *Performance requirements* are selected in accordance with the stakeholders' demands (e.g. owners, users, residents or maintenance team) and the requirements given by building codes and guidelines. These are established by means of *performance criteria*, i.e. quantitative limits defining the border between the desired and the adverse behaviour. In this regard, the concept of *limit state design* is generally applied. In a simplistic way, serviceability can be evaluated in terms of a deterministic verification. For instance, for a beam supporting a floor, a span-to-deflection ratio equal to 250 for a rare combination of self-weight and live loads represents the criteria that distinguishes a desired from an adverse situation.

As actions, environmental influences and structural properties may vary with time, the definition of a performance criteria must be constrained to the concepts of *design working life* and *reliability*. According to EN 1990 (2002), the design working life is the assumed period for which a structure is to be used for its intended purpose with anticipated maintenance but without major repair being necessary (in the case of existing buildings, Model Code 2010 uses the term *residual service life*). In turn, reliability refers to the acceptable probability of failure. In other words, the *performance requirements are satisfied if all relevant performance criteria are met during the working life at the required reliability level*. The importance of defining the working life is connected to the fact that every building suffers from natural decay and it is impossible (and economically unfeasible) to guarantee certain performance requirements indefinitely. Additionally, the likelihood of events with increasing magnitude gets higher with an extended working life. Accordingly, constraints related to reliability are specified by means of a *target reliability level*, generally expressed in terms of the target reliability index β or target probability of failure P_f. The latter is the probability of having the resistance R smaller than the effect of the loading E during the working life: synthetically P_f=Prob $(R \le E)$. Consequently, $1 - P_f$ is the probability of success. It is worth noticing that both resistance and loading are affected by random variations.

The choice of the acceptable failure probability is a delicate issue that should consider the possible consequences of failure in terms of loss of life or injury, potential economic losses and the degree of societal inconvenience. Additionally, the amount of expense and effort required to reduce the risk of failure should also be considered. It is evident that the costs involved when upgrading the performance of existing structures are usually higher than the costs of obtaining the same performance by structural design in the case of a new construction. Upgrading existing structures may also entail relocation of occupants and disruption of activities or influencing heritage values, which does not play a role in the case of the design of new structures. Accordingly, for existing structures the target reliability level may be chosen lower than that for new structures. From a reverse perspective, it is possible to keep the same target reliability level of a new structures but reducing the design working life.

In absence of well-founded analysis about the consequences of failure and of the costs of safety measures, for ultimate limit state analysis, Model Code 2010 (fib, 2013) suggests β in the range 3.1–3.8 (depending on the costs of safety measures for upgrading existing structures) for a design working life of 50 years. This means that the structure is considered safe if the probability of failure P_f in 50 years is below 10^{-3} to 10^{-4}, respectively. It is worth underlining that the given target reliability index is valid for ductile structural components or redundant systems for which a collapse is preceded by some kind of warning, which allows measures to be taken to avoid severe consequences. A structural component or structural system that would likely collapse suddenly without warning should be designed for a higher level of reliability than that recommended for ductile structural components.

Operatively, for all conditions that can occur during the working life, resistance R should be larger than the effect of the loading E. With this aim, according to the *semi-probabilistic method*, the performance requirements are not checked in a probabilistic way, but through a simple comparison of scalars: $R_d > E_d$. The analogy with the previous formula is evident, but the design values are now introduced (indicated with the subscript d). These are obtained according to an assumed probabilistic distribution of both resistance and action. The design values are calculated reducing the characteristic value of resistances and incrementing the characteristic value of actions. Whereas the latter can be still calculated according to the PSF (partial safety factor) method (for permanent load, live load, prestressing, etc.) in accordance with building codes, different methods are available for the determination of R_d.

For nonlinear analysis, these are presented next. In the *probabilistic method*, design resistance is explicitly evaluated for a required failure probability or reliability index. In the *global resistance method*, the design resistance is estimated based on a simplified probabilistic approach assuming significant approximations. This method is subdivided into two subcategories, namely, *global resistance factor* (GRF) and *estimation of a coefficient of variation* (ECOV) *of resistance methods*. In the *partial safety factor method*, the design resistance is determined directly using design values of random variables. The choice of a relevant method for the calculation of the design resistance depends on engineering judgement while taking into consideration the resistance model and possible uncertainties. In practical design, usage of at least two methods is recommended, which provides independent verifications of limit state.

It must be stressed that, despite their names, all methods consider the global resistance of the structure. In this regard, the term *global* is introduced to distinguish the check of safety of the entire structure vs. the local safety check with the PSF method based on linear analysis. The main difference between the two approaches is that the linear analysis ensures a safety check at the local material points, without providing information about the capacity of the structure as a whole, that may result underestimated. In turn, nonlinear analysis investigates the reserves of capacity that guarantee performance requirements even with local violations at the material level. Additionally, with the global approach, redundancy and robustness are implicitly included in the safety assessment.

4.5.3.2 Probabilistic method

The probabilistic method (sometimes referred to as *fully probabilistic design method*) is the most sophisticated tool for safety assessment and allows the analyst to explicitly include the reliability requirements in terms of reliability index β and reference period. The numerical simulation resembles real testing of a representative group of samples, statistically analysed for the assessment of safety. Basically, by assuming a random distribution of the input variables (material properties, dimensions, boundary conditions, etc.), the consequent distribution of resistance values implicitly considers all possible causes of failure. For instance, the method can reveal reserves of capacity or alternative force redistributions not easily identifiable with conventional methods. Certainly, the results are sensitive to the initial random distribution selected (in terms of function, average and standard deviation), marginally supported by a wide number of destructive and nondestructive tests. The latter aspect should be carefully evaluated for not compromising the authenticity of the building at hand. As a consequence, also due to its computational demands and the consequent economical effort, a full probabilistic analysis is suggested only in those cases where the consequences of failure justify the effort, or where real random properties need to be exploited.

4.5.3.3 Global resistance factor method (GRF)

The effects of uncertainties are integrated in a global design resistance and can also be expressed by a global safety factor. The design resistance is calculated according to Eq. (4.23), where the function r represents the nonlinear analysis. In this case, the analysis is performed considering the average input material parameters. It is worth reminding that Model Code 2010 (fib, 2013) suggests to use average values $f_c=0.85\, f_{ck}$ and $f_y=1.1\, f_{yk}$ for compressive strength of concrete and yield stress of reinforcing steel, respectively. It is important to highlight that the concrete strength is reduced to account for its higher variability when compared to reinforcing steel, and to treat both materials in a unified way. These values are needed to obtain the GRF consistent with the material variability. The other concrete parameters (such as fracture energy) can be derived from the average compressive strength by standard relations. Additionally, by considering the partial factor of resistance $\gamma_R=1.2$ and the model uncertainty factor for well-validated models $\gamma_{Rd}=1.06$, the global safety coefficient results $\gamma_{GRF}=1.27$. In other words, if P_u is the ultimate load obtained from the nonlinear analysis by inputting average mechanical properties, the design resistance is $R_d=P_u/1.27$.

$$R_d = \frac{r\left(f_{\text{average}}, \ldots\right)}{\gamma_R \gamma_{Rd}} = \frac{r\left(f_{\text{average}}, \ldots\right)}{\gamma_{GRF}} \tag{4.23}$$

Similar considerations of the GRF method may be adapted for structures made with other materials, such as masonry, timber and steel.

4.5.3.4 Method of estimation of a coefficient of variation of resistance (ECOV)

This method is based on the idea that the random distribution of resistance (described by the coefficient of variation V_R) can be estimated from average and characteristic values, R_a and R_k, respectively. In turn, these are estimated using two separate nonlinear analyses adopting average and characteristic values of input material parameters, $R_a=r$ $(f_{\text{average}}, \ldots)$ and $R_k=r$ $(f_k, \ldots)$, respectively. The underlying assumption of the method is that the random distribution of resistance follows a lognormal distribution, which is typical for structural resistance. In this case, assuming the probability for its characteristic value as 5%, the coefficient of variation can be expressed as shown in Eq. (4.24). The global resistance factor γ_R for the average is determined from Eq. (4.25), where α is a sensitivity (weight) factor for the reliability of resistance (generally $\alpha=0.8$) and β is the reliability index ($\beta=3.8$ means probability $P_f=0.01\%$). The design resistance R_d is then calculated from Eq. (4.26), where the model uncertainty is accounted for by means of the factor γ_{Rd}. For well-validated numerical models, $\gamma_{Rd}=1.06$, whereas higher values should be used for low-level validation.

$$V_R = \frac{1}{1.65} \ln\left(\frac{R_a}{R_k}\right) \tag{4.24}$$

$$\gamma_R = \exp\left(\alpha\beta V_R\right) \tag{4.25}$$

$$R_d = \frac{R_a}{\gamma_{Rd}\gamma_R} \tag{4.26}$$

As shown in the previous method, characteristic and average values of material parameters are used to evaluate the possible scatter of the resistance. This approach is preferable to the PSF method that is based on the design material values, usually rather low and far from the real behaviour of the material.

4.5.3.5 Partial safety factor method (PSF)

This is a simplified verification concept usually adopted for linear analysis of new constructions. It is based on past experience and calibrated in such a way that the general reliability requirements are satisfied with a sufficient margin during a defined period of time. By extending this method to nonlinear analysis, the design resistance R_d is calculated using the design values f_d as input parameters for the nonlinear analysis, namely, $R_d=r\,(f_d, ...)$. In this case the structural analysis is based on extremely low material parameters in all locations, which does not correspond to the probabilistic concept of simulation. This may cause deviations from reality in structural response, e.g. in failure mode or force redistribution. Accordingly, this method is not recommended for existing structures.

BIBLIOGRAPHY AND FURTHER READING

Almazán, F.J.A., Prieto, E.H., Martitegui, F.A., Richter, C., 2008. Comparison of the Spanish visual strength grading standard for structural sawn timber (UNE 56544) with the German one (DIN 4074) for Scots pine (Pinus sylvestris L.) from Germany. *Eur. J. Wood Wood Prod.* 66, 253–258. Doi: 10.1007/s00107-008-0241-9.

Amadio, C., Rajgelj, S., 1991. Shear behavior of brick-mortar joints. *Mason. Int.* 5, 19–22.

ANSI/AWC, 2018. *National design specifications for wood construction.* Leesburg, VA: American Wood Council.

AS 3700, 2018. Masonry structures. Standards Australia.

ASCE 41-06, 2007. *Seismic Rehabilitation of Existing Buildings.* Reston, VA: American Society of Civil Engineers. Doi: 10.1061/9780784408841.

ASCE 41-17, 2017. *Seismic Evaluation and Retrofit of Existing Buildings.* Reston, VA: American Society of Civil Engineers. Doi: 10.1061/9780784414859.

ASTM C270-19ae1, 2019. *Standard Specification for Mortar for Unit Masonry.* Doi: 10.1520/C0270-19AE01.

Atkinson, R.H., Amadei, B.P., Saeb, S., Sture, S., 1989. Response of masonry bed joints in direct shear. *J. Struct. Eng.* 115, 2276–2296. Doi: 10.1061/(ASCE)0733-9445(1989)115:9(2276).

Augenti, N., Parisi, F., 2011. Constitutive modelling of tuff masonry in direct shear. *Constr. Build. Mater.* 25, 1612–1620. Doi: 10.1016/j.conbuildmat.2010.10.002.

Backes, H.P., 1985. *On the Behavior of Masonry under Tension in the Direction of the Bed Joints.* PhD dissertation, Aachen University of Technology, Aachen.

Binda, L., Fontana, A., Frigerio, G., 1988. Mechanical behaviour of brick masonries derived from unit and mortar characteristics, in: *Proceedings of 8th International Brick and Block Masonry Conference*, Dublin, pp. 205–216.

Binda, L., Fontana, A., Mirabella, G., 1994b. Mechanical behavior and stress distribution in multiple-leaf stone walls, in: Shrive, N., Huizer, A. (Eds.), *10th International Brick and Block Masonry Conference*, Calgari, Canada, pp. 51–59.

Binda, L., Mirabella, G., Tiraboschi, C., Abbaneo, S., 1994a. Measuring masonry material properties, in: Abrams, D., Calvi, G.M. (Eds.), *U.S. Italy Workshop on "Guidelines for Seismic Evaluation and Rehabilitation of Unreinforced Masonry Buildings." Technical Report NCEER 94-0021*, Pavia, Italy, pp. 3–24.

Bisoffi-Sauve, M., Morel, S., Dubois, F., 2019. Modelling mixed mode fracture of mortar joints in masonry buildings. *Eng. Struct.* 182, 316–330. Doi: 10.1016/j.engstruct.2018.11.064.

Bodig, J., 2000. The process of NDE research for wood and wood composites, in: *12th International Symposium on Nondestructive Testing of Wood*, University of Western Hungary, Sopron, Hungary, pp. 1–17.

Borri, A., Corradi, M., Castori, G., De Maria, A., 2015. A method for the analysis and classification of historic masonry. *Bull. Earthq. Eng.* 13, 2647–2665. Doi: 10.1007/s10518-015-9731-4.

Borri, A., De Maria, A., 2019. Qualità muraria secondo il metodo IQM: aggiornamento alla Circolare esplicativa n. 7 del 2019 (in Italian). Struct. 222. Doi: 10.12917/STRU222.07.

Briceño, C., Azenha, M., Lourenço, P.B., 2019. *Systematic Report: Current Situation of the Influence of Masonry Mortar in the Eurocode 6 Part 1-1.* Report 2019-DEC/E-15. University of Minho, Guimarães, Portugal.

BSI, 2005. BS NA EN 1996-1-1: UK National Annex to Eurocode 6. Design of masonry structures. General rules for reinforced and unreinforced masonry structures. UK: British Standards Institution, p. 20.

Bussell, M.N., 1997. *Appraisal of Existing Iron and Steel Structures.* Ascot: Steel Construction Institute.

CEB-FIP, 1993. *Model Code 1990.* London: Thomas Telford Services Ltd.

Circolare, 2019. *Instruction for the Application of the Building Code for Constructions* (in Italian). Italy: Ministry of Infrastructure and Transport.

Dhanasekar, M., Page, A.W., Kleeman, P.W., 1985. The failure of brick masonry under biaxial stresses. *Proc. Inst. Civ. Eng.* 79, 295–313. Doi: 10.1680/iicep.1985.992.

DPCM, 2011. *Evaluation and Reduction of Seismic Risk of Cultural Heritage with Reference to the Building Code for Constructions Promulgated by the Ministry of Infrastructure and Transport on 2008 January 14th* (in Italian). Italy: Ministry for Cultural Heritage and Activities.

Drysdale, R.G., Vanderkyle, R., Hamid, A., 1979. Shear strength of brick masonry joints, in: *5th International Brick Masonry Conference*, Washington (DC), pp. 106–113.

EN 1990, 2002. *Basis of Structural Design (EN 1990:2002)*. Brussels (Belgium): CEN.

EN 1992-1-1, 2004. *Eurocode 2: Design of Concrete Structures. Part 1-1: General Rules and Rules for Buildings*. Brussels (Belgium): CEN.

EN 1993-1-1, 2014. *Eurocode 3: Design of Steel Structures - Part 1-1: General Rules and Rules for Buildings (EN 1993-1-1:2005/A1:2014)*. Brussels (Belgium): CEN.

EN 1995-1-1, 2014. *Eurocode 5: Design of Timber Structures. Part 1-1: General - Common Rules and Rules for Building (EN 1995-1-1:2004/A2:2014)*. Brussels (Belgium): CEN.

EN 1996-1-1, 2005. *Eurocode 6- Design of Masonry Structures. Part 1-1: General Rules for Reinforced and Unreinforced Masonry Structures*. Brussels (Belgium): CEN.

EN 1998-3, 2005. *Eurocode 8: Design of Structures for Earthquake Resistance - Part 3: Assessment and Retrofitting of Buildings*. Brussels (Belgium): CEN.

EN 10210-1, 2006. *Hot Finished Structural Hollow Sections of Non-Alloy and Fine Grain Steels - Part 1: Technical Delivery Conditions*. Brussels (Belgium): CEN.

EN 10025-2, 2019. *Hot Rolled Products of Structural Steels - Part 2: Technical Delivery Conditions for Non-Alloy Structural Steels*. Brussels (Belgium): CEN.

EN ISO 6892-1, 2019. *Metallic Materials - Tensile Testing - Part 1: Method of Test at Room Temperature*. Brussels (Belgium): CEN.

EN 14081-1, 2005. *Timber Structures – Strength Graded Structural Timber with Rectangular Cross Section – Part 1: General Requirements*. Brussels (Belgium): CEN.

EN 384, 2018. *Structural Timber. Determination of Characteristic Values of Mechanical Properties and Density (EN 384:2016+A1:2018)*. Brussels (Belgium): CEN.

EN 338, 2016. *Structural Timber-Strength Classes*. Brussels (Belgium): CEN.

EN 1912, 2012. *Structural Timber. Strength Classes. Assignment of Visual Grades and Species*. Brussels (Belgium): CEN.

EN 408, 2012. *Timber Structures. Structural Timber and Glued Laminated Timber. Determination of Some physical and Mechanical Properties (EN 408:2010/A1:2012)*. Brussels (Belgium): CEN.

fib, 2013. *Model Code for Concrete Structures 2010*. Berlin: Wilhelm Ernst & Sohn,.

fib Bulletin 45, 2008. Practitioners' guide to finite element modelling of reinforced concrete structures, fib Bulletins. *Int. Federat. Struct. Concrete*. Doi: 10.35789/fib.BULL.0045.

Gardner, L., Yun, X., 2018. Description of stress-strain curves for cold-formed steels. *Constr. Build. Mater.* 189, 527–538. Doi: 10.1016/j.conbuildmat.2018.08.195.

Giuffrè, A., 1996. A mechanical model for statics and dynamics of historical masonry buildings, in: Petrini, V., Save, M. (Eds.), *Protection of the Architectural Heritage Against Earthquakes*. Springer, Vienna, pp. 71–152. Doi: 10.1007/978-3-7091-2656-1_4.

Glos, P., 1978. *Zur bestimmung des festigkeitsverhaltens von brettschichtholz bei druckbeanspruchung aus werkstoff- und einwirkungsgrössen* (in German). PhD dissertation, TU Munich, Germany.

Hegemeier, G.A., Arya, S.K., Krishnamoorthy, G., Nachbar, W., Furgerson, R., 1978. On the behaviour of joints in concrete masonry, in: *North American Masonry Conference*, Boulder, CO.

Hendriks, M.A.N., de Boer, A., Belletti, B., 2017. *Guidelines for Nonlinear Finite Element Analysis of Concrete Structures - RTD:1016-1:2017*. Rijkswaterstaat Centre for Infrastructure, The Netherlands.

Hilsdorf, H.K., 1969. Investigation into the failure mechanism of brick masonry loaded in axial compression, in: Johnson, F.H. (Ed.), *Designing, Engineering and Constructing with Masonry Products*. Houston, TX: Gulf Publishing Company, pp. 34–41.

Hoffmann, G., Schubert, P., 1994. Compressive strength of masonry parallel to the bed joints, in: Shrive, N., Huizer, A. (Eds.), *10th International Brick and Block Masonry Conference*. Calgari, Canada, pp. 1453–1462.

Huber, H.A., McMillin, C.W., McKinney, J.P., 1985. Lumber defect detection abilities of furniture rough mill employees. *For. Prod. J.* 35, 79–82.

ICOMOS / ISCARSAH, 2003. *Recommendations for the Analysis, Conservation and Structural Restoration of Architectural Heritage*.

Kasal, B., Anthony, R.W., 2004. Advances in situ evaluation of timber structures. *Prog. Struct. Eng. Mater.* 6, 94–103. Doi: 10.1002/pse.170.

Kaushik, H.B., Durgesh, R.C., Sudhir, J.K., 2007. Stress-strain characteristics of clay brick masonry under uniaxial compression. *J. Mater. Civ. Eng.* 19, 728–739. Doi: 10.1061/(ASCE)0899-1561(2007)19:9(728).

Kržan, M., Gostič, S., Cattari, S., Bosiljkov, V., 2015. Acquiring reference parameters of masonry for the structural performance analysis of historical buildings. *Bull. Earthq. Eng.* 13, 203–236. Doi: 10.1007/s10518-014-9686-x.

Lourenço, P.B., 1996. *Computational Strategies for Masonry Structures*. PhD dissertation, Delft University of Technology.

Lourenço, P.B., Barros, J.O., Almeida, J.C., 2002. *Characterization of Masonry under Uniaxial Tension* - Report 02-DEC/E–12. Guimarães, Portugal: University of Minho.

Lourenço, P.B., Pereira, J.M., 2018. *Recommendations for Advanced Modeling of Historic Earthen Sites: Seismic Retrofitting Project Research Report*. Getty Conservation Institute, Los Angeles, CA.

Lourenço, P.B., Ramos, L.F., 2004. Characterization of cyclic behavior of dry masonry joints. *J. Struct. Eng.* 130, 779–786. Doi: 10.1061/(ASCE)0733-9445(2004)130:5(779).

Lourenço, P.B., Rots, J.G., 1997. Multisurface interface model for analysis of masonry structures. *J. Eng. Mech.* 123, 660–668. Doi: 10.1061/(ASCE)0733-9399(1997)123:7(660).

Magenes, G., 1992. *Seismic Behavior of Brick Masonry: Strength and Failure Mechanisms* (in Italian). PhD dissertation, University of Pavia, Pavia.

Mander, J.B., Priestley, M.J.N., Park, R., 1988. Theoretical stress-strain model for confined concrete. *J. Struct. Eng.* 114, 1804–1826. Doi:10.1061/(ASCE)0733-9445(1988)114:8(1804).

Mann, W., Müller, H., 1982. Failure shear-stressed masonry – an enlarged theory, tests and application to shear walls. *Br. Ceram. Soc.* 30, 223–235.

Muñoz, G.R., Gete, A.R., Saavedra, F.P., 2011. Implications in the design of a method for visual grading and mechanical testing of hardwood structural timber for designation within the European strength classes. *For. Syst. INIA* 20, 235–244. Doi: 10.5424/fs/2011202-9771.

NTC, 2018. *Italian Building Code for Constructions* (in Italian). Italy: Ministry of Infrastructure and Transport.

NZSEE, 2006. *Assessment and Improvement of the Structural Performance of Buildings in Earthquake.* Wellington, NZ: New Zealand Society for Earthquake Engineering.

NZSEE, 2017. *Seismic Assessment of Existing Buildings (New Zealand Technical Guidelines for Engineering Assessments) - Part C - Unreinforced Masonry Buildings.* Wellington, NZ: New Zealand Society for Earthquake Engineering.

Oudjene, M., Khelifa, M., 2009a. Elasto-plastic constitutive law for wood behaviour under compressive loadings. *Constr. Build. Mater.* 23, 3359–3366. Doi: 10.1016/j.conbuildmat.2009.06.034.

Oudjene, M., Khelifa, M., 2009b. Finite element modelling of wooden structures at large deformations and brittle failure prediction. *Mater. Des.* 30, 4081–4087. Doi: 10.1016/j.matdes.2009.05.024.

Page, A.W., 1981. The biaxial compressive strength of brick masonry. *Proc. Inst. Civ. Eng.* 71, 893–906. Doi:10.1680/iicep.1981.1825.

Page, A.W., 1983. The strength of brick masonry under biaxial compression-tension. *Int. J. Mason. Constr.* 3, 26–31.

Ridley-Ellis, D., Stapel, P., Baño, V., 2016. Strength grading of sawn timber in Europe: an explanation for engineers and researchers. *Eur. J. Wood Wood Prod.* 74, 291–306. Doi: 10.1007/s00107-016-1034-1.

Roberti, G.M., Binda, L., Cardani, G., 1997. Numerical modeling of shear bond tests on small brick-masonry assemblages, in: Pande, G.N., Middleton, J., Kralj, B. (Eds.), *Computer Methods in Structural Masonry -4*, London, UK: CRC Press. pp. 145–152.

Roca, P., Lourenço, P.B., Gaetani, A., 2019. *Historic Construction and Conservation:Materials, Systems and Damage*, 1st ed. New York: Routledge. Doi: 10.1201/9780429052767.

Rots, J.G. (Ed.), 1997. *Structural Masonry: An Experimental/ Numerical Basis for Practical Design Rules (CUR report 171).* Rotterdam: A.A. Balkema Publishers.

SCI, 2006. *Design Manual for Structural Stainless Steel*, 3rd ed. Ascot, UK: The Steel Construction Institute. ISBN: 2879972043.

Selby, R.G., Vecchio, F.J., 1993. *Three-Dimensional Constitutive Relations for Reinforced Concrete - Tech.* Rep. 93-02. Toronto, Canada: University of Toronto.

Stöckl, S., Hofmann, P., 1986. Tests on the shear-bond behaviour in the bed-joints of masonry. *Mason. Int.* 1–15.

Thomas, G.H., 1995. Overview of non-destructive evaluation technologies, in: *SPIE 2457- Nondestructive Evaluation of Aging Bridges and Highways*, Oakland, CA. Doi: 10.1117/12.209375.

TMS, 2016. *Building Code Requirements and Specification for Masonry Structures (TMS 402/602-16).* Longmont, CO: The Masonry Society.

Tomaževič, M., 1999. *Earthquake-Resistant Design of Masonry Buildings*. London: Imperial College Press.

UNI 11119, 2004. *Cultural Heritage - Wooden Artifacts - Load-Bearing Structures - On Site Inspections for the Diagnosis of Timber Members* (in Italian). Italy: Ente Nazionale Italiano di Unificazione.

UNI 11035-2, 2010. *Structural Timber - Visual Strength Grading for Structural Timbers - Part 2: Visual Strength Grading Rules and Characteristics Values for Structural Timber Population* (in Italian). Italy: Ente Nazionale Italiano di Unificazione.

Van der Pluijm, R., 1992. Material properties of masonry and its components under tension and shear, in: Neis, V.V. (Ed.), *6th Canadian Masonry Symposium*, University of Saskatchewan, Saskatoon, Canada, pp. 675–686.

Van der Pluijm, R., 1999. *Out-of-Plane Bending of Masonry: Behaviour and Strength*. PhD dissertation, Eindhoven, the Netherlands: Eindhoven University of Technology.

Van der Pluijm, R., 1993. Shear behavior of bed joints, in: Hamid, A.A., Harris, H.G. (Eds.), *6th North American Masonry Conference*, Philadelphia, PA, pp. 125–136.

Vasconcelos, G., 2005. *Experimental Investigations on the Mechanics of Stone Masonry: Characterization of Granites and Behavior of Ancient Masonry Shear Walls*. PhD dissertation, University of Minho, Guimaraes, Portugal.

Vecchio, F.J., Collins, M.P., 1993. Compression response of cracked reinforced concrete. *J. Struct. Eng.* 119, 3590–3610. Doi: 10.1061/(ASCE)0733-9445(1993)119:12(3590).

Chapter 5

Guidelines for practical use of nonlinear finite element analysis

INTRODUCTION

Besides the mathematical discussion of the previous chapters, any numerical analysis relies on the judicious understanding of the analyst who is asked to make assumptions and interpret the results. Dealing with the nonlinear behaviour of structures, the possibilities offered by finite element method software codes are vast and it is not easy to select the appropriate modelling strategy for a given problem. Typical examples are the understanding of the existing damage, the prediction of load carrying capacity or the definition of the failure mechanism of a single element, portion of the structure, or the entire building. Throughout the analysis, a heuristic approach is thus needed on a case-by-case basis, supported by experimental or existing evidence, further numerical analyses, simple calculations or structural intuition. The present chapter focuses on how to build an efficient model and how to avoid typical pitfalls and mistakes. The input sequences (e.g. geometry, type of elements, mesh and loads) are discussed independently, supported by illustrative examples. A final section addresses the numerical analysis of steel, concrete and masonry structural members according to theoretical insight and experimental evidence.

5.1 OVERVIEW

A finite element model is a numerical simulation of the reality. Being an abstraction of a physical phenomenon, it is based on assumptions, generalizations and idealizations. On the one hand, abstraction refers to the *mechanical model*, namely, which parts and details of the structure are involved and worth considering, as well as boundary conditions and significant loads. Basically, according to experience, hand calculations, experimental evidence or a possible pilot analysis, the analyst should be able to anticipate relevant characteristics of the results (a so-called *educated guess*) and understand which simplifications are possible. On the other hand, abstraction refers to the *discretization* of these properties, i.e. how to

DOI: 10.1201/9780429341564-5

attach the necessary attributes (e.g. material models, constraints and loads) to finite elements. With no doubt, a simulation cannot be more precise than its input allows: once assumptions are made, the results are the mere mathematical consequence. Accordingly, in the abstraction process, *model optimization* is essential.

Model optimization is a recursive procedure whose outline is illustrated in Figure 5.1 (the starting point is the grey-shaded box). Concisely, the numerical analysis can be subdivided into three phases (bold-frame boxes). The first is the *preprocessing*, which is the recipient of the abstraction process. Basically, after a careful evaluation of the physical phenomenon and the goal of the analysis, simplifying assumptions are made about the input data. As a matter of fact, every simplification influences the results to some extent, an issue the analyst should ponder carefully. During the second

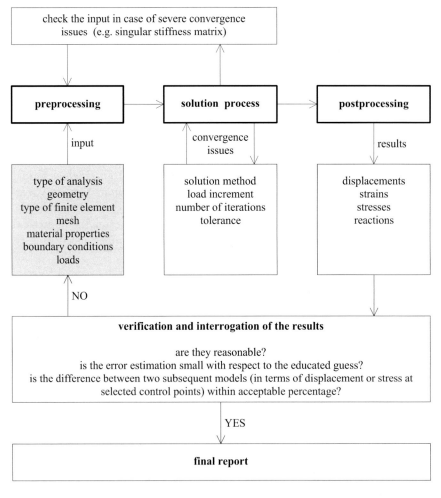

Figure 5.1 Outline of finite element analysis and model optimization.

phase, the solution is sought according to the methods/algorithms available in the software code and the relative parameters. As shown in Chapters 2 and 3, this solution process is iterative in itself, and its complexity is a result of the nonlinearities considered. Errors may be detected in this phase, such as a singular stiffness matrix due to boundary conditions insufficient to prevent rigid movements. The third phase of *postprocessing* must include the *verification* of the output of the analysis such as displacements, stresses and strains to assess the correctness of the model and its efficacy in meeting the initial goals. This process implicitly provides data to refine hypotheses and mesh, until a balance is found between accuracy of the solution and computational effort (thus the name *optimization*). As stressed in this chapter, revisions of the model assumptions should not be regarded as attempts to correct previous failures, but as necessary steps to move towards more adequate hypotheses, thus to more accurate results. The lack of verification checks and interrogation of the results may lead to severe miscalculations.

The verification process relies on the analyst's judgement that is based on theoretical insight and experience. However, it is not uncommon that analysts use advanced structural analysis software codes with little understanding of the theoretical background of the same, and with little understanding of the mechanical properties required. As simple examples, the fracture energy is a mesh-dependent quantity, whose miscalculation may provide the model with unrealistic capacity, whereas the value of the flexural tensile strength in concrete-like materials is different from the value of uniaxial tensile strength, and these cannot be used interchangeably. Nevertheless, even when aware of the theoretical background and with a good knowledge of the mechanics of the materials involved in the model, the analyst may be not aware of the engineering significance of the results.

In the case of lacking experience with the software code or when dealing with new topics, it is highly recommended to start with simple models, for which the result is known or can be approximated by analytical calculations. This learn-by-doing approach is unfeasible with complex models. Additionally, many possibilities may be available to solve the same problem and the analyst should be aware of them, understanding their suitability and effects. Different parts or different nonlinearities can be added successively in order to understand their influence and isolate possible errors. These may be difficult to find in a large and complex model even by long-experience analysts. In this regard, it is worth underlining that the outcome of an impressive simulation using a large model is just the tip of the iceberg of the work made by experts, where the *assumptions-and-checks* procedure was reiterated many times before reaching the final result.

Another process that must be introduced is *calibration*. This regards the case when a few input parameters are not known due to either lack of information or their large scatter. In this regard, it is possible to run several analyses varying (thus calibrating) the missing parameters until the simulation matches experimental data or existing damage for a known load set or load level (the former being associated to laboratory or in situ tests, and the latter being associated to a given crack pattern or structural deficiency). In other words, the input parameters are obtained according to the known results, paving the way to the so-called *inverse parameter identification*. Numerical optimization methods can be also adopted, although being efficient only in the case of linear elastic analysis.

Careful attention should be paid to software procedures. Even if not directly selected, the software code assumes default values or strategies that may need to be checked by the analyst. For instance, in presence of moderate load increments, implicit solution methods (e.g. Newton–Raphson family) are expected not to affect the final solution but they may determine the convergence and the numerical efficiency in terms of computational costs and time. Differently, conceptual features such as material properties, element types or explicit solution methods may lead to inaccurate (if not definitely wrong) answers. Software manuals and technical support assistance, together with accredited benchmark solutions, represent an appropriate reference. In this regard, a major role is also played by fellow experienced engineers that can help identifying problems or errors in the simulation. It is important that the analyst becomes aware of his/her own technical capabilities and of those of the selected software code. This also helps in knowing the resources, timescale and, thus, costs required for the analysis, avoiding underestimations that will result in counterproductive time-pressure.

Common simple and avoidable errors when performing structural analysis are usually related to: (1) units, which have no limitations as long as they are used consistently; (2) reference system, e.g. gravity is downward, but also orientation of finite elements such as interface axes; (3) incorrect element data, e.g. wrong thickness, beam cross-section and cross-section dimensions; (4) supports wrong in location, type, direction or stiffness; (5) loads wrong in location, type, direction or magnitude; (6) duplicated entities, either force or geometry; (7) connections not working as intended, e.g. a beam element connected to a plane stress/strain element does not transfer any moment.

Finally, although not directly pertinent to structural analysis, it is advisable to adopt a clear organization of files and folders. This can be made either with synthetic names (e.g. acronyms or according to a predefined user-defined terminology) or by means of an ancillary text file describing the differences and the model evolution. This would be of much help in the case a recap is needed, or when the model is used again after some time or by different analysts.

5.2 PREPROCESSING

The term *heuristic*, previously mentioned, refers to the process of improving the current simulation thanks to the subsequent feedbacks (or self-educating evaluations) of the previous simulations. In this regard, moving from a simple model, the general idea is to refine it step-by-step or to add new features as needed. Simpler models should be preferred over more complex models, where "simpler" does not have a term of comparison, but it indicates the point where further complexities do not significantly improve the accuracy of the simulation and reduce the focus of the results. In other words, before including further details in the model, the analyst should question the needfulness, being aware of the consequent simplification in the case the addition is discarded. Along this line, the definitions *sufficient* and *efficient* are also rather common.

Dealing with existing structures, this process is complicated by the intrinsic uncertainties involved in the abstraction phase. These are related to two main aspects, namely, *epistemic* and *aleatory*. Epistemic is related to the knowledge process of the construction that may be not exhaustive or may be affected by inaccuracies. This reflects itself in the structural modelling, being generally based on modelling decisions difficult to be mathematically characterized. Aleatory is related to the uncertainties of loads and material properties, included with a mathematical framework in the safety format of design and assessment. In this scenario, the analyst is asked to understand how the overall behaviour of the structure is affected by such uncertainties. This goal can be achieved thanks to a sensitivity analysis, i.e. the evaluation of how the structural response is affected by the variations of selected factors and model assumptions. The sensitivity analysis may also shed light on the need of further investigations in the case a material parameter or modelling decision is found to be crucial.

Regarding the preprocessing, whose main points are discussed independently next, the knowledge of the following aspects may be incomplete and should be evaluated on a case-by-case basis:

- geometry and morphology (such as structural form and elements, internal composition or connections between structural elements) exacerbated by possible long construction periods or subsequent (unknown) alterations along the history;
- material properties due to the possible large scatter;
- actions, either mechanical, physical (e.g. thermal or hydric) or chemical, which may have changed along the construction history, also affecting the actual load distribution;
- existing alterations and damage (such as cracks, construction mistakes, disconnections, crushing or leanings) not detailed by the survey or within the structural elements;
- interaction with surroundings (such as soil, fluids and nonstructural elements).

5.2.1 Type and main aspects of analysis

The analyst should select the appropriate type of the analysis in agreement with the goal of the simulation. However, as described throughout the book, the discussion is here limited to quasi-static nonlinear analysis. The selection of analysis type should be performed in the case of: (1) new contacts, i.e. constraints detected along the analysis; (2) phased or sequential analysis (e.g. boundary conditions added or removed, and geometry or load variations during the construction process); (3) material nonlinearities (e.g. plasticity, damage and cracking) and (4) geometrical nonlinearities (namely, large strains, large displacements and nonconservative forces).

More in detail, contacts are related to finite element analyses such as a rocking block on a fixed base or metal forming, in which new contacts are enforced. Phased analysis has long been used in structural engineering for specific structural typologies, such as bridge design, in which temporary supports in the form of construction ties or columns exist. A typical example is the balanced cantilever method, according to which the superstructure of the bridge is built by sequentially joining the segments to form a span by posttensioning and balancing them left and right from each pier. Therefore, mechanical characteristics, constraints and loads can be considered variable during the different construction phases. From an inverse perspective (which is the analysis, not the design), the same concept applies to historical constructions, although for a much-dilated time lapse. Historical analysis of a building may shed light on valuable events useful to interpret the current state. Material nonlinearities are extensively discussed in Chapters 3 and 4, to which the reader is referred.

In turn, geometrical nonlinearities deserve additional information. First, the difference between *large* (or *finite*) *strains* and *large displacements* must be highlighted. Without entering the merit of the discussion, small strains can be calculated as the first derivative of displacements, being accurate when strains do not exceed indicatively 1%–2%. Larger strains and corresponding stresses cannot be defined linearly as "displacement over initial length" or "force over initial area", respectively. The consequent measures are nonlinear adding complexity to the analysis. This is evident in the different definition of constitutive relationships for linear and nonlinear strain and stress measures. In turn, large displacements induce a change of geometrical configuration and, possibly, of external loads, with respect to the original/undeformed one. Stability problems fall in this group, such as compressed slender columns, shallow arches, slender footbridges or long-span cable-stayed bridges. As instability may occur, geometrical nonlinearities cannot be neglected (see first part of Chapter 2). However, the hypothesis of small strains can be still considered valid, thus, limiting the computational effort of nonlinear strain measures. The contemporary presence of small strains and large displacements can be idealized by means of a long cantilever beam: small strains throughout the element may produce a large deflection at the free end.

The typical formulations used to deal with geometrical nonlinearities are termed *Total Lagrangian* and *Updated Lagrangian*. The difference between them regards the reference system according to which kinematic variables are evaluated. In the total formulation, the displacement field is described according to the original coordinate system of the undeformed configuration, which is assumed fixed throughout the analysis. In turn, in the updated formulation, the coordinate system is updated continuously and the previous known configuration becomes the reference configuration. The main advantage of total formulation is that all derivatives regarding the spatial coordinates are calculated with respect to the original configuration and therefore can be precomputed. The disadvantage of this approach is the complicated description of finite strains. As a matter of fact, at every step of the analysis, displacements are the sum of the previous (known) displacements and the current increment. Regarding the updated formulation, its main advantage is the simplicity in evaluating the increment of strains. Basically, by shifting the reference configuration, there is no initial displacement effect but only increments. In turn, all derivatives with spatial coordinates must be recomputed in each load step. According to continuum mechanics, if both formulations are applied to the strong form of the equilibrium equation, provided that the appropriate constitutive relationships are employed, identical numerical results can be obtained. In turn, approximations and linearization of the equilibrium conditions used in the FEM may lead to some discrepancies.

As a general recommendation, if nonlinear geometrical effects are neglected, it should be verified whether strains and displacements are sufficiently small to support this assumption and if instability is not controlling the response. In the case of doubt, geometrically linear and nonlinear simulations should be compared to check the differences.

5.2.2 Geometry

One of the first simplifications for a finite element model regards the spatial domain, namely, the possibility of performing either two- or three-dimensional analyses. This simplification is aimed at focusing the analysis, avoiding superfluous computations. For instance, if the scope of the analysis is the in-plane behaviour of a masonry or concrete wall, the stress and strain distribution through the thickness may be assumed constant and plane-stress conditions may be adopted.

Regarding the geometry of the elements, most FEM programs are furnished with a CAD (Computer-Aided Design) interface for the preprocessing phase. However, geometry can also be generated using popular modelling packages, and then imported in the finite element program. In this case, the analyst should check the correctness of the import, avoiding missing parts. The scale factor and dimension units should be carefully examined. The imported drawing should have only the bare essentials, avoiding constructions lines, fillets, holes, grooves, chamfers and all details not significant for the analysis at hand. Special care should be taken for boundary conditions or geometrical details in the loading area. In those cases, details may be significant and impact the structural response. Duplicate nodes or duplicate elements should be also checked, as they could lead to overlapping finite elements or to unwanted connections (especially if meshing is done automatically by the software code). In turn, nodes outside of a certain geometrical tolerance may be assumed not coincident, leading to a lack of connection (see Section 5.2.4 for meshing aspects and element connectivity).

Geometry idealization also plays a crucial role in the definition of the model. On the one hand, idealization refers to the *simplification* of real-world geometry (especially when too many geometric details are present); on the other hand, idealization refers to the *type of elements* to be considered. As shown in Sections 5.2.3 and 5.2.5, each finite element has its own basic "understanding" of what constitutes a structural response, from a simple truss element (provided with only axial stiffness) to complex three-dimensional elements.

A typical example of geometrical simplification is shown in Figure 5.2, where a steel mechanical component is considered. Figure 5.2a shows the three-dimensional model, the most detailed one as it accounts for fillets of the perimetral edge of the bottom disk, the plate-to-tube welding fillets and chamfer at the top of the tube and holes. The first simplification regards the axial symmetry of the geometry and the loading (if axial pulling load is assumed), see Figure 5.2b. Subsequently, it is possible to reduce the mentioned geometrical details, together with the slot along the external tooth of the disk, as illustrated in Figure 5.2c. At this stage, the reader should notice that the plate-to-tube welding fillets were not deleted to avoid stress

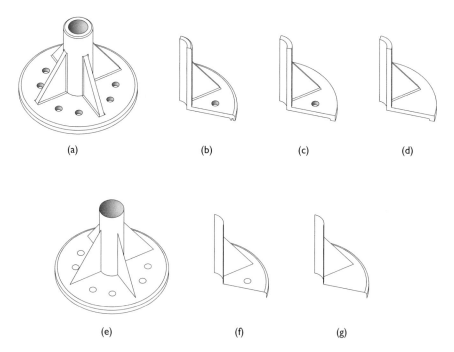

(a)   (b)   (c)   (d)

(e)   (f)   (g)

Figure 5.2 Geometry simplification of a steel mechanical component. Solid elements: (a) entire model, (b) simplification due to axial symmetry, (c) reduction of geometrical details and (d) model without holes. Shell (for the base) and plate (three-dimensional but plane elements, for tube and stiffeners) elements: (e) entire model, (f) simplification due to axial symmetry and (g) model without holes.

singularities. Additionally, according to the geometry of the element and the goal of the analysis, the holes may or may not be relevant. For instance, a horizontal action may produce local damage due to bearing pressure. If this aspect is of no interest (e.g. small hole diameter compared to the distance from the closest edge), a further simplification is possible (Figure 5.2d).

Regarding the type of elements, a large number of solid elements are needed to capture the stress (and strain) distribution through the element thickness. A different approach is shown in Figure 5.2e where the mechanical component is modelled with shell and plate elements. In this case, discarding the actual geometry of the structure, the finite elements are placed on the middle plane of the structural members (with a significant simplification). Figure 5.2f and Figure 5.2g follow the reasoning provided earlier. In general, looking at all the models of Figure 5.2, it is not possible to state which model is the best. As discussed throughout this chapter, accuracy is a relative concept interconnected with the goal of the analysis. The most suited option is based on the analyst's experience, resources available and insight into the problem at hand.

Another example regarding the choice of element types is shown in Figure 5.3 where a steel column and concrete foundation connection are illustrated. The first model is the most detailed one as it accounts even for the web-to-flange fillets and the fillets due to welding (needed to avoid stress singularities). In this case too, a large number of solid elements are needed to capture the stress (and strain) distribution through the element thickness, and interface elements must be placed carefully wherever two different objects interact e.g. between lower part of the nut and its projection on the horizontal plate, between this and the concrete foundation, along the shaft of the bolt. On top of this, the constitutive relationship of the interface elements must be considered carefully.

Given the consolidated experience on this type of connection, a detailed and time-consuming modelling is recommended only when a critical situation is evident (e.g. existing damage that may compromise the structural capacity) or for a detailed analysis of the damage evolution up to the

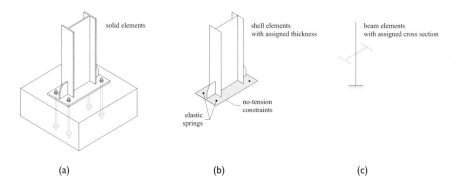

(a)   (b)   (c)

Figure 5.3 Geometry idealization of a steel column-concrete foundation connection in terms of type of elements. (a) Solid (three-dimensional), (b) shell and plate and (c) beam elements (one-dimensional, or line, representation).

collapse. Figure 5.3b shows a simplified approach made, again, of shell and plate elements. The greater easiness in building the model is evident from both geometric and constitutive relations viewpoints, still maintaining a good accuracy. Finally, Figure 5.3c shows the simplest model of a canti-lever beam, where the cross-sectional properties are assigned to the beam finite element. The geometry of the problem is idealized even more by simply accounting for the axis of the structural member. This approach, although useful for modelling large structures, focuses only on the column, disregard-ing the behaviour of the steel plates at the connection, that may be designed afterwards (or will be assessed independently using an analytical model).

In civil engineering, simplification of geometry is often needed for the case of masonry structures, whose walls are often provided with niches and openings, or their middle plane is not aligned with the one of the adjoint elements, either in plan or in elevation. The analyst should pay proper atten-tion to cases in which simplifications may lead to erroneous results. For instance, whereas walls may be able to accommodate small imperfections in carrying in-plane actions, the same may not apply for instability where small eccentricities or misalignment may foster failure. Also, in the case of multileaf walls, misalignment may exacerbate the leaf separation with disproportionate effects. In the case of concrete columns and beams with some eccentricity, it is recommended to consider the real eccentricity and use rigid elements to account for parasitic bending or torque moments.

In the case of large models, it may be possible to focus only on selected parts of the structure, although it is not easy to discuss which parts are valuable and how to manage the interaction (and interconnection) with adjacent parts. In the case the model does not incorporate the entire building, the missing parts may be substituted by equivalent boundary conditions through interface, con-tact or spring elements. This is usually encountered with the so-called build-ing aggregates, i.e. a group of existing masonry buildings in historic centres that will mutually interact under seismic actions, as shown in Figure 5.4a.

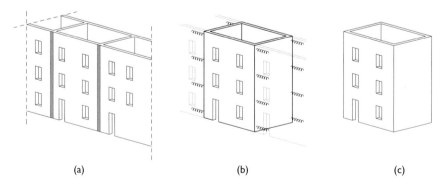

(a)　　　　　　　　　(b)　　　　　　　　　(c)

Figure 5.4 Geometry idealization of a masonry building within an aggregate. (a) Full block, (b) individual building with springs to account for interaction with adja-cent constructions and (c) isolated building.

In particular, the grey stripes mark the common walls between adjacent buildings. Two modelling alternatives are also shown, namely, the individual building modelled with springs of stiffness equivalent to the surrounding (Figure 5.4b), and the isolated building configuration (Figure 5.4c).

Attention should also be paid to volume and stiffness while modelling. For instance, if a curved boundary is modelled as a polygon or a faceted surface, this simplification should be done in a way that preserves the correct volume/area/weight of the structure. In turn, shell, plate and beam elements provide a lower stiffness in the case of an intersection. As an illustrative purpose, it is possible to consider again a masonry building. Besides the stiffness of each wall, it is known that, in correspondence of an intersection, each wall provides stiffness to the adjacent one, working as a sort of flange. If masonry walls are modelled with shell or plates, as their intersection is a line, their interaction is limited, leading thus to a smaller stiffness of the node. This affects the global response of the structure due to its intrinsic multidimensional behaviour, especially in presence of thick walls (compared to their length or height). To account for the contribution of transversal walls (otherwise neglected), it is possible to consider an enlargement of the shell thickness, as shown in Figure 5.5a. The same applies to equivalent frames with beam elements where rigid links are often used to account for the larger stiffness of joints (Figure 5.5b).

As a final remark on geometry idealization, it may happen that the structure under consideration already has a visible crack pattern. Again, it is not easy to discuss how to model this aspect in FEM. Different alternatives are usually available, such as reducing the element thickness, the tensile strength, the fracture energy or Young's modulus of the original material in

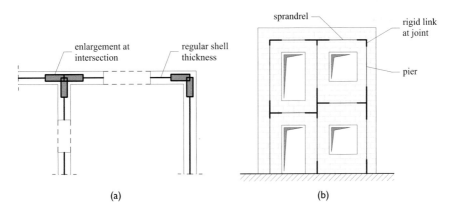

Figure 5.5 Two examples of geometry idealization of a masonry building. (a) Shell thickness enlargements at the intersection with transversal walls (plan); (b) equivalent frame model of the façade (elevation) using beam elements and rigid links at the nodes.

that part of the model. The evident goal is to lower the mechanical capacity and enforce larger flexibility, which is the implicit effect of cracks. However, the linear and nonlinear consequences of the modelling decision must be carefully considered. As an example, reducing Young's modulus may induce the force flow towards a different path, with the weakest part resulting as unloaded and undamaged.

5.2.3 Type of element

As shown in Figure 5.3, the choice of the type of finite element is problem-dependent, i.e. it is based on the goals of the analysis. In this regard, suitable kinematic assumptions must be made about the finite elements that better fit the analysis, such as truss, beam, plate, shell, solid or interface. Choosing the element type and the relative shape function is crucial as the analyst is selecting the discretization/distribution of the field variables. Additionally, this decision influences the complexity of the numerical model, together with the type of analysis and results, as well as the consequent processing time and the result interpretation.

Although already described in Chapter 1, the main finite elements are synthesized in Table 5.1, where the typical uses are also highlighted. This summary is purely indicative because different strategies for the same structural member are still possible (see Figure 5.3) and the combination of different element types in the same model is not uncommon. For instance, a masonry wall can be modelled through solid elements in a full-size complex monumental building, through flat shell elements in the case of in-plane and out-of-plane actions, through membrane elements if the in-plane response is predominant, and, finally, even as beam elements with proper constitutive laws in terms of shear and bending moment for simplified full modern building representation. Again, to define the most adequate elements, a heuristic approach is beneficial. Different models can be tried and compared to verify the element type selection and, sometimes, combining various element types (e.g. beam and plate) may be efficient for building the model as a whole. As a matter of fact, reality behaves without considering which is the model to follow (e.g. beam, membrane, plate or solid elements), but the analyst should select the most appropriate description of the real behaviour. Section 5.2.5 provides examples regarding this topic, with different types of elements compared.

Plate and shell elements deserve additional considerations. Although two-dimensional (they ideally represent the middle plane of the structural element), plate and shell elements can bear out-of-plane actions by exhibiting a variation of stresses along the thickness. Concisely, shell and plate elements are geometrically two-dimensional, but their structural behaviour is three-dimensional. This stress variation is evaluated by means of integration points (through the thickness) that should be carefully examined. Accordingly, bottom, middle and top planes have different stress conditions,

Table 5.1 Summary of typically available finite element types

Geometry	Element	Stiffness	Typical uses
Node	External spring	Translational or rotational stiffness	Nonrigid constraints (e.g. Winkler foundation model) or to replace structural parts with their equivalent stiffness
Line	Truss	Axial stiffness	Bar structures with hinged connections, such as ties, stiffeners in frames or roof trusses in which loads may be considered in the nodes only
	Beam	Axial and bending stiffness	Trusses with distributed loads (requiring possibly hinges on nodes), two- and three-dimensional frames, slab stiffeners, columns, lintels or arches (curved lines)
	Rigid link or MPC (multipoint constraint)	Rigid connection	Constraints between selected nodes (master and slave-type dependency), such as precluding rotation at the end of a specimen or enforcing equal horizontal displacement on top of a wall
Plane	Plane stress	Stiffness in plane (membrane). Free out-of-plane displacement	Structural elements with small thickness and loads acting in the plane, e.g. masonry walls and reinforced concrete deep beams
	Plane strain	Stiffness in plane (membrane), but with zero out-of-plane displacement	Cross-sectional elements of long structures for which the longitudinal deformation can be considered null, e.g. tunnels, retaining walls or dams.
	Plate bending	Out-of-plane bending stiffness	Building slabs or walls subjects to wind loading
	Flat/curved shell	Combination of plane stress and plate bending elements	Building shafts, walls subjected to a combination of vertical and horizontal loading, vaults or domes
Solid	Solid	Stiffness along all axes	Voluminous structures, such as thick walls, concrete foundations or soil masses
Joint between parts or boundary	Interface	Translational stiffness between connected elements. New contacts can hardly be considered	Discrete cracking, joints in masonry, no-tension bedding, bond-slip along reinforcement, friction between surfaces, soil–structure interaction or interaction between buildings
	Contact	Rigid contacts (no interpenetration). New contacts can usually be considered	

and the normal to the element must be carefully defined, as it defines the orientation of the surface.

It is also interesting to compare plane and solid elements. Whereas solid elements may represent the natural choice for three-dimensional elements, the relative calculation and analysis of the results may be computational costly and time-consuming. Solid elements also fail to provide internal forces in a way that designers are used to, such as axial force, shear force or bending moment (so-called generalized stresses). In turn, the discretization of structural members by means of plane element provides generalized stresses and strains, but their location and geometry deserve careful attention. For instance, in the case plate bending elements are used to model slabs, their location is not univocal with respect to the actual thickness of the slab, affecting the response of adjoining vertical elements. Similarly, if shells are used for vaults or domes, the stiffness of the connections can be misrepresented at the connection with the vertical walls.

After selecting the element type, attention should be paid to the adopted shape functions. These are defined as mathematical interpolation functions that describe the displacement field within the given element according to its nodal values. Being usually polynomials, the crucial parameter that directly affects the accuracy of the solution is their order of approximation. Shape functions are typically linear or quadratic, and these represent the variations of displacement along the element axes. As both the selection of finite elements and shape functions are crucial steps of modelling, some recommendations follow.

Linear elements require a finer mesh to approximate complex displacement fields, when compared with quadratic elements. Figure 5.6 shows the discretization of a quadratic displacement field by means of four different distributions of linear elements, with increasing refinement. It is evident that a unique quadratic element would be sufficient, and different linear

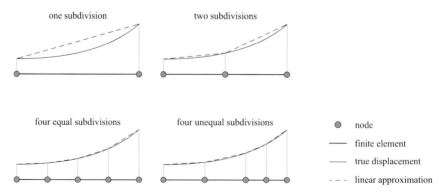

Figure 5.6 Approximation of a quadratic displacement distribution by means of linear interpolation.

discretizations lead to more or less accurate solutions. Analogous approach can be assumed in the case of high stress or strain gradients, i.e. strong variations within a small region. It is worth recalling that, as shown in Chapter 2, the univocal relation between stresses and strains is lost in the case of nonlinear analysis, meaning that high stress gradients are *not* necessarily associated with high strain gradients through the analysis process (e.g. due to yielding or softening behaviour).

In general, when complex nonlinear analysis is used, the robustness of the finite element type should be preferred to the order of approximation. Linear elements are less prone to spurious displacements than quadratic elements. This may be beneficial when dealing with very low stiffness values in one or more integration points due to inelastic behaviour. In this regard, linear elements with a sufficiently refined mesh are advisable in the case of nonlinear physical analysis of concrete-like materials. In turn, linear elements are more sensitive to error in the case of element distortion as the quadratic function is more "accommodating", and they also represent poorly out-of-plane bending, requiring careful discretization through the thickness.

Shape functions and integration points (also named optimal points) are two connected features of finite elements, as shown in Chapter 1. The careful analysis of these features may be of help in interpreting *shear locking*, *volumetric locking* and *hourglassing* spurious behaviours occurring in numerical analysis. Without entering the merit of the discussion, shear locking is the phenomenon that occurs when quadrilateral elements are subject to bending or equivalent deformation, as they appear stiffer than they actually should be (thus the term *locking*). According to the theory, the expected displacement field of a bent element is shown in Figure 5.7a where transversal lines are kept perpendicular to the longitudinal curved ones. If a four-node isoparametric linear element is considered (Figure 5.7b), the respect of linear shape functions imposes the longitudinal lines not to curve

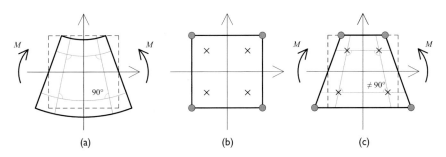

Figure 5.7 Shear locking effect (integration points marked with crosses). (a) Theoretical displacement field in bending; (b) isoparametric linear finite element at rest and (c) deformed.

and to remain straight. A shear distortion is thus evident (i.e. the angle between the lines is not 90°, see Figure 5.7c), with a shear stress arising unintentionally. Due to this, the element becomes too stiff (energy is spent for both bending and shear deformations) with a consequent underestimation of the overall deflection. The phenomenon is even more relevant in the case of distorted elements. In general, a mesh refinement or a change in the interpolation functions can limit shear locking.

Volumetric locking is related to an almost incompressible material (i.e. Poisson's ratio ≈ 0.5) such as some rubbers. However, it may occur also in the case of fully developed plastic flow when the deformation is iso-volumetric (i.e. constant-volume process) and a parasitic stiffness is found, while it should be null. Without going extensively into this topic, linear triangle and quadrilateral elements in a plane strain or axisymmetric stress state are usually more prone to volumetric locking, particularly when a coarse mesh is used. Volumetric locking is also possible in constrained parts of three-dimensional models, but it will not occur in plane stress models, as the third dimension is unconstrained. In the case volumetric locking is present, the simulation may significantly overestimate the collapse load and a mesh convergence check may be needed (mesh convergence is discussed in Sections 5.2.4.1 and 5.2.5).

Hourglassing represents another spurious behaviour occurring when elements with a single integration point are used (either plane or solid). Figure 5.8a shows a plane element at rest where the integration point is located at its centre. Figure 5.8b displays a certain distortion that does not alter the distance between the integration point and external sides, nor its location. More in detail, none of the thin solid internal lines has changed in length, and the angle between them is also unchanged, which means that all components of stress at the element's single integration point are null. This "bending" deformation mode is thus a zero-energy mode because no strain energy is generated by the displacement field. This means the element has zero stiffness in this mode and the finite element is not able to counteract this deformation. In coarse meshes, this pattern can grow unbounded and produce meaningless results (Figure 5.8c). Although most programs use numerical techniques to suppress hourglass modes, exaggerating the displacements may help in visualizing hourglassing, especially if localized in the interior of a three-dimensional model. It is worth noticing that three-dimensional elements may show more complex hourglassing modes.

Linear triangular and tetrahedral elements do not suffer hourglassing issues. However, in some applications, their behaviour is stiffer than quadrilateral or hexahedral elements, and the final displacements may be underestimated, especially when bending occurs. The reason lays on the fact that these elements have constant strain, thus stress.

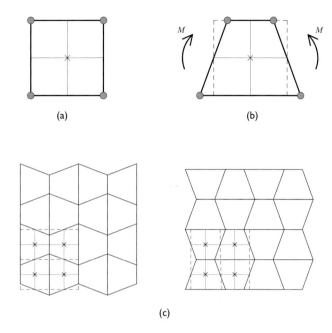

(a)   (b)

(c)

Figure 5.8 Hourglassing effect (integration points marked with crosses). (a) Isoparametric linear finite element with single integration point; (b) hourglassing
mode with null strain and (c) consequent propagation.

Consequently, equilibrium is satisfied by means of underestimated nodal
displacements, with consequences on the assessment of strain and stress.
The shortcoming is usually avoided by using multiple elements that better approximate the expected stress distribution. It is worth reminding
that averaging or smoothing the results may help visualizing the otherwise piecewise constant results. However, as shown in Section 5.4, it is
recommended to compare averaged and unaveraged results to look for
inaccuracies, together with a check of mesh convergence (see Section
5.2.4). Consequently, triangular (two-) and tetrahedral (three-dimensional) meshing (often done automatically by software codes) should be
compensated by a larger number of elements needed, that in turn affects
the computational costs. The analyst should consider these two aspects
when choosing quadrilateral or triangular, hexahedral or tetrahedral
elements. If possible, quadrilateral or hexahedral elements should be
preferred, in the case of linear analysis. However, in the case of large
models with complex geometry, a fine mesh with triangular or tetrahedral elements is likely to be the best option.

5.2.4 Mesh

After the decision about which element(s) to use, two aspects must be examined, namely, *size* and *shape*. The size of finite elements is strongly related to their amount and their distribution in the model. Starting from a preliminary mesh, the so-called *mesh refinement* is usually enforced to approach accurate results with adequate computational efforts. As a rough estimation, when fully modelling a large masonry building, a mesh dimension of 0.30–0.50 m may be appropriate, while for a detailed study of a steel component, a mesh dimension at mm scale may be needed.

Meshes are generally classified as structured, following some uniform pattern and, typically, orthogonal quadrilateral (two-) or hexahedral (three-dimensional) elements, or unstructured, with arbitrary structure and, typically, with triangles (two-) and tetrahedra (three-dimensional). Automatic meshing often uses unstructured meshes, which usually complies only with the maximum element size criterion, the location of loads and boundary conditions. The result may be poor and the analyst is asked to check it carefully. To overcome this issue, it is possible to adopt local mesh refinements, different element sizes in different parts of the model and different mapped areas in which the number of divisions is predetermined, among other techniques to ensure a high-quality mesh. Two examples of structured and unstructured meshes are shown in Figure 5.9.

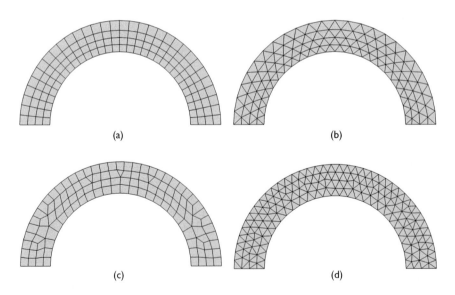

(a)   (b)

(c)   (d)

Figure 5.9 Examples of automatic meshing. (a and b) Structured and (c and d) unstructured quadrilateral and triangle elements.

5.2.4.1 Mesh refinement

The mesh density (i.e. finer or coarser distribution of elements) is problem-dependent because it depends on the requested accuracy of the physical quantities investigated. For example, displacements are often calculated more precisely than strains (or stresses) because strains involve derivatives. Accordingly, if deformation of a portion of the structure is negligible, coarser elements are adequate. Conversely, a fine mesh is needed where large gradients of stress and strain are expected. As shown in Chapter 1, strain and stress fields are evaluated at the integration points. Thus, mesh regions undergoing plastic deformation or cracking require a reasonable number of integration points in order to properly capture the distribution of such nonlinearities. The critical judgement should guide the analyst in adopting the best compromise between the higher accuracy of the solution (fine mesh) and the efficient use of computer resources (coarse mesh). Increasing the number of nodes in a model dramatically increases the amount of time it takes to solve the model. Even using modern computers, large models can take hours or days to perform the calculations needed to find the solution.

With this purpose, a preliminary simulation with a coarse mesh may help identifying the regions where a better discretization is required. Subsequent mesh refinements can be performed iteratively until appropriate accuracy is reached. The effects of the mesh refinement can be assessed by a so-called *mesh convergence study*. This is aimed at quantitatively evaluating the gain in accuracy between subsequent simulations according to selected outcomes at given control points. For instance, the gain of maximum stress at the given cross-section can be adopted as the subject of the convergence study, or the displacement at a certain point in the model. The refinement becomes inefficient when the gain of accuracy is below a certain threshold (e.g. 1%–2%). An example in Section 5.2.5 will make this concept clearer. However, the analyst needs to judge the effect of the refinement as a larger error may be assumed suitable (with due acknowledgment) if a higher accuracy is overly complex or the impact of the structural part is limited.

Regarding mesh refinement, some considerations must be highlighted. The first is quite logical: when comparing different meshes, the variation in the mesh density should be sufficiently large to make sure that the effects are noticeable. In this case, when elements get smaller in a certain part of the model, attention should be paid to the transition between elements of different size. Coarse elements are usually stiffer than fine ones, and a very large change of size may lead to spurious stresses or to an ill-conditioned stiffness matrix. Furthermore, a coarser/stiffer region may work as a constraint for the finer one. As a practical rule, it is recommended that adjacent elements do not differ by more than a factor of 2–3 in their area (or volume). Along this line, although not related to mesh refinement, it is worth underlining that a large difference in terms of stiffness naturally happens when elements crack or crush, as their stiffness becomes zero.

As mentioned in Chapter 1, the accuracy of finite elements can be improved either by reducing the dimension (or size) *h* of the elements (i.e. *h-refinement*) or increasing the polynomial order *p* of the shape function (i.e. *p-refinement*). The former maintains the same element type but increases the number of elements. No connectivity problems arise, but transition elements towards coarse regions must be examined. The latter keeps the element size constant, but increases the number of nodes per element, thus of the number of degrees of freedom DOFs (for instance, from linear four-node to quadratic eight-node quadrilateral elements). Special care should be taken when elements with different orders of approximation are used in the same model, as discussed below (see Figure 5.15).

In general *p-refinement* produces exponential decay of the discretization errors for sufficiently smooth solutions, as typically encountered in linear elastic problems. However, it is well known that *p-refinement* does not perform well in the vicinity of boundaries, corners or stress singularities and should be avoided. *h-refinement* is less susceptible to such issues and it is to be preferred if elements distort strongly in the case of nonlinear physical analysis. It must be noted that automatic *hp-refinement* methods (also known as adaptive meshing) are rather effective, even if they remain to large extent an academic advanced tool. Basically, a high polynomial degree *p* is selected wherever the solution is sufficiently smooth, while the size *h* is reduced at places wherever the solution lacks regularity.

Regarding mesh refinement, it is interesting to provide an estimate of time and computational costs. In this regard, it must be noted that it is not easy to discuss computational complexity in nonlinear models, naturally more expensive than their linear counterparts. Whereas linear problems define one set of simultaneous equations, which can be solved in a single iteration, the inclusion of geometric and material nonlinearities involves iterative solution methods that may dramatically increase the computational costs. Limiting the discussion to linear analysis, when *h-refinement* is considered, the number of nodes, thus of the DOFs, increases non linearly with the number of elements. If mesh refinement is assumed to halve the element size, these operations approximately double the number of line elements, quadruple the number of plane elements and increase the number of solid elements by a factor of eight. Shortly, the DOFs increase proportionally to a factor equal to the ratio of refinement (e.g. 2 in the case of halving the element size) to the power of the space dimension. As experience shows, for many problems the computational cost is roughly proportional to the square of the number of DOFs. Accordingly, the computational cost of dividing the element size to the half increases about $(2^2)^2 = 16$ in two- and $(2^3)^2 = 64$ in three-dimensional problems. The disk space and memory requirements increase in the same manner, although the actual extent is difficult to predict. As a general recommendation, for many problems, the mesh only needs to be refined in the areas of interest and/or in areas with high stress (and strain) gradients. Thus, local mesh refinement is usually more economical and as effective as global mesh refinement.

In turn, the relatively fast convergence rate of *p-refinement* is achieved at the expenses of significantly increased computational cost. The higher polynomial order results in an increase of the total DOFs of the system, and even larger increase of the operation related to formulation and matrix manipulations. These aspects usually require longer solver times (although still dependent on selected shape functions and the linear solver). As a rough estimation for large problems, ignoring the time required in the additional mathematical complexity with *p-refinement*, keeping fixed the number of elements, when moving from linear to quadratic elements, the number of DOFs increases 3 and 4 times for the two- and three-dimensional space, respectively. This means that the computational cost related to the increase of DOFs magnifies about $3^2 = 9$ times in two- and $4^2 = 16$ times in three-dimensional problems, a computational cost that must be added to those related to the formulation and matrix manipulations stated above.

For the sake of completeness (see also Chapter 1), stress singularities represent a concentration of stresses due to point loads or restraints, sharp edges, re-entrant corner (with zero fillet radius), openings, cut-outs or sudden change on cross-sectional areas or material properties. Upon refinement, conversely to the paradigm of FEM, stress value is unbounded and converges to infinity. In a simplistic way, if stress is assumed as the ratio between force and area, as the area tends to a null value the stress becomes infinite. It is rather evident that real structures do not contain such stress singularities that follow from simplifying assumptions of the mathematical model. However, as explained below, stress singularities are not influential after a certain distance. Additionally, in presence of nonlinear analysis the singularity is bounded by the strength of the material.

An illustrative example is shown in Figure 5.10, where a simply supported steel beam with cross-section $300 \times 300\,mm^2$, 5.0 m long (net span 4.7 m), subjected to a distributed downward vertical load of 100 kN/m is analysed (for the sake of simplicity, the self-weight is set null). The element size is defined according to the height of the beam, with two, four and eight subdivisions: $h = 150$, 75, and 37.5 mm for Figure 5.10a–c, respectively. A linear quadratic element is selected with 2×2 integration scheme. The three illustrations show that the vertical normal stress component SYY at the support increases upon mesh refinement, from about 10 to 30 MPa. For the sake of clarity, the same results are shown in terms of stresses extrapolated to the nodes (top) and averaged (bottom) through the mesh.

Another example is shown in Figure 5.11, where the free end of a cantilever beam subjected to a point load is analysed. The example is extensively discussed in Section 5.2.5, to which the reader is referred for further data. Here the goal is to show how the stress below the point load evolves with the mesh refinement. The three illustrations show the extreme values of the

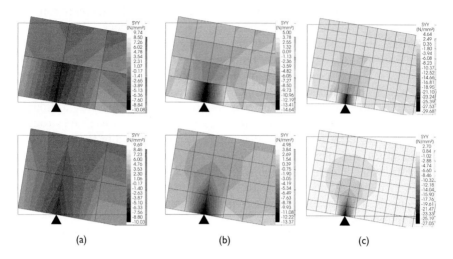

Figure 5.10 Stress singularity at concentrated support: the case of a simply supported beam under distributed load. (a–c) Normal vertical stress extrapolated to the nodes (top) and averaged (bottom) according to mesh refinement.

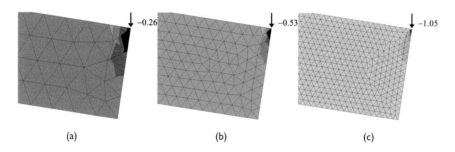

Figure 5.11 Stress singularity at the free end of a cantilever beam under point load (values in MPa). Discretization by means of quadratic triangular elements with (a) four, (b) eight and (c) sixteen elements through the height at the free end.

vertical stress considering quadratic triangular elements. Looking at the three figures, the stress almost doubles each time the mesh size is halved, and none of them is correct, as the stress cannot be locally defined. However, all the three models correctly evaluate the maximum deflection at the free end equal to 0.29 mm (not reported here), which shows that the stress singularity does not influence the global behaviour of the beam. In other words, spurious gradients caused by stress singularities do not propagate.

This is in line with the Saint-Venant's principle, according to which the effects of two different but statically equivalent loads become very small at sufficiently large distances from the load. Of course, this approach does not apply to the cases in which local effects (such as punching) are critical.

Accordingly, if the stress distribution around a point load is of no interest, provided the surrounding mesh is satisfactory, larger elements may better accommodate the loads. This would avoid softening and failure of small elements close to the singularities, as they do not provide appreciable insight to the understanding of the problem and, in most cases, deviate from reality. Obviously, boundary conditions may lead to the same stress singularities if they replicate isolated reactions.

Sharp corners represent an additional source of a stress singularity. Figure 5.12a shows the case of a compressed concrete cube (side 100 mm) with a cubic hole (25 mm side length) subjected to a distributed vertical load of 10 MPa. Starting from eight linear brick elements per side of the cube (Figure 5.12b), the refinement consisted into two subsequent size halving. Consequently, Figure 5.12b–d has 63 elements (300 DOFs), 504 (1,938 DOFs) and 4,032 (13,764 DOFs), respectively. As shown, the refinements provide a better discretization (the displacement field is more accurate), but the absolute values of stress increase: the maximum compressed vertical normal stress SZZ moves from −19.86 to −22.88 up to −27.30 MPa for the last case shown (almost three times as big as the external load pressure).

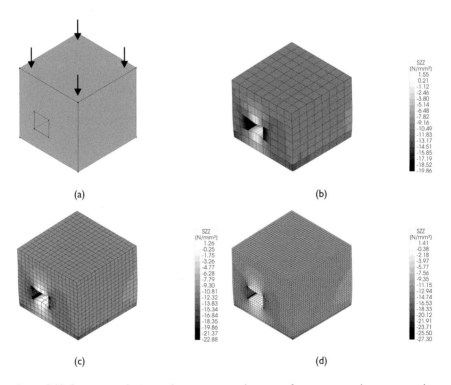

Figure 5.12 Stress singularity at sharp corners: the case of a compressed concrete cube with a cavity. (a) Overall structure; (b–d) normal vertical stress (extrapolated) according to mesh refinement.

Again, none of these values is correct and they are symptomatic of the stress singularity.

Besides mesh refinement, it must be noted that an increment of integration points may also improve the accuracy in the case of nonlinear analysis. For instance, in the case of smeared crack models, it has been seen that the one-dimensional constitutive law is enforced at every integration point independently. This is particularly relevant through the thickness of line and plane elements to account for out-of-plane behaviour. Usually, 7–11 points per cross-sectional axis are suitable options to integrate the nonlinear constitutive relations in the cross-section. It is recommended not to choose less than 5 integration points in the case of coarser meshes. In perfect analogy, the same number of linear solid elements must be put in place across the thickness of wall, slab or vault, if provided with a single integration point (this may be reduced to 3 elements, if provided with two integration points).

5.2.4.2 Mesh quality and controls

The element shape, due to geometry or meshing processes, may result distorted from their basic regular shape. With the aim of assessing the degree of distortion, mesh quality controls are usually performed. The importance of these controls relies on the fact that, in the calculation of structural stiffness, each element is evaluated numerically. The accuracy of the element stiffness (thus, the accuracy of the model) depends on how well this evaluation is carried out. In general, elements distorted from their basic shape may be less accurate, and the higher is the distortion, the greater is the error. In presence of nonlinear analysis, this problem usually is more severe and may lead to unexpected responses. As a general recommendation, the analyst should never accept a mesh without checking it.

Whether a coarse or fine mesh is used, several measures are available to quantitatively check the regularity of the elements adopted in the model, i.e. to assess the degree of conformity of a finite element to the regular shape of its type (i.e. equilateral triangle, square, regular tetrahedron and cube). These controls are usually made on screen as the software code displays the finite elements coloured according to a given control measure or predefined threshold value. The analyst should be aware that poor elements may be in the interior of the model and may not be directly visible.

However, building an entire model with only regular elements is usually impossible and a compromise must be found according to the analyst's judgement, also based on the importance of the part of the model considered. Poorly shaped elements should be avoided in regions where high gradients are expected and in regions of interest. In this regard, it is possible to define *meshing blocks*, which are subdivisions of the structure suitable for local meshing process, either by changing the reference mesh size or by mapping the divisions directly.

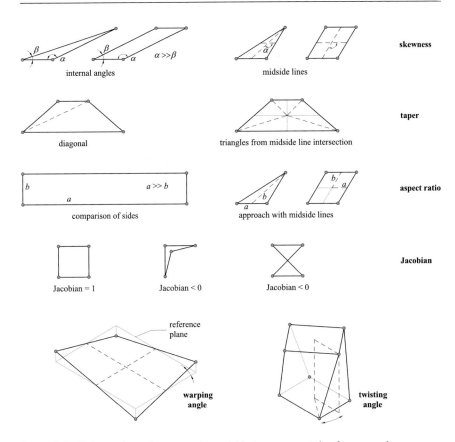

Figure 5.13 Main mesh quality controls available in commercial software codes.

The following checks are usually available in a software code (the exact definitions and even the designation may differ), with explicative drawings shown in Figure 5.13. It is noted that an irregular element may have more than one check infringed, and the software code being used may give the possibility to verify only a few of the mentioned checks:

- **Skewness** indicates how much the element shape digresses from the regular one of its type, e.g. a parallelogram vs. a square. Indicatively, optimal quadrilateral faces have internal angles close to 90°, whereas triangular elements have internal angles close to 60°. Different approaches are proposed to calculate skewness, all being referred to angles between internal lines. For quadrilateral elements, a possible evaluation regards the minimum angle between two lines joining opposite midsides of the element. Then, skewness is 90° minus the minimum angle found (although other strategies are available). In this case, a rectangle provides a skewness value of $0°$ and acceptable values for skewness are smaller than $45°$.

- **Taper** regards the ratio of the areas of two triangles separated by a diagonal within a quadrilateral face, defining the limits of an acute trapezoid or a rectangle. Another approach regards the comparison of the areas of the triangles connecting the nodes and the midside line intersection.
- **Aspect ratio** is a measure of the stretching of an element face. A very stretched rectangular element is optimal from the skewness and taper points of view, but it has a poor aspect ratio. An acceptable element should have the ratio between largest and smallest dimensions less than 3, and values larger than 5 are indicative of less accurate elements;
- **Jacobian ratio** (often referred to as *Jacobian* only), in this context, measures the deviation of a given element from an ideally shaped element. This is defined as the ratio between the smallest and largest determinant of the Jacobian matrix in an element. The value ranges from -1.0 to 1.0, where 1.0 is representative of a regular shape (see also Chapter 1). Not convex or much distorted elements (e.g. edges crossing over each other) are marked with negative values and they may cause the analysis to quit. Acceptable Jacobian ratio values are larger than 0.6.
- **Warping** indicates the departure of a quadrilateral face from being planar. A distortion less than $10°$ can be considered acceptable.
- **Twist** indicates how much two opposite element faces are twisted. Recommendations for warping hold here too.
- **Minimum angle** is the minimum value of the angle between two adjacent sides of an element face. Acceptable values for quadrilateral and triangular elements are larger than $45°$ and larger than $20°$, respectively.
- **Length, area and volume** return the length of elements or edges, the area of mesh faces and the volume of solid elements, respectively. They usually help the analyst in evaluating the transition between a coarse mesh and a fine mesh.

Although not related to the quality of the elements, other controls are recommended to check the accuracy of the mesh as a whole (even inside the model). In particular:

- **Free nodes** are nodes not connected to any mesh element, perhaps after an element deletion or erroneous copy-paste, reflection or translation process.
- **Free edges** indicate borders not shared among elements.
- **Free faces**, analogously to free edges, indicate that either nodes are not connected properly or mismatching elements are present.
- **Duplicate nodes** are coincident with other nodes (or within a given tolerance). Operations like copy-paste, translation, orientation or

reflection, and even dividing neighbouring meshing blocks, can result in duplicate nodes at a common edge. Different tolerances when importing the geometry from other software codes may also foster duplicate nodes.

- **Duplicate elements** are elements based on the same nodes, perhaps due to erroneous copy-paste, reflections or translations. These duplicate elements do not cause error or warning messages in the computational phase, but they double the stiffness and strength of that part of model, affecting displacements, strains and stresses.
- **Shell normals** is a check for inspecting the local coordinate system in shell elements. For each structural member, it is recommended to have all shell normals aligned in the same direction. Whereas no problem will arise during the solution process, discrepancies will be evident when results are displayed. For instance, if shell elements used for a slab are not properly aligned, top surfaces (of the shell element) will be either at the top or at the bottom of the slab, displaying misleading results in the postprocessing.

Another simple check to evaluate the mesh connectivity is by shrinking the elements by a factor slightly lower than the unit value. The approach is shown in Figure 5.14: the shrunk view on the right highlights an element missing, not evident from the wireframe view of the mesh on the left side. If this error is not detected in the preprocessing phase, the structural response (similar to what was shown in Figure 5.12) would, possibly, reveal the inconsistency of the result in the postprocessing phase. In the case of large and complex three-dimensional structures, a missing element in the internal part of the mesh may be not evident in the postprocessing phase.

Other issues about element connectivity follow. In general, although possible, it is not recommended to mix elements with different orders of approximation. First, low-order elements are stiffer and may lead to spurious stress or strain discontinuities. Second, mixing different-order elements may cause overlaps or gaps between elements (usually not displayed on screen). Figure 5.15a clarifies the case of possible overlaps when quadratic and linear elements are connected. Despite all nodes are shared between adjacent elements,

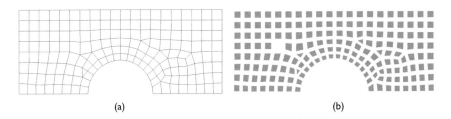

(a)   (b)

Figure 5.14 Check of connectivity conformity by means of element shrinking. (a) Wireframe and (b) shrunk views of the mesh (note the missing element).

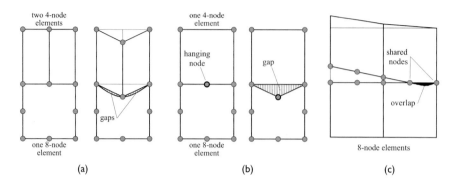

Figure 5.15 Connectivity issues. Mixing elements with quadratic and linear displacement approximation: (a) without and (b) with hanging node; (c) possible overlap in the case of contact/interface elements.

the difference in terms of order of approximation is evident in gaps in the displacement field (parabolic vs. linear approximations). This lack of compatibility is exacerbated in the case of *hanging nodes* (i.e. vertices of a certain element not shared with their other neighbouring elements, being thus a sort of "free nodes"). Figure 5.15b displays the case of a hanging node, whose displacement is erroneous. Hanging nodes can be suppressed by either remeshing the neighbouring elements or by imposing fictitious constraints on the system (to maintain the kinematic compatibility). Figure 5.15c addresses a different connectivity issue, related to the case of interface/contact for which only a few nodes are in contact (e.g. to model a possible discontinuity such as a crack). As shown, the parabolic interpolation leads to overlaps, whereas linear elements would not suffer this issue (i.e. they do not have midside nodes).

Figure 5.16a shows the case of incorrect meshing with hanging nodes. The structure is symmetric with dimensions $1 \times 1 \times 0.1\,\text{m}^3$ with a $0.25 \times 0.50\,\text{m}^2$ hole in the middle. It is subjected to self-weight only (mass density equal to $1,800\,\text{kg/m}^3$) and it is supported at the bottom. Additionally, Young's modulus is equal to $2,100\,\text{MPa}$ and the Poisson's ratio is equal to 0.25. Quadrilateral elements were used, with average size equal to $80\,\text{mm}$, with linear interpolation everywhere (four-node element) except for the left side pier, where quadratic interpolation is used instead (eight-node element). Figure 5.16b shows the total displacement field. Although symmetric in terms of geometry, loads and supports, the wall shows an evident lack of symmetry in the results. The contour plot over the hole is unbalanced towards the left side, whereas the "quadratic" pier displays a highly nonuniform displacement field if compared with its counterpart on the right. Looking at Figure 5.16c, the shear stress SXY distribution in correspondence of the nodes connecting the two mesh sets shows unexpected gradients, being not uniform if compared with the right side of the hole. By

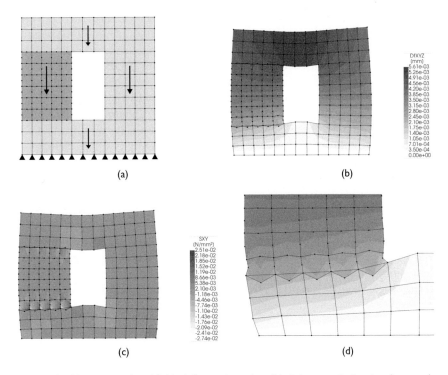

Figure 5.16 Hanging nodes. (a) Undeformed mesh, with light grey indicating four-node elements and dark grey indicating eight-node elements; (b) total vertical displacement; (c) shear stress distribution (extrapolated to the nodes, not averaged) and (d) enlargement and magnification of the displacement field.

magnifying the total displacements, as illustrated in Figure 5.16d, it is possible to detect the presence of hanging nodes that invalidate the results.

As shown, to check element connectivity in terms of discontinuity or overlaps, magnifying the displacements (with mesh visible) is of much help. Along the same line, exaggerating a specific load component from the full load set may also contribute. Finally, even though not tackled in the present text, a simple modal analysis, regarding the dynamic properties of a structural system and providing its vibration modes, gives precious information about possible connectivity issues.

5.2.5 Examples of element types and mesh refinement

Let us consider a cantilever 5 m long with a vertical force of 1 kN downward at its end. The cross-section is $0.30 \times 0.30\,\text{m}^2$ with the material properties given by $E = 210,000\,\text{MPa}$ and $\nu = 0.3$. For the sake of simplicity, null mass density is adopted. According to basics of Bernoulli–Euler beam theory, a linear elastic analysis leads to a maximum deflection of 0.29 mm, a maximum bending moment of 5 kNm and a maximum normal stress of ± 1.11 MPa. The neutral axis is located at the centre line and a linear normal (axial) stress distribution through the height is expected. Finally, the vertical displacement distribution along the beam length is a polynomial of third order (no distributed loads are present, as self-weight is disregarded). Different element types and mesh refinements are discussed next to tackle this structural problem.

The first simulation involves a beam element. Only a single element is necessary if the interpolation polynomial for the displacements is of the third order. Two integration points are sufficient to describe the moment and normal stress distribution, both varying linearly throughout the cantilever. The results shown in Figure 5.17 confirm the accuracy and efficiency of a single beam high-order element.

If a coarse mesh (maximum element size 500 mm) of linear plane stress triangular elements with a single integration point is used, see Figure 5.18a, the results are unexpected and not even close to the theoretical ones. Due to the single integration point, the element and the structure behave much

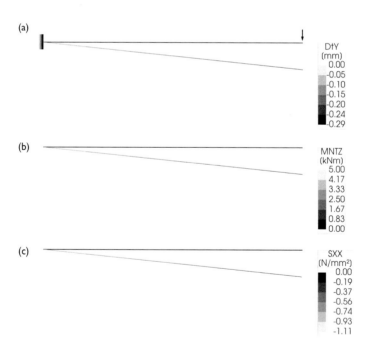

Figure 5.17 Cantilever structure modelled with one third-order-polynomial beam element. (a) Vertical displacement, (b) bending moment and (c) positive normal stress distributions.

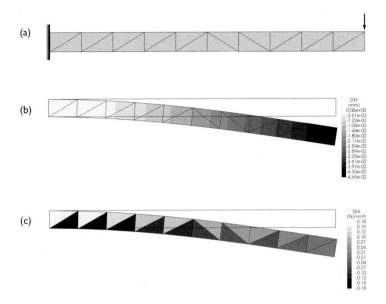

Figure 5.18 Cantilever structure modelled with one quadratic triangular element through the height. (a) Mesh, (b) vertical displacement and (c) normal stress distribution.

stiffer than expected with a maximum deflection six times smaller than the theoretical value, see Figure 5.18b. Normal stresses are also severely underestimated. As a linear displacement distribution is considered within the element, constant stress is implicitly assumed, paving the way to the patched diagram of Figure 5.18c.

Upon mesh refinement, i.e. with two triangular elements through the height (maximum element size 150 mm), the results are better, but still with poor accuracy. They are shown in Figure 5.19 and the relative description is in line with what was shown earlier.

With the aim of a *mesh sensitivity study*, the results of further simulations are collected in Table 5.2. As it is possible to notice, only a discretization with maximum element size equal to 40 mm provides values of displacement at the free end of the beam and stress at the support with a change lower than 2% with respect to the previous analysis. For the sake of simplicity, the results are plotted in the graphs of Figure 5.20, where the theoretical results are also shown. Looking at the plots, similar results from different mesh refinements provide a certain confidence on their accuracy. From the reverse perspective, large and unexpected change of results may be associated to mistakes. It is worth highlighting that, for verification purposes, the use of unaveraged results is recommended.

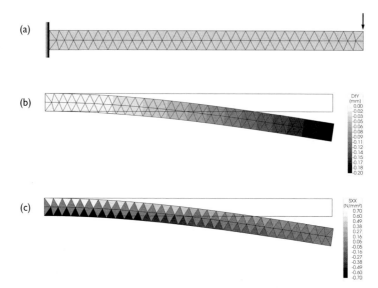

Figure 5.19 Cantilever structure modelled with two linear triangular elements through the height. (a) Mesh, (b) vertical displacement and (c) normal stress distribution (unaveraged in the nodes).

Table 5.2 Synthesis of the numerical simulations made with linear triangular elements.

Max element size [mm] (total number)	Max deflection [mm]	Increment of deflection [%]	Stress at the support [MPa]	Increment of stress [%]
500 (20)	0.0469	-	0.1785	-
200 (102)	0.1940	314	0.6925	288
120 (256)	0.2382	23	0.9113	32
80 (508)	0.2617	10	1.0042	10
60 (996)	0.2788	7	1.0940	9
50 (1402)	0.2828	1	1.1200	3
40 (2020)	0.2856	1	1.1303	1

For the sake of completeness, six-node triangular elements are also discussed for this problem (i.e. *p-refinement*), with a polynomial interpolation function that yields an approximately linear strain variation in both directions. As it is possible to notice from Figure 5.21, the results are in full agreement with the expected results. The values of stress at the integration points (three per element) are shown in Figure 5.21c. These values

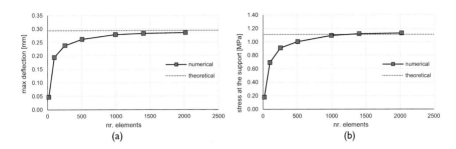

(a)   (b)

Figure 5.20 Synthesis of the mesh sensitivity study about linear triangular elements in terms of (a) deflection and (b) normal stress at the support vs. the total number of elements of the beam model.

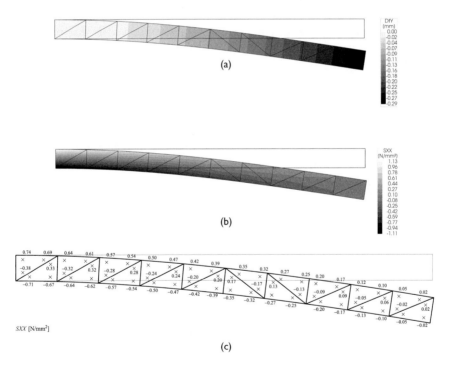

SXX [N/mm²]

(c)

Figure 5.21 Cantilever structure modelled with one quadratic triangular element through the height (quadratic shape function and three integration points). (a) Vertical displacement; (b) and (c) normal stress distribution at nodes (extrapolated but unaveraged in the nodes) and values at integration points, respectively.

are those calculated at specific locations within the element (see Chapter 1 for further insight). If the shape function coincides with the theoretical/expected one, these values are the correct ones (only in correspondence of their location, as nodal values are extrapolated). Sometimes, instead of punctual values, software codes may output a contour plot where the integration point results are assigned to the closest node, to allow mapping the results in a visual output. The analyst should be aware of this computational aspect.

In the case of quadrilateral elements, the effects of *h*- and *p-refinement* are also evaluated. Figure 5.22 shows the case of linear elements (maximum element size 500 mm) with one integration point, thus constant stress. The results are evidently incorrect: besides hourglassing issues and incorrect order of magnitude of displacements and stresses, the maximum normal stress is in the middle of the cantilever (white in Figure 5.22c).

Let us consider the same geometrical discretization, but with 2×2 integration scheme. The results are shown in Figure 5.23 and a much better accuracy is evident. The larger stiffness observed is due to small shear locking (see Figure 5.7c) and the consequent deflection depicted in

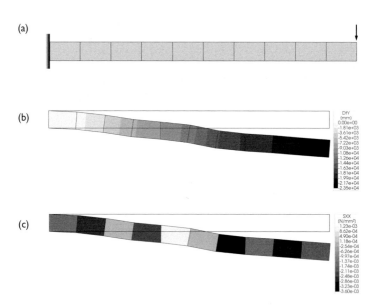

Figure 5.22 Cantilever structure modelled with one linear quadrilateral element through the height (one integration point). (a) Mesh, (b) vertical displacement and (c) normal stress distribution (unaveraged in the nodes).

Figure 5.23a is slightly smaller than the theoretical one, namely, 0.27 vs. 0.29 mm. The nodal stresses SXX are extrapolated from the integration points and they represent a good estimation of the theoretical values (Figure 5.23b). The stress values at the integration points are shown in Figure 5.23c. As it is possible to notice from the last two graphs, the distribution of SwX is constant in x direction and varies linearly in y direction. This is due to the choice of the selected shape function (associated to the selected finite element). From the theoretical point of view, a linear distribution is expected along both the beam axis and through the height. Although not in line with the theoretical results, the approximation is acceptable.

It is interesting to adopt the same geometrical discretization and increase the number of integration points to 3×3. Although not reported here, the results are identical to what is shown in Figure 5.23 for obvious reasons: full integration is considered with 2×2 scheme and over-integration

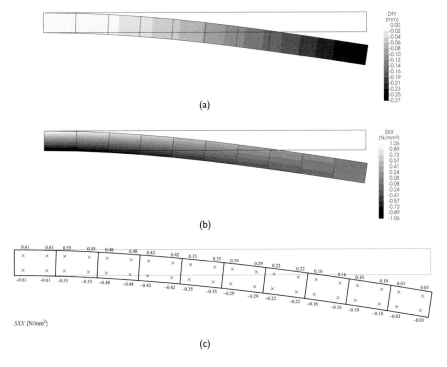

Figure 5.23 Cantilever structure modelled with one linear quadrilateral element through the height and 2 × 2 integration scheme. (a) Vertical displacement; (b) and (c) normal stress distribution at nodes (extrapolated but unaveraged in the nodes) and values at integration points, respectively.

produces no change in the case of linear elastic analysis. The addition of one integration point within the beam height does not add any information to a stress distribution that is linear (note that the response would be different in the case of nonlinear behaviour). The same applies along the beam axis where a constant stress distribution is enforced for the given finite element. However, the maximum stress at the integration point is equal to 0.82 MPa, a higher value justified by the location closer to the beam intrados and extrados.

For the sake of completeness, the *h-refinement* with 2 × 2 integration scheme is shown in Table 5.3. As it is possible to see, very few elements are needed to achieve accurate results. More in detail, a maximum element size of 120 mm (which corresponds to a discretization of the height with three elements) is able to guarantee a change of displacement and stress of about 1% and 2%, respectively, when compared to the previous coarser simulation, achieving results in line with the theoretical expected ones. It is noted that a finer discretization leads to a maximum stress at the support slightly higher than theoretically expected due to the approximation of the finite element approach at the boundary.

If quadratic quadrilateral elements are selected, the case is different. A typical element has 8 nodes, 2 × 2 integration scheme and a polynomial function for displacement interpolation such as the normal stress varies linearly in longitudinal direction and quadratically in transversal direction. The last feature is even excessive (for linear analysis) as a linear distribution is expected. The consideration of possible curved edges, besides a better discretization of the beam deformation, also prevents shear locking. The results are shown in Figure 5.24, with perfect agreement with the theoretical expected results. As it is possible to notice from Figure 5.23c, as the linear shape function provides a constant value of *SXX* along the beam length, this value approximates the average of those in Figure 5.24c (see beam intrados and extrados).

Table 5.3 Synthesis of the numerical simulations made with linear quadrilateral elements and 2 × 2 integration scheme

Max element size [mm] (total number)	Max deflection [mm]	Increment of deflection [%]	Stress at the support [MPa]	Increment of stress [%]
500 (10)	0.2675	-	1.0556	-
150 (66)	0.2870	7	1.1434	8
120 (126)	0.2908	1	1.1697	2
80 (252)	0.2925	1	1.1967	2
60 (415)	0.2932	0	1.2112	1

Dealing with mesh refinement, a selective procedure can be adopted to locally refine the discretization. In general, to reduce computational time and costs, it is advisable to use more elements only at locations of interest. This is the case shown in Figure 5.25 where a masonry wall with overall dimensions $5.0 \times 2.5 \times 0.5\,\mathrm{m}^3$ is considered. The bottom of the structure is constrained against the two translations and the left pier is subjected to a vertical downward settlement of 50 mm. The material has $E = 2,100\,\mathrm{MPa}$, $\nu = 0.25$ and mass density of $1,800\,\mathrm{kg/m}^3$. A coarse and unstructured mesh is used to provide the first insight for a better discretization. In particular, quadratic finite elements with maximum size of around 250 mm are used for the left side of Figure 5.25a, where the averaged vertical and horizontal normal stresses obtained with a structural linear elastic analysis are shown

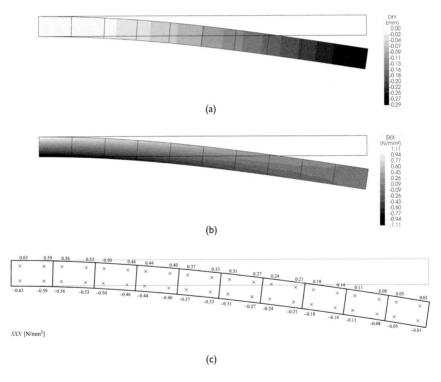

(a)

(b)

(c)

SXX [N/mm²]

Figure 5.24 Cantilever structure modelled with one quadratic quadrilateral element through the height and 2 × 2 integration scheme. (a) Vertical displacement; (b) and (c) normal stress distribution at nodes (extrapolated but unaveraged in the nodes) and values at integration points, respectively.

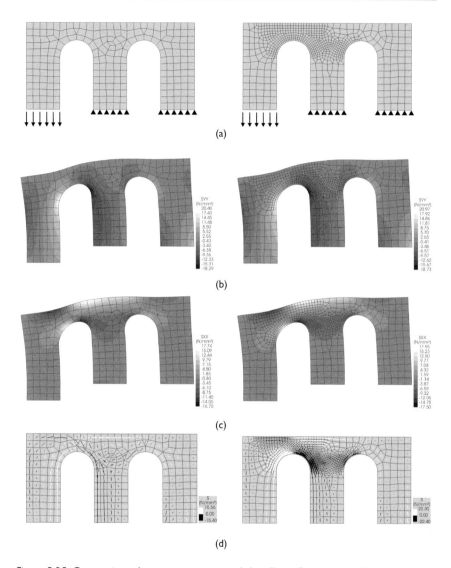

Figure 5.25 Comparison between coarse and locally refined mesh for a masonry wall subjected to vertical downward settlement at one pier. (a) Overall mesh; (b) and (c) vertical and horizontal averaged normal stresses, respectively and (d) vector plots of the principal stresses.

in Figure 5.25b and c. These are used to quickly locate the regions of the model where high stress values and gradients are expected, namely, where a better discretization is needed. A locally refined mesh and the consequent stress distributions are shown on the right side of Figure 5.25a–c. The concept of selective discretization is evident as only the areas with larger gradients are now covered by a finer mesh, leaving the rest of the model with the coarse mesh.

For the sake of completeness, the principal stresses are also discussed as they better describe the local stress level and make clear the stress flow through the structure. The comparison of the two mentioned models is reported in Figure 5.25d, where the principal stresses are shown in terms of vectors (whose length indicates the stress magnitude). Limiting the discussion to the left opening, on both coarse and refined meshes, tensile and compressive stresses flow on the left and right side of the arch, respectively, highlighting important shear on top of the keystone. The comparison in terms of principal stresses shows a significant increment on both compressive and tensile values because these are mesh-dependent. Even though linear elasticity was assumed for this example (as it is possible to notice, the stress values are far beyond the masonry capacity), this mesh refinement approach is efficient also for nonlinear analysis, although a possible different stress flow will occur, with stress redistribution. It is noted that principal stresses are computed in correspondence of each integration point (whose location depends on mesh size: the smaller is the size, the closer are integration points to intrados). Regarding Figure 5.25d (and in the rest of this chapter), the principal directions and stresses are plotted at the centre of each finite element.

5.2.6 Material properties

Idealization of the material properties needs to consider experimental evidence, either requested by the analyst for the specific case study or available in literature. As described in Chapter 4, the more complex is the constitutive law, the more detailed the experiments must be, possibly with a large number of parameters required. This dramatically differs from linear elastic analysis where isotropic materials are fully described thanks to only two mechanical constants, namely, Young's modulus and Poisson's ratio. Additionally, it is not easy to precisely describe the material behaviour through simple mathematical functions and further simplifications are also needed (e.g. hardening and softening branches as parabolic or exponential curves, respectively).

In general, when approaching a nonlinear analysis, some parameters can be directly investigated while others are adopted according to codes of practice or literature. If not directly investigated, the analyst should pay due attention when referring to identical materials (or within a certain degree of similitude). For instance, dealing with masonry, mechanical properties differ according to the overall morphology of the structural element and to the nature of components. Additionally, mechanical parameters can be derived from regression curves or tables, the applicability of which is related to a specific range of input parameters. In other words, *interpolation* (within the original range) is generally allowed, but careful attention should be paid to *extrapolation* (beyond the original range). If the material behaviour cannot be determined accurately, varying the mechanical parameters within a plausible interval can help understanding their influence in the structural response. In the case important sensitivity to the input is found, further experimental or in situ investigations on the parameters are advisable. If this is not possible, higher degree of caution may need to be adopted when making conclusions from the analysis results.

Two additional considerations regarding mesh details and material properties are of interest. When multiple materials are considered, a boundary such as contact or interface elements may need to be placed in between. This may be the case of masonry buildings where foundations and superstructure are made with different materials (e.g. stone and adobe, respectively). Along the same line, large difference in terms of stiffness between neighbouring elements may lead to numerical difficulties (e.g. large and stiff elements being connected to small and less stiff elements). The consequent large variations of magnitude for the stiffness terms may lead to a global stiffness matrix susceptible to ill-conditioning, as in the case of dramatic differences in terms of element size. For detecting these issues, it is recommended to check the ratio between the largest and smallest diagonal terms of the stiffness matrix (usually available in the output file). Finally, strain localization will occur in the case of softening and the response typically exhibits moderate mesh size and orientation dependency, as well as some dependency on the order of approximation. A technique to avoid mesh size sensitivity is to make the stress–strain relation a function of the element size. See Chapter 3 for further details.

Regarding typical material behaviour and observed structural response, three examples are presented in the conclusive section of this chapter to which the reader is referred.

5.2.7 Constraints and loads

Constraints are usually idealized as fully rigid or ideally hinged. However, actual supports show an intermediate behaviour between these extreme conditions, which, due to insufficient data, is not easy to represent adequately in a numerical model. An example is shown in Figure 5.26a, where a propped cantilever beam is considered. A possible representation is illustrated in Figure 5.26b, with a fixed support in A and a spring in B, whose stiffness represents the axial stiffness of the post. In the case the post is assumed axially rigid, the scheme of Figure 5.26b may be used. Moving to the wall, the actual support may allow some rotation, which may be included in the model by introducing a set of surrounding finite elements, such as the ones of the wall mesh (Figure 5.26c). Another intermediate situation may be with a rotational spring in A, not shown in the figure.

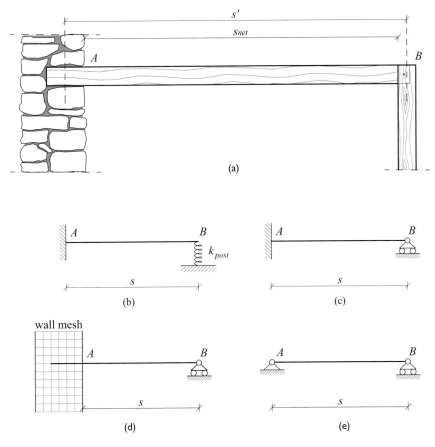

Figure 5.26 Propped cantilever beam. (a) Actual structure and (b–e) more or less refined structural models.

Finally, as an extreme condition to bound possible outcomes, the case of a hinge in *A* is advisable, which leads to the maximum deflection and bending moment for the beam (Figure 5.26d). Along the same line, the span length to be used in calculation is questionable as it is not clear where the ideal constraint starts working. In this regard, the analyst may make assumptions about the span *s* as an intermediate value between the centre-to-centre distance and the net span, *s′* and s_{net} in Figure 5.26a, respectively.

Dealing with an idealized constraint, Figure 5.27a shows the detail of a clamped beam modelled with plane elements. As illustrated, hinges are applied to all the nodes of the edge (it is noted that a rotational constraint, typical of a clamped support, would result meaningless for an element not provided with this degree of freedom). The selected constraint distribution prevents the Poisson's effect as vertical strains are prohibited, meaning that

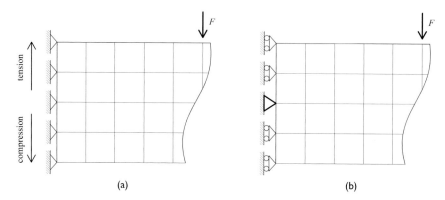

Figure 5.27 Two alternatives for modelling a full support. (a) Fully fixed conditions to all nodes, which introduce a fully clamped support and prevent expansion transversal to the beam axis, thus vertical reactions in the supports, and (b) fully fixed conditions in correspondence of the neutral axis and vertical rollers elsewhere.

vertical tension and compression stresses arise. A possible alternative is shown in Figure 5.27b where one hinge is placed in correspondence of the neutral axis and free transversal deformation to the beam axis is permitted. According to Saint-Venant's principle, the two alternatives do not influence the global behaviour of the beam at a certain distance from the support. However, spurious stresses in nonlinear analysis may influence the local response.

In general, external constraints are needed to avoid rigid-body movements of the entire model or of part of it. Additional constraints may be needed to set a certain relation between the displacements of two or more nodes. Typical examples are a rigid-body behaviour, where nodes translate and rotate together in a rigid connection, or an equal-displacement behaviour, where displacements are set equal only for certain DOFs. It is worth recalling that no solution is possible if a singular stiffness matrix is detected, i.e. rigid movements are not prevented. Typical mistakes are made when a planar problem (e.g. a simply supported beam) is studied through a tri-dimensional analysis without constraining the out-of-plane displacement variables (translation along the depth axis, out-of-plane bending and torsion). Quite opposite is the case when constraints or loads are applied to elements lacking such degree of freedom. For instance, if a truss element is considered, applying a bending moment to a node or constraining its rotation becomes meaningless. The same applies for rotational constraint or bending moment for nodes of a membrane element.

External constraints may also be useful in the case of the contemporary presence of geometrical, material, loading and support symmetries. Selecting the proper constraints along the line or plane of symmetry may significantly reduce the displacement variables involved, thus the computational effort of the analysis (see Section 5.6.1). Structures may have more than one plane of symmetry (e.g. masonry vaults) but care is necessary to check if the apparent geometrical symmetry is accompanied by symmetry of loads, materials and supports. It is noted that, in the case of brittle failure, symmetry is usually lost in the response (e.g. shear failure in a four-point loaded reinforced concrete beam), as only on side of the structure fails.

Dealing with loads, it is suggested to control their application points, direction and magnitude before running the analysis. If loads are not known precisely, conservative estimates should be adopted. A parametric analysis within a reasonable range may shed light on the sensitivity of the structure to a certain load type or magnitude. This applies also to combinations of loads. Usually, self-weight, other permanent loads and, if appropriate, prestressing are applied to the structure first. The subsequent loading sequence depends on the goal of the analysis. Different is the case of load incrementation, which depends on the case at hand and on the level of nonlinearities, as shown ahead.

5.3 SOLUTION PROCESS

Before approaching a nonlinear analysis, it is recommended to perform a linear elastic analysis. Many of the issues discussed in the previous sections can be identified this way. It is advisable to check the overall behaviour of the model such as the deformed shape together with the stress or strain fields (the two are related univocally by the elastic constitute law), and compare them with expected results (i.e. the educated guess) or look for anomalous responses.

In the case of quasi-static nonlinear simulations, the main goals usually regard the ultimate capacity in terms of carried loads (i.e. the level of safety, capacity or performance in terms of force and displacement), and how the structure fails (e.g. ductility and structural components involved in the failure mechanism). Failure is often associated with loss of stability due to material or geometrical nonlinearities (i.e. negative stiffness) and the solution process may be a challenging task. Other aspect of relevance in the nonlinear analysis may be the serviceability conditions, such as cracking or deformation in service. The analyst's patience, persistence, experience and theoretical insights are needed to investigate in full the behaviour of a building or a structural component.

The solution methods for nonlinear analysis are discussed in Chapters 2 and 3. In particular, two main convergence problems may arise and lead the solution method to diverge: (1) failure of the (local) iterative solution method for evaluating the stress update at the integration points from the strain increment and (2) failure of the (global) iterative solution method to obtain equilibrium. In both cases, the solution method is not able to reach the requested tolerance, or it reaches the maximum number of iterations. It is noted that the problem of the global iterative solution method is usually the most complex to resolve. In general, convergence issues are related to a large load increment, an inappropriate solution method, an inadequate control of the solution in the case of using the arc-length method or a poor implementation in the software code. Examples of difficulties are given by a structural negative tangent stiffness combined with a load-controlled approach, for which no solution exists, or a singular problem for the given load increment with multiple competing solutions. In the case of convergence issues, no unique solution procedure is suitable for solving all nonlinear problems. The approach to handle convergence issues is based on hypotheses, such as why the simulation fails to converge, variation of the solution procedure or the load step size and evaluation of the consequent outcome.

In general, if convergence is not reached with smaller load steps, the model may be unstable. Possible solutions are the change of the iterative method (e.g. changing from Newton–Raphson to secant), the change of the nodes involved in arc-length control or even using a larger load increment to go over the momentary instability of the model. Alternatively, it is possible to consider small viscous artificial effects, unrelated to real time dependency of the analysis, as used in many software codes to introduce stabilizing damping forces in the system.

Other suggestions regard using simpler constitutive laws (e.g. considering either tensile failure or compressive failure, but not both simultaneously, to identify the source of problems), removing geometrical nonlinearities (if considered), small variations of the mechanical properties values (e.g. increasing the fracture energy value in a reasonably way or reducing the tensile strength to increase ductility, noting that the latter is a conservative approach), the change of the type of finite elements used (e.g. moving from quadratic to linear approximation), alterations in the integration scheme (e.g. from full to reduced), the sequential removal of external loads (to individuate possible errors or incorrectly used units), the sequential consideration of only selected model parts (to identify model parts leading to problems) or the sequential definition of parts of the model as elastic (again, to identify model parts leading to problems). If the problem persists, the model may contain other inaccuracies related to the overall setting. This may be the case when new features of the software code are used by the analyst for the first time, e.g. contact elements or interface elements. In this case, it is suggested to create simple examples for which the result is known or can be approximated by analytical calculations, with the aim of verification of data used.

At the end of any analysis, it is advisable to check the output file provided by the software code, if available, and any warnings encountered during the analysis, especially related to number of steps or not converged load increments. In general, even in absence of convergence problems, tentatively harmless warnings (i.e. the ones that do not compromise the solution process, such as distorted elements) can affect the overall accuracy of the results and should be considered as suggestions to improve the model. Furthermore, being an incremental analysis, it is possible to study subsequent blocks of steps with different approaches. For instance, if regular Newton–Raphson fails due to a nonsymmetric stiffness matrix, the secant method may be used in its place to circumvent this problem. If the analysis is load-controlled, switching to displacement-control may help in studying unstable paths (e.g. snap-through). Additionally, arc-length and line search control methods can improve the convergence of the analysis with a limited computational effort.

If available, the comparison between the results of the final iteration of non-converged step and the last converged step may also help. Frequently, large changes in a given quantity may indicate the location of the problem. In this regard, the incremental values of displacements (i.e. not cumulative, but the values relative only to the current step) are fundamental (see also the examples in Section 5.6). It is recommended to subdivide large variations of loads in small steps to smoothly allow the model to accommodate a different configuration. In this regard, according to the adopted software code, several options may be available to tune the solution process, and the software manual should be the main reference. A few options about load increments follow.

The simplest way to define a step-by-step procedure is by explicitly setting the value of the load increments and applying them subsequently. Blocks of analysis can be performed with different values, either with the same load set or with a different load set (e.g. the application of the gravitational loading followed by other loads), but the general idea is that the load increments are set beforehand and linked to some physical representative quantity (e.g. the gravitational loading is applied in 10 steps of 10% of the total load to reach the desired value). It is worth reminding that arc-length control does not result in a fixed load increment, but still a certain value must be set a priori. It may also happen that, approaching an unstable path (e.g. snap-through), the load increment becomes negative instead of positive, indicating a postpeak response. To improve the load increments procedure, software codes usually provide different methods (so-called *automatic* or *adapting loading*) aimed at varying the size of the subsequent increments based on the information of the previous one. Examples of loading methods are:

- **Iteration-based:** By fixing a desired number of iterations, the software code increases (or decreases) the current load increment size if the number of iterations used in the previous increment was smaller (or larger) than the desired number.

- **Energy-based (in combination with arc-length method)**: Considering the current step, the load increment is calculated in such a way that the energy of the first iteration of the new load step equals the energy of the previous step.
- **Cutback-based automatic incremental loading**: It is aimed at taking as few load steps as possible, with a constant focus on limiting the number of iterations in the iterative procedure. First, the full loading (or the amount of interest) is applied in a single step. If the iterative procedure fails to converge (or it does not converge within a maximum number of iterations), the load step is scaled down by a cutback factor (e.g. 0.25) and the calculation is restarted. Basically, the main advantage of the method is that it recovers from not convergent or slow solution processes (i.e. with large number of iterations) and restarts with smaller (and hypothetically more efficient) load increments. If, after successive failures, the load step becomes smaller than a user-specified value, the automatic load controller quits. On the other hand, if the iterative procedure converges fast, i.e. the number of iterations is less than or equal to a certain amount, the step size may also be increased. This last feature is similar to the iteration-based method.

Although appealing, a unique solution process does not exist for all non-linear problems. The experience and judgement of the analyst provide guidance in evaluating the outcome or the failure of the adopted methods. For instance, it may happen that the algorithm gets stalled in proximity of an equilibrium point where many points in the structure are expected to become inelastic at the same time. Finding the "exact" sequence of e.g. fracture propagation or yielding may be complex and not needed in most engineering application. In turn, setting a larger load increment or loosening slightly the convergence criteria would allow to find all the points already cracked, so that the solution process can proceed.

Regarding tolerance, it is aimed at checking the discrepancy between external and internal forces and gives a quantitative insight on how close the approximated solution is to the true one (see Chapter 2). Typical tolerance values for the convergence criteria are $10^{-3} - 10^{-2}$ for the force norm, $10^{-4} - 10^{-3}$ for the displacement norm and $10^{-4} - 10^{-3}$ for the energy norm. If more than one criterion is enforced, load increments in which at least one of the norms is satisfied can be usually considered as converged, but any jump in the global load–displacement diagram requires careful inspection, particularly if the force convergence criterion is violated (as this may indicate that equilibrium is not ensured). Additionally, the solution methods are iterated until the tolerance criterion is met in absolute terms, when e.g. $0.99 < 1$. However, it is evident that a small deviation from the given threshold does not affect much the accuracy of the results. In other words, if not mathematically, the criterion can be considered satisfied from the engineering point of view. Load steps that do not fully comply with the convergence criteria may be still admissible if they are followed by converged load steps, the response is plausible and the convergence error is not too large.

Finally, a strong nonlinear behaviour may cause highly distorted meshes that prevent convergence. In this case, remeshing may improve the shape of the elements and the solution quality.

5.4 POSTPROCESSING

Nonlinear analysis typically produces a large amount of data for each load step, such as nodal displacements, reactions at the supports and stresses and strains at each integration point of each element. The analyst should have a certain awareness of what to investigate and how to interpret the results. Keeping in mind that the results should not be assumed correct without questioning them, *verification* is needed. This is the process that investigates the correctness of the model in terms of input (e.g. boundary conditions, loads or material properties) and if the analysis meets the initial goals in terms of output (e.g. displacement, stress and strain). Checks have been described in the previous sections and overall recommendations in the postprocessing phase follow.

The verification phase should address the model as a whole, as it may happen that the focus on a specific objective makes the analyst overlook mistakes elsewhere. The verification can be done by visual inspection of the deformed mesh, provided the representation scale is sufficiently large (even much exaggerated) to appreciate small details. Mesh should be also checked with respect to distorted elements, appropriate density in regions with high gradient of stresses (and strains) and possible connectivity issues. In this regard, it is worth noticing that, in the case the model underwent numerous and severe modifications, the final version may not meet the initial requirements anymore and should be verified again.

The outcome of the analysis should be compared with educated guesses. In the case of deviation, further controls should be made to check the correctness of either the intuition of the analyst or the numerical analysis. In the case of existing buildings, the current condition may be useful for evaluating the adequacy of the numerical model: damage and deformation under existing or past load should be reflected by the numerical results. The same applies to experimental data. However, if both cases are used to calibrate the model (i.e. to adapt unknown parameters), the consequent results are only verified but not validated. In other words, the model can accurately describe that given physical phenomenon, but caution should be used when taking the same model to assess the structural response under different circumstances.

Reactions can help identifying errors related to external loads and mass density (mistaken sign in load application or incorrect units are common). In this regard, reactions should be consistent with the expected results and it is recommended to check each load case independently by means of a linear elastic analysis. As far as stresses are concerned (the same considerations apply for strains), it is suggested to refer to the values in correspondence of the integration points or extrapolated at the nodes. It is worth noticing that averaging is also referred to as stress recovery or smoothing. For both cases, either extrapolated or averaged, a significant difference between manipulated values and the ones at the integration points is indicative of poor mesh density. In turn, averaged values should be used with caution especially in proximity of: (1) internal boundaries between different materials, where the stress and strain fields are discontinuous in reality, (2) outer boundaries (or internal discontinuum boundaries with interface or contact elements), where extrapolation should be preferred and (3) singularities that may be flatten by the averaging process (see also Section 5.2.4.1).

The stress at a free surface should be zero. Additionally, if the variation of strains/stresses between neighbouring elements is not smooth, connectivity issues or abrupt changes in stiffness should be investigated. In the case of three-dimensional elements, stresses should be investigated also in the interior of the model (to check for possible inaccuracies), by means of sections (or cuts). In the case of interface elements, also normal and tangential stresses should be continuous. Moreover, in the case principal stresses and strains are considered, it should be clear to the analyst that these values are in correspondence with principal directions, which are vectors, even if a scalar contour plot on magnitude may be shown.

As a final recommendation, the results of the analysis should be checked to assess the possibility of development of a different, and often dangerous, brittle failure mode such as shear failure, punching failure or buckling.

Tables 5.4–5.6 are meant to recap the concepts discussed earlier, aiming at a successful analysis and good-quality postprocessing.

Table 5.4 Synthesis of the main input in finite element analysis

Input
Analysis type (e.g. linear or nonlinear)
Reference system and units
Dimensions (e.g. plane stress or three-dimensional)
Geometrical simplification
Possible symmetry
Element type and shape function
Integration scheme
Element size/mapping
Material data (e.g. fracture energy is size-dependent)
Existing damage
Boundary conditions and possible internal constraints
Load cases and combinations
Solution method and tolerance
Load discretization

Table 5.5 Synthesis of the main results in finite element analysis

Results
Displacement history
Evolution of nonlinear phenomena (e.g. cracking, crushing, yielding and damage)
Reactions
Deformed shape
Stresses and strains (even generalized, such as axial force and bending moment)
Distribution and integration of results through selected lines
Stress and strain equivalent measures (e.g. von Mises stress)

Table 5.6 Synthesis of the main verifications to be made in finite element analysis

Input	Results
Mesh quality	Software warnings
Free nodes/edges/faces	Convergence issues (stress update and global
Internal edges	equilibrium)
Lacking or duplicated elements	Singular or ill-conditioned stiffness matrix
Correctness of input data	Excessively small or large strains (and displacements)
Mesh quality	Stresses continuity between elements
Connectivity issues	Null stresses at free faces
	Stress singularities and spurious behaviour
	Critical interpretation of results
	Agreement with an educated guess
	Incremental displacements
	Mesh refinement
	Possible sensitivity to input data
	Possible occurrence of brittle failure

5.5 STRUCTURAL ANALYSIS REPORT

As the final act of the analysis, a report should describe the basic hypotheses adopted in modelling, the degree of approximation embedded in the decisions made at modelling stage, the careful judgement of proposed interventions (if any), the significance of the results and the sensitivity analysis made. In this regard, it must be stressed that finite element analysis usually produces a huge amount of data that must be sorted out and presented in an easy-to-understand way. This report will help other persons involved in the process (or the analyst at a subsequent stage) in checking the analysis. When reporting a finite element analysis, the structural analysis report may contain at least the information mentioned in Table 5.7.

Table 5.7 Minimum information for the structural analysis report

Overall specification	• objectives of the analysis • type of analysis • software code used, with version and date of the release • building codes and the safety assessment approach adopted • units for geometry, loads, etc.
Model description	• detailed explanation of the decisions made, and simplifications introduced. This may include a discussion about additional modelling used to reach the final version of the model • type, number and integration scheme of elements, as well as the plot of the finite element mesh • material models and parameters, with possible correlation with experimental data or literature • in the case the response obtained with the used materials models is not obvious, a report with the analysis results of single element tests with a well-defined displacement history is recommended • description and plot of the boundary conditions and loading, including details of loading areas and locations • miscellaneous data necessary to re-analyse the model, if necessary
Analysis	• information about the model (number of elements, nodes and degrees of freedom) • loading scheme, and time used for the analysis (if significant) • condition of the stiffness matrix by comparing the ratio between smallest and largest diagonal terms (if known and significant) • discussion about warnings issued by the program and motivation of why these can be ignored • convergence behaviour, preferably iteration and variation of the norm in a graphical fashion • number of cracking points, crushing points or yield points, as applicable, at the most significant steps in the loading history
Results	• contour plots of field variables (such as displacements, or stress and strain values) for the most relevant load steps, highlighting the extreme values. This is particularly important for load steps close to change of stiffness in the global response, maximum load, ultimate load (in which the analysis was stopped), snap-throughs and snap-backs, as applicable • deformed configurations • possible evolution of field variables along the load history at significant points in the structure • reactions • vector plots (such as displacements, principal stresses or principal strains). It is important to highlight the nature of principal stress and strain, associated to principal directions but in some cases displayed in a contour plot • line plots of field variables along selected lines or sections • discussion of the validity of the results both in qualitative and quantitative sense (e.g. by using hand calculation, by comparing the results with the expected outcome or by comparing additional models)

(Continued)

Table 5.7 (Continued) Minimum information for the structural analysis report

> • failure mechanisms, i.e. the conditions that bring the structure to collapse
> • regions where damage is more pronounced; for instance, where the minimum strain of concrete or masonry is larger in absolute value than 0.3%, where fully open cracks are present or where damage above a threshold of 0.98 is present

5.6 EXAMPLES

The following examples discuss three paradigmatic case studies about the use of nonlinear analysis with steel, concrete and masonry structural components (the last, with both macro- and micro-modelling approaches). The analyses are not meant to *calibrate* the numerical model according to literature or experimental outcomes: the goal is to recap the insights of the previous chapters based on examples. As stressed, in absence of experimental evidence, the mechanical parameters discussed in Chapter 4 are meant to be, in general, conservative and the results are expected to be on the safe side. Finally, it is noted that the analyses are performed with the software code DIANA FEA v. 10.4 (Delft, The Netherlands).

5.6.1 Steel plate

Let us consider the structural response of a steel plate with a hole in the centre subjected to tension. The plate is $1,000 \times 800 \times 50 \, mm^3$ and the hole has a radius of 160 mm. Thanks to symmetry of geometry, constraints and loads, the numerical model regards only half of structural element, as shown in Figure 5.28a. The mesh is unstructured and built with quadratic quadrilateral elements with 2×2 integration scheme (except for only four quadratic triangular elements used to meet the current geometry, furnished with three integration points each). The external load is 1,000 N/mm applied on the right side of the plate and, according to the plate thickness, this load is equivalent to an external tensile stress of 20 MPa. The plate is fully fixed on the left side, by restraining all nodes in the horizontal and vertical direction, while the symmetry conditions are introduced with horizontal rollers in all nodes along the symmetry axis. Finally, the material is mild steel with $E = 210,000 \, MPa$ and $\nu = 0.3$. For the sake of simplicity, the self-weight is neglected. By performing a linear elastic analysis, the deformed configuration (mirrored along the horizontal axis) is shown in Figure 5.28b. The distribution of horizontal normal stresses (extrapolated and averaged) *SXX* is shown in Figure 5.28c together with the stress distribution along the most significant vertical sections, namely, at the support and in correspondence of the maximum horizontal stress (i.e. at the smallest cross-section). As it is possible to notice, an external traction of 20 MPa leads to a maximum

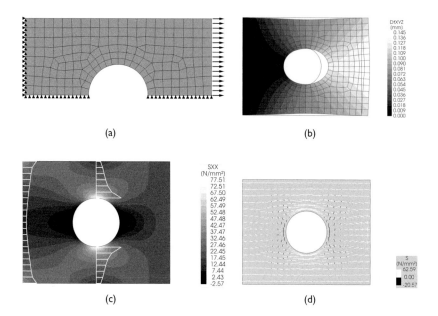

(a)   (b)

(c)   (d)

Figure 5.28 Linear static analysis of a plate with a hole under tension. (a) Half of the
model thanks to symmetry considerations; (b) displacement distribution
(magnified 480 times); (c) horizontal normal stresses (averaged from values at the integration points) for the entire mesh and along selected lines,
respectively, and (d) principal stresses and directions.

horizontal tensile stress of 77.51 MPa at the hole. Finally, Figure 5.28d illustrates the principal stress components and principal directions, with the
stress flow evident around the hole.

More in detail, it is interesting to discuss further the horizontal normal
stress value at the hole (77.51 MPa) and to compare this to the theoretical
one. By knowing that the cross-section is reduced by 320 mm in height (i.e.
40% of the full plate), an average stress equal to 33.33 MPa is expected in
this line, with a maximum in proximity of the hole. This phenomenon is
usually named *stress concentration* and the *stress concentration factor* is
generally used to evaluate the expected stress. This factor depends on the
ratio between hole diameter and plate width, namely, 320/800 = 0.4 for
the case at hand, which corresponds to a factor equal to 2.24. Accordingly,
the theoretical expected stress is 33.33 MPa × 2.24 = 74.67 MPa, which is
in good agreement with the numerical result.

Before performing a nonlinear analysis, a few considerations are worthy.
According to the stress distribution of Figure 5.28c, it is possible to calculate the load multiplier that leads the first element to yield. Considering a
yield stress $\sigma_y = 200$ MPa and assuming a uniaxial stress condition at this
point, 200/77.51 = 2.6 is the elastic threshold, i.e. up to this load factor

the structural response is elastic (note that this value is mesh-dependent because it considers the extrapolation at the nodes and not the values at the integration point, where the constitutive law is actually enforced). In turn, the load multiplier attains the maximum possible value when the smallest cross-section is fully yielded. An approximation of this maximum can be obtained by a simple uniaxial calculation in the framework of limit analysis, leading to a value of 6.0 (concisely, the height of the plate is reduced by 40% in correspondence of the hole, whereas the external applied stress and the yield stress are related by a factor of 10, thus the results of $10 \times (100\%-40\%)$). The nonlinear analysis can be performed with 20 load steps equal to 1.0 with arc-length control; it is worth reminding that this load history would be impossible with a force-controlled approach, as divergence would occur for a total load factor larger than 6.0. Additionally, it is a good practice to include the line search control algorithm to ensure global convergence of the solution process.

The load multiplier vs. horizontal displacement diagram is shown in Figure 5.29a, where the maximum displacement in the free edge is considered as control node. Considering step 20 with a load multiplier equal to 6.03 (vs. 6.0 provided by limit analysis), the displacement distribution is shown in Figure 5.29b. As it is possible to see, the maximum displacement is equal to 2.58 mm. The effect of the material nonlinearity is evident when this figure is compared to Figure 5.28b. Basically, there is no proportion between load (six times larger) and displacement (almost twenty times larger). This is also clear from the graph of Figure 5.29a where the linear elastic response is depicted. The magnification factor of displacements in Figure 5.28b helps the comparison: the analysis being linear elastic, displacements are proportional to the load.

With the aim of localizing the source of nonlinearity, the analysis of incremental displacements is recommended. It is stressed that this does not represent a physical set of displacements and depends on the step size, indicating what occurred in the structure in the last load step. Still, incremental displacements usually allow to clearly understand the nonlinear processes involved in the response. It is worth reminding that portions of the structure with equal incremental displacements are indicative of rigid movements (i.e. the mutual displacement between their parts is null). Figure 5.29c shows the contour plot for the same load step seen earlier. Three main areas with rigid movements are evident (i.e. areas with uniform colours): (1) the left part, which undergoes almost null incremental displacement, (2) the right part, with a rigid movement rightward, and (3) the inverted triangle above (and symmetrically below) the hole. It is across these two last regions that strain occurs, and the closer are the regions (i.e. the sharper is the change of colour), the larger strains are expected. For the case shown in the figure, the larger deformations are mainly localized along the inclined bands that move from the hole. This is confirmed by the vector plot of the principal strains shown in Figure 5.29d, with larger values in correspondence of the mentioned bands. Additionally, it is interesting to compare the hole deformation in Figure 5.29b and c (note that the magnification factors are not the same). The former accounts for the entire displacement history (i.e. total deformation) and the circular hole became a sort of ellipse with a more pronounced deformation on the right side. The latter, indicative of localized plastic deformations, shows a different incremental mechanism according to which the right semi-circular part of the hole is rigidly pulled.

Figure 5.29e confirms this outcome and provides more insight on the distribution of the plastic strain within the steel plate, totally localized along two inclined bands. It is worth noticing that, in the fashion of the criteria

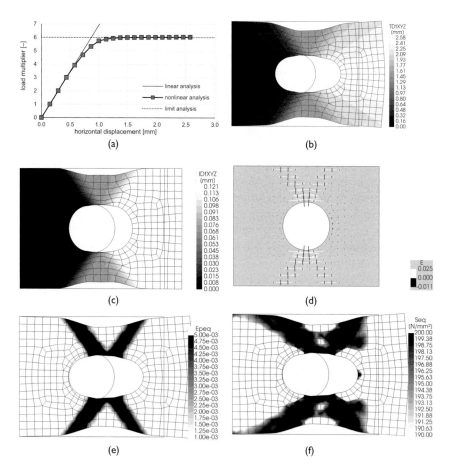

Figure 5.29 Nonlinear static analysis of a holed plate under tension. (a) Load multiplier–displacement diagram; (b) total displacements (magnified 80 times); (c) incremental displacements (magnified 1,000 times); (d) principal strains and directions; (e) plastic strains and (f) von Mises equivalent stresses (at the integration points). Note that results b-d are shown at the last load step.

seen in Chapter 3 for stresses, this measure furnishes a single scalar that synthesizes the six strain measures (three in this case of plane analysis). In particular, the legend is arranged such as the values below 10^{-3} are discarded, until 5×10^{-3} are grey-shaded, and above this value are fully black. Accordingly, the contour plot clearly shows where plastic deformation concentrates. In turn, Figure 5.29f shows the von Mises stress distribution, also in this case, arranged such as the values equal or larger than 200 MPa are shown in black. Basically, considering von Mises's formula, yielding occurs when the stress combination approaches the yielding uniaxial stress (see Chapter 3). As it is possible to notice, at the load step number 20, yielding combinations localize in the V-shaped bands above (and symmetrically below) the hole. It is noted that values of equivalent stress larger than the yield stress 200 MPa are possible at the nodes, although not allowed by the constitutive law, due to the extrapolation and averaging (see Section 5.4).

At this point it is worth recalling the difference between current plastic strain and current stress. With ideal plasticity (and discarding cyclic loading), once a portion of the structure yields, there is no possibility for it to recover the consequent plastic deformation. Basically, the strains shown in Figure 5.29e represent the cumulative effects of all previous load steps: subsequent ones can only stabilize (elastic unloading) or increment these values (in the case of subsequent pulling actions). In turn, Figure 5.29f gives a picture of the current situation. With reference to the concept of yield surface with ideal plasticity, a plastic stress state can either keep yielding or exhibit elastic response. Recalling the strain decomposition, elastic unloading recovers the elastic part of strain, but not the plastic one.

For the sake of clarity, Figure 5.30 illustrates the results of the plate after a total unloading (i.e. null external load). Extensive plastic deformation and residual stresses in the plate are evident. Although the external load is null, the X-shaped yielded region around the hole is still subjected to tensile strain (Figure 5.30a). This does not mean that part is subjected to

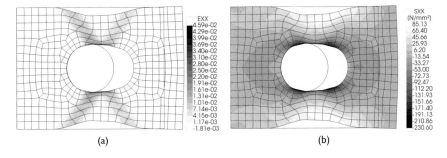

(a)   (b)

Figure 5.30 Elastic unloading of the holed plate with null external load (total displacements magnified 80 times). (a) Total strain EXX and (b) stress SXX along the horizontal direction (extrapolated and averaged at the nodes).

tensile stress but, with respect to the original configuration, this region is elongated (as shown in the figure and intuitively grasped). In turn, the stress distribution of Figure 5.30b shows large residual values, as the unloading produced a significant compression around the hole edge (upon the yield process), leaving the rest of the central section in tension (with maximum value on the external edges).

Going back to the loading phase, for the same load discussed earlier (step 20), it is also interesting to evaluate the horizontal normal stress distribution and the relative values along two selected lines, as shown in Figure 5.31a. As it is possible to notice, the maximum values at the nodes are higher than the yielding value (228.89 vs. 200 MPa), which occurs mostly not due to node extrapolation, but because failure under biaxial stresses provides higher normal stress values than the uniaxial strength (see Figure 3.16c in Section 3.2.2). For the sake of comparison, the results of the elastic analysis for the same load factor are shown in Figure 5.31b (six times larger than what was shown in Figure 5.28). The maximum stress is 465.07 MPa (i.e. 77.51 MPa × 6), which is not admissible for the given ideal plasticity law (the reader should not be misled by the grey shades, arranged according to the different scale limits). Due to subsequent yielding and consequent

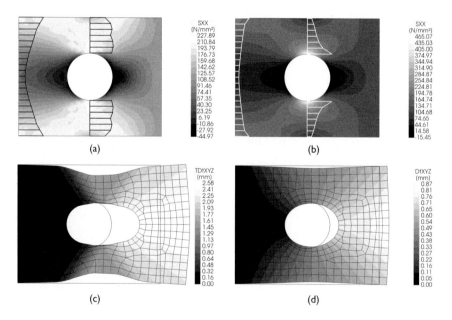

Figure 5.31 Comparison between nonlinear and linear static analysis of a plate with a hole under tension, for the same external load (linear distributed load equal to 6,000 N/mm). Horizontal normal stress for (a) nonlinear and (b) linear analysis; displacement distribution for (c) nonlinear and (d) linear analysis (total displacements magnified 80 times).

stress redistribution, in fact, the stress distribution is dramatically different in nonlinear analysis. This is clear observing the stresses along the middle line (in correspondence of the hole). In turn, as the material close to the support still behaves elastically and is only marginally affected by the yielding of the middle part of the plate, the relative stress distribution at the support is in line with the outcome of the elastic analysis.

The comparison between nonlinear and linear analyses is extended to displacements, as reported in Figure 5.31c and d. The nonlinear response localizes the larger displacements on the right side of the plate with strong plastic deformations in the middle part, which is the weakest one. The linear response, with a theoretically infinite strength, shows a more uniform distribution of displacements. This is clear also if the horizontal top and bottom sides of the plate and the shape of the hole are compared. Finally, it is also important to stress that, except for Figure 5.30, all plots shown here regard the same load step, and more investigation is usually needed in complex structures to evaluate other steps of the analysis.

5.6.2 Concrete beam

The following example regards a reinforced concrete beam subjected to a point load in the midspan according to the experimental campaign carried out by Vecchio and Shim (2004). The beams tested in the experimental campaign were provided with heavy amounts of flexural reinforcement to make them shear critical, which is more demanding for a numerical simulation. The consequent failure generally exhibits small ductility at peak-load level before a sudden drop in load capacity. Additionally, test observations revealed that the behaviour of the beams was highly influenced by crushing of concrete beneath and adjacent to the loading plates. Flexure cracks in the midspan regions were relatively insignificant, with crack widths generally in the range of 0.5–1.0 mm. The goals of the numerical analysis presented in this section are the evaluation of the force–displacement diagram (thus ultimate load and deflection) and the failure mode.

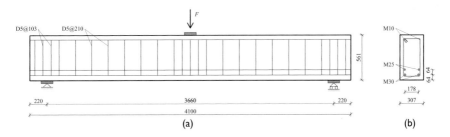

Figure 5.32 Beam specimen AI from Vecchio and Shim (2004), dimensions in mm. (a) Elevation and (b) cross-section details.

Table 5.8 Synthesis of the mechanical data for the concrete beam of Figure 5.32

Concrete					
Young's modulus [MPa]	Poisson's ratio [–]	Compressive strength [MPa]	Fracture energy compression [N/mm]	Tensile strength [MPa]	Fracture energy tension [N/mm]
36,500	0.2	22.6	21.0	1.79	0.13

Reinforcement				
Bar	Cross-section area [mm²]	Young's modulus [MPa]	Yield strength [MPa]	Ultimate strength [MPa]
M10	100	200,000	315	460
M25	500	220,000	445	680
M30	700	200,000	436	700
D5	32	200,000	600	649

Shaded grey cells indicate estimated values from Chapter 4.

The geometrical data of the beam originally named A1 are shown in Figure 5.32, whereas the material properties are synthesized in Table 5.8 according to the information provided in the mentioned paper and the indications of Chapter 4 (the latter in grey-shaded cells). In particular, for the reinforcement bars and stirrups, a hardening plasticity model was used, with linear hardening and ultimate strain equal to 5%. In turn, the concrete model is based on the total strain rotating crack model, exponential softening in tension and parabolic behaviour in compression. Additionally, the reduction of compressive strength of concrete due to lateral cracking was considered according to Vecchio and Collins (1993) with a lower limit equal to 0.4. Regarding the effect of confinement, instead, the model proposed by Selby and Vecchio (1993) was selected. For the steel plates considered in correspondence of the support and the load application, a linear elastic model was considered with Young's modulus equal to 200,000 MPa and Poisson's ratio equal to 0.3. Assuming minor contribution of these small plates for the overall behaviour of the beam, no interface was considered between the two materials. Finally, the self-weight was neglected, given its small contribution to failure, and tension stiffening was also neglected, given the small element size adopted.

The finite model is shown in Figure 5.33a, where linear triangular elements (thus constant strain and stress) and one integration point are chosen. The use of a structured mesh using regular quadrilateral elements is more normal in academic context but, given the engineering purpose of the present book and the extended use of automatic meshing procedures based on triangles and tetrahedrons for complex geometries, a triangular mesh was adopted. Due to symmetry, only half beam was modelled. Additionally, due to its crucial role in the overall response of the beam, special care is needed

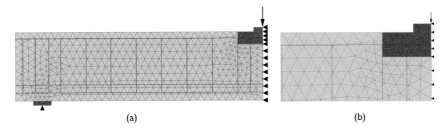

Figure 5.33 Finite element model of the beam shown in Figure 5.32. (a) Overall model (with constraints and vertical load) and (b) enlargement of the loading zone.

for the concrete beneath and adjacent to the loading plate. In this regard, the numerical model in Vecchio and Shim (2004) was provided with an extra out-of-plane reinforcement in the neighbouring elements. As explicitly stated by the authors, this choice resulted in strength enhancement and considerable ductility increase.

A different approach was selected for the present case, where the fracture energy value was doubled for the zone underneath the loading plate with respect to the rest of the beam. The former is $206 \times 100 \, mm^2$ for half of the beam (enough to incorporate the stirrups at the midspan) and is highlighted in Figure 5.33b, with a mesh maximum size equal to 40 mm, which is the same discretization of the steel plates ($75 \times 35 \, mm^2$). For this zone, a null reduction of the compressive strength due to lateral cracking was also assumed. The rest of the beam has maximum mesh size equal to 60 mm. Regarding steel bars and stirrups, embedded reinforcement elements were considered. Figure 5.33 shows also the boundary conditions and the imposed force on top of the loading plate (darker). Although scaled throughout the analysis by means of the load multiplier, a value of 2 kN was considered as reference load (1 kN in the half beam modelled).

The analysis was performed considering the linearized form of arc-length control (namely, updated normal plane) decreasing the load multiplier, from 20 to 1, as progressively approaching the peak, the minimum value was used also for the subsequent softening phase. The arc-length control was set to monitor the vertical displacement of the node where the force is applied (i.e. indirect displacement control). Finally, regular Newton–Raphson method was selected for the equilibrium iterations, checked according to the simultaneous satisfaction of energy and force criteria (10^{-4} and 10^{-2}, respectively), with line search control activated.

The comparison between experimental and numerical results is shown in Figure 5.34 in terms of load–deflection diagram, using the vertical displacement in the mid of the beam as control point. As it is possible to notice, the numerical analysis is stiffer in the first branch of the curve, with an accurate match for the peak load, ultimate deflection and quasi-brittle failure

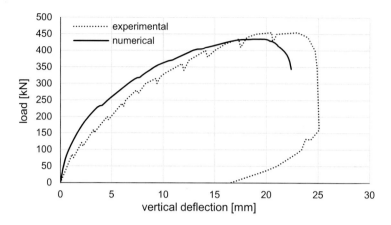

Figure 5.34 Load–deflection diagram. Comparison between experimental test and numerical analysis of the concrete beam.

mode. In particular, the numerical simulation achieved a maximum load of 435 vs. 455 kN experimentally reached (an underestimation of about 4%), while the ultimate deflection is 22.4 vs. 25.1 mm (an underestimation of about 10%). The following description sheds light on the damage evolution and failure modality.

Figure 5.35a–d shows the stages of the numerical analysis with the principal crack width evolution. The scale is arranged such that dark elements are representative of cracks width larger than 0.1 mm, whereas 2.73 mm is the maximum value achieved during the analysis. In turn, Figure 5.35e shows the principal strains and directions, where the colour and the symbol size are indicative of the strain sign and magnitude, respectively. As it is possible to notice, white symbols are more pronounced across the cracks, while black symbols are denser close to the damaged region in compression below the loading plate. The last drawing (Figure 5.35f) is about the experimental results where the area interested by the concrete crushing beneath the loading plate is shaded. Although the overall behaviour could be characterized as shear-flexural in nature, the experimental beam exhibited severe diagonal-tension cracks during later load stages, achieving a crack width as large as 2.0 mm (bold in Figure 5.35f). The real damage is not symmetrical (as normally happens in real-word), and the numerical results are more in line with the right side where failure finally occurs, with a prominent crack (although steeper than the real one) and a diffused damage along the reinforcement bars in tension. In line with what stated in Chapter 4, smeared crack models are not meant to reproduce localized (and sharp) cracks but their effects on the structural behaviour.

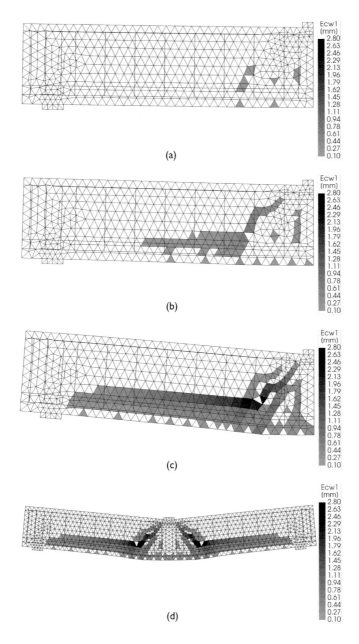

Figure 5.35 Comparison between numerical and experimental crack pattern. (a–d) Subsequent stages of the analysis (displacement field magnified eight times) with load equal to 230, 300, 435 and 366 kN (in the softening branch), respectively.

(Continued)

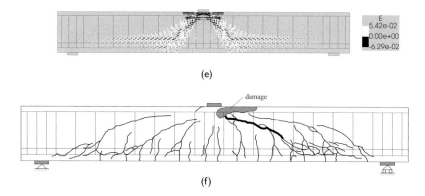

(e)

(f)

Figure 5.35 (Continued) (e) principal strains and directions and (f) experimental results where the thicker line is representative of a crack width of about 2.0 mm.

The effects of the major crack are also evident if the increments of displacements are considered. These are shown in Figure 5.36 by means of contour plots on top of incremental deformations. It is worth reminding that this deformation is not representative of any state of the test, but it illustrates the difference between the previous and the current step of the numerical analysis. While the beam displays an overall response during the first stage of the loading process (Figure 5.36a), deflection starts to localize at the midspan wedge as soon as cracks occur (Figure 5.36b). This process is even more evident once the peak load is reached and beyond (Figure 5.36c and d), when the area localized between the two main cracks is almost independent (see the extrados line). Besides small differences of shades, this part is coloured the same, which means that, at ultimate condition, the central wedged behaves as a rigid body.

The stress distribution at peak can be visualized by means of principal stress components and directions, as shown in Figure 5.37. This is in line with the expected flexural behaviour of the beam at the ultimate state, where the stress block in the compressive zone is highlighted by the black horizontal lines. The presence of cracks is marked by inclined principal compressive stresses beneath the loading plate. According to continuum mechanics, principal directions are perpendicular, so cracks (due to tension) open perpendicular to compressive directions. Looking at the magnitudes, the stress values are contained within the concrete compressive and tensile strengths, namely, −22.6 and 1.79 MPa, respectively (compare with Table 5.8). Larger values of compressive stress (lower in absolute value) are likely due to the positive contribution of the confinement stress.

Looking at the behaviour of the longitudinal bars at the beam bottom and of the stirrups, the relative stress distributions are shown in Figure 5.38. All the bars are extensively yielded, with longitudinal bars with yield strength about 450 MPa and ultimate strength about 700 MPa, and stirrups with

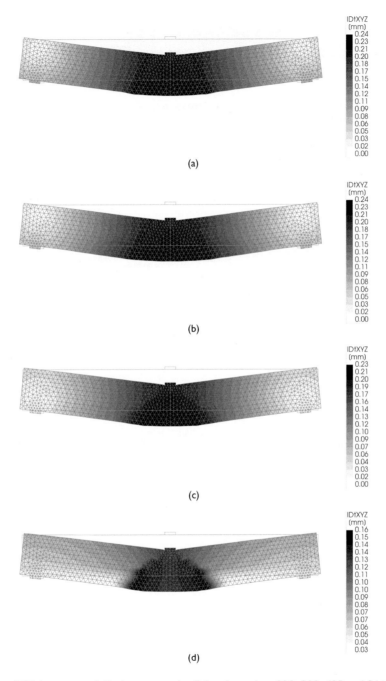

Figure 5.36 Incremental displacements. (a–d) Load equal to 230, 300, 435 and 366 kN (in the softening branch), respectively. The deformations of the four figures are not in the same scale and the absolute values depend on the step size of the incremental analysis.

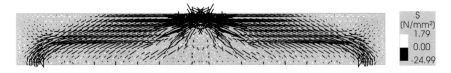

Figure 5.37 Principal stress directions and magnitudes at peak load (external load equal to 435 kN).

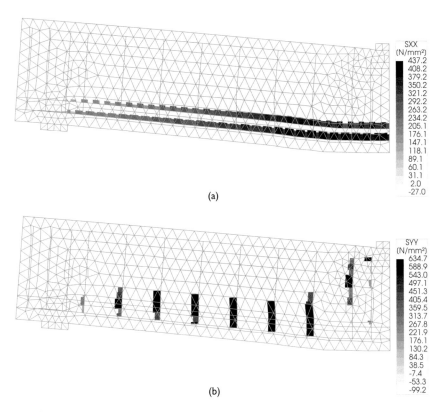

(a)

(b)

Figure 5.38 Longitudinal stresses along the reinforcement bars for (a) longitudinal bars and (b) stirrups at peak load (external load equal to 435 kN).

yield strength of 600 MPa and ultimate strength of 650 MPa (see Table 5.8). Considering the perfect bond with concrete (due to the assumption of embedded reinforcement), the larger stresses in the stirrups follow the crack presence. The analogies with Figure 5.35c and d are evident. In this regard it is important to remind that reinforcement bars are assumed to be one-dimensional line elements without transverse shear stiffness or flexural rigidity. This means that the dowel effect is not considered.

5.6.3 Masonry shear wall

The example discussed in the present section is taken from Anthoine, Magenes and Magonette (1994) and regards the masonry wall named "Low wall" subjected to a constant vertical load and cyclic in-plane shear force at its top (with consequent bending moment and shear). As for the concrete beam example, the goals of the numerical analysis presented in this section are the evaluation of the force–displacement diagram (thus ultimate load and deflection) and the failure mode. The comparison is made between the cyclic (experimental) and the quasi-static (numerical) behaviour of the wall, as shown ahead. The wall specimen was constructed with brick units and hydraulic lime mortar arranged in two-leave thickness English bond pattern. The brick dimension is $55 \times 120 \times 250 \, \text{mm}^3$, whereas the mortar joint is $10 \, \text{mm}$ thick. The geometrical data and the testing setup are shown in Figure 5.39. In particular, the wall is $1,350 \, \text{mm}$ high and $1,000 \, \text{mm}$ wide, leading to a height/width ratio equal to 1.35. The thickness is $250 \, \text{mm}$ and the wall is subjected to a constant vertical stress of $0.6 \, \text{MPa}$ (i.e. a vertical force equal to $F_v = 150 \, \text{kN}$). Upon testing, the wall exhibited a moderately brittle failure by diagonal cracking at a drift of 0.2% for a maximum horizontal force of $F_h = 84 \, \text{kN}$.

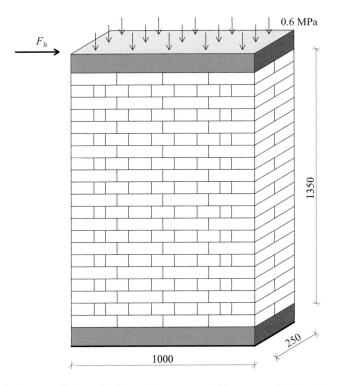

Figure 5.39 "Low wall" from Anthoine, Magenes and Magonette (1994), dimensions in mm. Schematic view.

As far as the mechanical properties are concerned, the indications in the mentioned paper are about the constitutive relationship adopted for the macro-modelling approach. In particular, the original authors carried out a numerical analysis with compressive strength and Young's modulus (calculated up to 70% of the maximum strength) of masonry equal to 6 and 2,100 MPa, respectively, with a consequent ratio equal to 350. It must be noted that there is no evidence whether these values correspond to the actual masonry properties, but they are also adopted in the present example. The rest of the properties have been evaluated according to Chapter 4, as synthesized in Table 5.9. Given the scarcity of available data, the first

Table 5.9 Synthesis of the mechanical data for the masonry wall of Figure 5.39

Macro-modelling					
Young's modulus [MPa]	Poisson's ratio [–]	Compressive strength [MPa]	Fracture energy compression [N/mm]	Bond tensile strength [MPa]	Fracture energy tension [N/mm]
2,100	0.2	6	1	0.13	0.02

Mass density = 1,800 kg/m³
Total strain rotating crack model, exponential softening in tension and parabolic behaviour in compression

Micro-modelling (brick)			
Young's modulus [MPa]	Poisson's ratio [–]	Compressive strength[a] [MPa]	Equivalent brick height = actual brick height + joint thickness [mm]
20,000	0.2	25.0	65 = 55 + 10

Mass density = 1,800 kg/m³

Micro-modelling (linear material properties)	
Normal stiffness [N/mm³]	Shear stiffness [N/mm³]
36.1	14.4

Micro-modelling (cracking–shearing)					
Bond tensile strength [MPa]	Mode I fracture energy [N/mm]	Cohesion [MPa]	Friction coefficient (angle)	Dilatancy coefficient	Mode II fracture energy [N/mm]
0.1	0.01	0.14	0.75 (37°)	0	0.014

Micro-modelling (crushing and compressive inelastic law): bed and head joints			
Compressive strength [MPa]	Factor C_s [–]	Fracture energy compression [N/mm]	Equivalent plastic relative displacement [mm]
6	9	1	0.015

Shaded grey cells indicate estimated values from Chapter 4.

[a] Not useful for the model but needed to evaluate other parameters.

step was to assess the possible values of brick and mortar compressive strengths. Although not useful for either micro- or macro-approach, this provides a general overview of the structure at hand and helps evaluating other parameters. More in detail, starting with the compressive strength of brick, set equal to 25 MPa, to comply with the macroscopic compressive strength according to Eurocode formula, mortar strength should be around 1.5 MPa, which is in line with aerial lime mortar. Young's modulus of brick can be assumed as $800\, f_c = 20{,}000\,\text{MPa}$.

Regarding the compressive fracture energy, as damage was detected throughout the cyclic behaviour, to compare the quasi-static analysis with the envelope of the corresponding experimental curves, a value of $G_c = 1\,\text{N/mm}$ was selected. It is reminded that the total energy dissipated by the wall is given by the area defined by the different cycles in the force–displacement diagram. A reduced fracture energy in quasi-static analysis considers, to some extent, the energy dissipated by the wall during the multiple cycles. Additionally, as far as the bond tensile strength is concerned, a lower bound value of 0.1 MPa was suggested in Chapter 4. However, in order to better compare the macro- and micro-modelling approaches, Eq. (4.11) is adopted here to calculate the tensile strength along the direction parallel to the bed joints. Assuming that failure does not involve cracking of the units, the tensile strength can be related to the cohesion ($c = 0.14\,\text{MPa}$), the units overlapping length ($l_o = 62.5\,\text{mm}$) and the unit height (h_b is considered equal to 65 mm, including the mortar thickness of 10 mm), leading to 0.13 MPa, which is adopted because the failure mechanism is mostly controlled by vertical cracking.

Moving to micro-modelling, the interface normal and shear stiffnesses were calculated according to the formulas of Chapter 4 leading to 36.1 and $14.4\,\text{N/mm}^3$, respectively. These values were obtained considering Young's modulus of masonry (as a composite material) and of the brick, with an equivalent height of the brick equal to 65 mm. It is noted that the bond tensile strength is here assumed equal to 0.1 MPa. Regarding the nonlinear parameters of the interface, they agree with Chapter 4, except for fracture energy in compression, for which the considerations adopted for macro-modelling hold here too. As far as geometry is concerned, units are expanded to include the mortar joint, achieving dimensions equal to around $65 \times 130 \times 250\,\text{mm}^3$. Operatively, the actual dimensions of the wall ($1{,}000 \times 1{,}350\,\text{mm}^2$) were divided according to the number of courses and bricks along vertical and horizontal directions, respectively. Given the low stiffness of the wall, the equivalent plastic relative displacement κ_p was calculated so that the total masonry strain at the end of the hardening phase is equal to 0.3%.

Finally, for the steel beams considered in correspondence of the support and the load application, a linear elastic model was considered with Young's modulus equal to 200,000 MPa and Poisson's ratio equal to 0.3. The interface between steel and masonry was considered as perfectly rigid given the existing boundary conditions.

5.6.3.1 Macro-modelling

The finite model is shown in Figure 5.40, where a structured mesh with linear quadrilateral elements with 2×2 integration scheme is chosen, with a maximum element size of 35 mm. The boundary conditions at the bottom of the steel element are also shown, together with the distributed load on top and the horizontal force. Although scaled throughout the analysis by means of the load multiplier, a reference load value of 1 kN was considered.

The analysis was performed considering the linearized form of arc-length control (namely, updated normal plane) decreasing load multipliers, from 5 to 1 as progressively approaching the peak. In the softening phase, different load steps have been used to comply with the complexity of the analysis. The arc-length control was set to monitor the horizontal displacement of the node where the horizontal force is applied (i.e. indirect displacement control). Finally, the regular Newton–Raphson method was selected for equilibrium iteration, checked according to the simultaneous satisfaction of energy and force criteria (10^{-4} and 10^{-2}, respectively), with line search control activated.

The comparison between experimental and numerical results is shown in Figure 5.41 in terms of load–displacement diagram. A good match with the

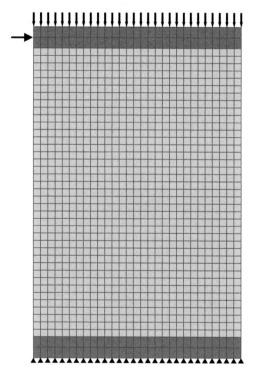

Figure 5.40 Macro-modelling approach to masonry. Finite element model of the wall shown in Figure 5.39: overall model, with vertical distributed load and horizontal point load.

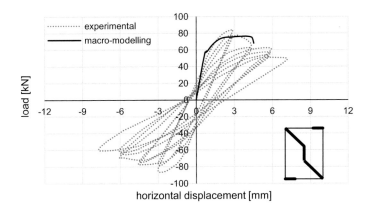

Figure 5.41 Load–displacement diagram. Comparison between experimental test and numerical analysis according to the macro-modelling approach of the masonry wall.

experimental results is found until around 4.6 mm of horizontal displacement, with a maximum load equal to 76 vs. 84 kN experimentally reached, an underestimation of around 9%. The following description sheds light on the damage evolution and failure mode. The only experimental outcome is represented by the illustration on the right side of Figure 5.41, where the solid lines represent the shape of the wall and the bold ones the crack locations.

First of all, it is interesting to understand the behaviour of the wall at the end of the elastic phase, i.e. around 58 kN. This is shown in Figure 5.42 where, after a first detachment of the diagonal corners, the wall behaviour is dominated by the presence of a significant crack in the middle part of the wall. Due to the change of scale, the horizontal bending cracks at the upper right and lower left corners become less evident if compared with

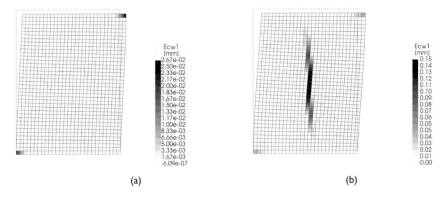

Figure 5.42 Crack width at the end of the elastic branch (deformation magnified 100 times). Horizontal load equal to (a and b) 50 and 58 kN.

the vertical crack at the middle, which controls the final failure mode. This aspect is in line with the experimental behaviour. The same load steps are considered in Figure 5.43 where the incremental displacements are represented instead. After the initial cracks at the corners (highlighted with large incremental displacements in Figure 5.43a), the entire deformation is concentrated in the middle part of the wall where the crack sides move apart (Figure 5.43b).

The sudden drop of capacity after reaching the maximum load is due to the occurrence of an additional pseudo-vertical crack (Figure 5.44). This was not detected during the experimental campaign. This may be due to the implicit assumptions of macro-modelling approach about a homogenized isotropic continuum with no difference between units and mortar joints. In this specific wall, a pseudo-vertical straight crack would have requested cracking of units for several courses in sequence. Further damage close to the corners is also evident in Figure 5.44.

The significant damage experienced by the wall leads the stress flow to pass between the cracks, in the fashion of isolated struts. This is clear if principal stresses are analysed, as shown in Figure 5.45. However, the increasing stress concentration at the bottom right and upper left corners of the wall is indicative of the larger demand for these areas. In the end, the wall is about to fail due to the disintegration of the central area and additional crushing of the corners.

5.6.3.2 Micro-modelling

The finite model with the independent representation of masonry units and joints is shown in Figure 5.46, using a structured mesh with quadratic quadrilateral elements with 2×2 integration scheme, with a maximum size of 35 mm. Regarding the head and bed joints, each interface is modelled with quadratic line elements with four-point Newton–Cotes integration scheme.

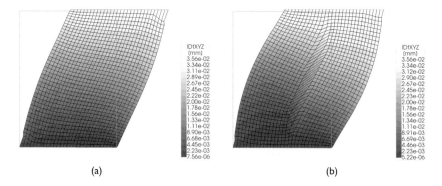

(a)   (b)

Figure 5.43 Incremental displacement at the end of the elastic branch (deformation magnified 15,000 times). Horizontal load equal to (a and b) 50 and 58 kN.

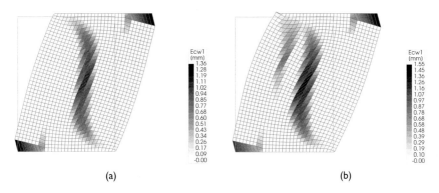

Figure 5.44 Crack width at load peak and postpeak branch (deformation magnified 100 times). Horizontal load equal to (a and b) 76 and 66 kN.

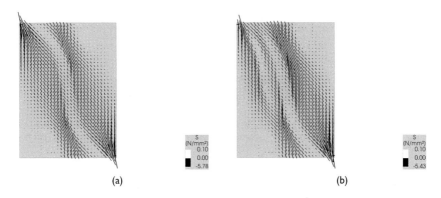

Figure 5.45 Principal stresses and directions at load peak and postpeak branch. Horizontal load equal to (a and b) 71 and 66 kN.

The rest of the model is in line with what was described in the previous subsection. The analysis was performed again with the updated normal plane of the arc-length method, load multiplier equal to 5. In this case too, the arc-length control was set to monitor the horizontal displacement of the node where the horizontal force is applied (i.e. indirect displacement control). Finally, regular Newton–Raphson method was selected for equilibrium iteration, checked according to the simultaneous satisfaction of energy and force criteria (10^{-4} and 10^{-2}, respectively), with line search control activated.

The comparison between experimental and numerical results is shown in Figure 5.47 in terms of load–displacement diagram. The numerical simulation describes correctly the overall behaviour of the wall, approaching accurately the quasi-brittle softening branch. The maximum load is 80 vs. 84 kN experimentally reached (an underestimation of about 5%). From the diagram it is possible to see that the micro- and macro-approach are almost coincident until a horizontal displacement of 4.2 mm.

Figure 5.46 Micro-modelling approach to masonry. Finite element model of the wall shown in Figure 5.39: overall model, with vertical distributed load and horizontal point load.

As it is possible to notice from Figure 5.47, the softening branch can be subdivided into three parts, each characterized by a plateau and a sudden drop. This means that, after a mechanism is stabilized, which guarantees a constant load capacity, quick alterations make the wall jump to other mechanisms. In this respect, the left column of Figure 5.48 shows the total displacements at peak and at the steps before the drops of the softening branch, while the right column shows the related incremental displacements. The first two rows of the results show a mechanism compatible with what was detected with macro-modelling (compare with Figure 5.44), with a main vertical crack in the middle part of the wall (and minor ones next to it), together with the support flexural cracks. However, the presence of linear elastic masonry units forces the crack to develop through the joints. As shown in Chapter 4, this aspect finds correspondence in actual experimental tests, provided the low level of compressive stress. Starting from Figure 5.48b, the wall failure is localized at the upper left and lower right corners, the main responsible of the loss of capacity (so-called toe-crushing). Overall, comparing the micro- and macro-approach, the former gives a better representation of the real failure mode because the dimension of the bricks (with respect to the dimension of the specimen) makes the wall behaviour markedly anisotropic.

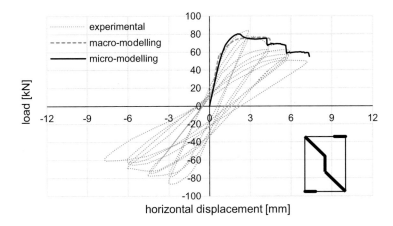

Figure 5.47 Load–displacement diagram. Comparison between experimental test and numerical analysis according to the micro-modelling approach of the masonry wall.

The same behaviour can be understood in Figure 5.49 where the principal stresses and directions of the units are depicted for the same load levels. It is noted that these values are not coincident with the interfaces' and, being linear elastic, may experience larger values of tensile stress (especially due to bending). However, the principal stresses provide a qualitative behaviour of how the wall reacts. In detail: (a) an initial homogeneous stress flow is present along the diagonal, (b) the wall splits and two individual struts appear (compare with Figure 5.45a); (c) the upper left corner crushes and the location of the maximum (in absolute value) compressive stresses move towards the centre and (d) similar crushing occurs at the lower right corner. The last two illustrations are indicative of toe-crushing.

Moving to the normal and shear stresses at the interface, the relative diagrams are shown in Figure 5.50 for the same levels of external load discussed earlier. First, it is noted that head joints have a marginal impact on the overall behaviour of the wall. Second, except for the first row of results, the stress distribution is rather discontinuous along the individual interface, a consequence of the reduced contact area between adjacent units. Looking at the normal stress (left side), it is possible to notice the evolution of stress due to toe-crushing. Basically, the larger is the horizontal displacement at the top of the wall, the larger is the compressive stress in this area. Due to the softening of the constitutive law in compression, the maximum stress migrates towards the centre of the wall in the supports, meaning that the area around the corner is crushing providing a vanishing resistance to the wall. Looking at the shear stress (right side), it is possible to notice the large shear in the interface responsible for increasing the horizontal equivalent tensile strength in a macro-model, in the compressed struts, with the subsequent unloading as a full crack develops.

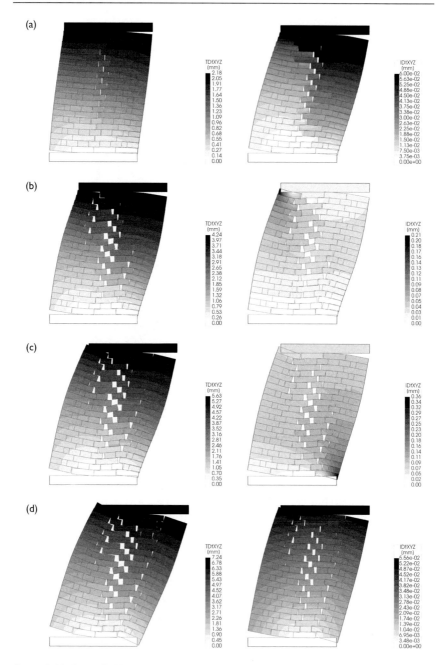

Figure 5.48 Total (left) and incremental (right) displacements (magnification factor for total displacements equal to 80, and variable for incremental ones). Horizontal load equal to (a–d) 81, 75, 67 and 59 kN (the last three values in the softening branch).

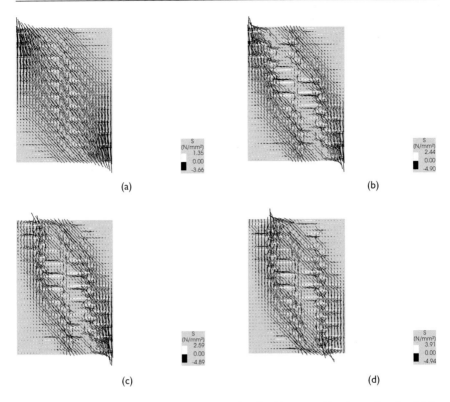

(a)   (b)

(c)   (d)

Figure 5.49 Principal stresses for linear elastic bricks. Horizontal load equal to (a–d) 81, 75, 67 and 59 kN (the last three values in the softening branch).

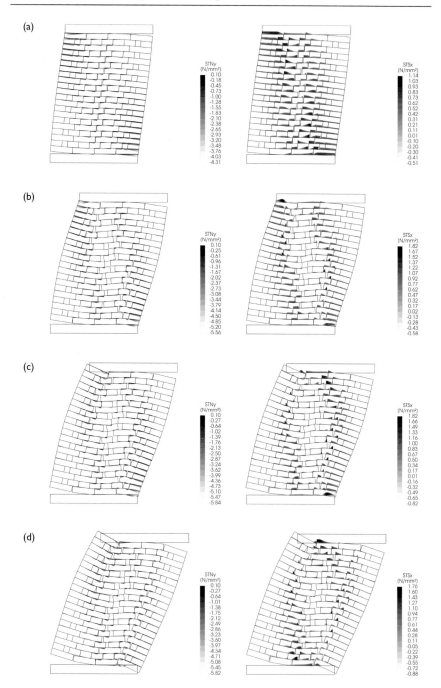

Figure 5.50 Normal (left) and shear stresses (right) at the interface (magnification factor for total displacements equal to 80). Horizontal load equal to (a–d) 81 kN, 75 kN, 67 kN and 59 kN (the last three values in the softening branch).

BIBLIOGRAPHY AND FURTHER READING

Anthoine, A., Magenes, G. and Magonette, G. (1994) 'Shear compression testing and analysis of brick masonry walls', in GERALDDUMA, Central Institute for Meteorology and Geodynamics. (ed.) *10th European Conference on Earthquake Engineering*, Vienna, Austria, pp. 1–6.

Baker, M. K. (2018) How to get meaningful and correct results from your finite element model.

Cook, R. D., Malkus, D. S. and Plesha, M. E. (1989) *Concepts and Applications of Finite Element Analysis*. New York, N.Y.: John Wiley & Sons, Inc.

Hendriks, M. A. N., de Boer, A. and Belletti, B. (2017) *Guidelines for Nonlinear Finite Element Analysis of Concrete Structures - RTD:1016-1:2017*. The Netherland: Rijkswaterstaat Centre for Infrastructure.

Lourenço, P. B. and Pereira, J. M. (2018) *Recommendations for Advanced Modeling of Historic Earthen Sites: Seismic Retrofitting Project Research Report*. Los Angeles, CA: Getty Conservation Institute.

Rajadurai, S. et al. (2014) FEA best practices approach. *Int. J. Recent Develop. Eng. Technol.* 2(3), 58–66.

Selby, R. G. and Vecchio, F. J. (1993) *Three-Dimensional Constitutive Relations for Reinforced Concrete - Tech. Rep. 93-02*. Toronto: University of Toronto.

TNO DIANA BV. (2019) *DIANA Finite Element Analysis; Documentation Release 10.3*. Edited by F. Denise and M. Jonna. Delft: DIANA FEA BV.

Vecchio, F. J. and Collins, M. P. (1993) Compression response of cracked reinforced concrete. *J. Struct. Eng.* 119(12), 3590–3610. doi: 10.1061/(ASCE)0733-9445(1993)119:12(3590).

Vecchio, F. J. and Shim, W. (2004) Experimental and analytical reexamination of classic concrete beam tests. *J. Struct. Eng.* 130(3), 460–469. doi: 10.1061/(ASCE)0733-9445(2004)130:3(460).

Index

Note: **Bold** page numbers refer to tables and *italic* page numbers refer to figures.